ENVIRONMENTAL BIOGEOCHEMISTRY

Volume 1

Carbon, Nitrogen, Phosphorus, Sulfur and Selenium Cycles

ENVIRONMENTAL BIOGEOCHEMISTRY

Volume 1

Carbon, Nitrogen, Phosphorus, Sulfur and Selenium Cycles

Proceedings of the 2nd International Symposium on Environmental Biogeochemistry organized by the Canada Centre for Inland Waters and co-sponsored by UNESCO, International Association of Geochemistry and Cosmochemistry, the Geochemical Society and the Canadian Society of Microbiologists.

edited by

JEROME O. NRIAGU

Research Scientist
Department of the Environment
Canada Centre for Inland Waters
Burlington, Ontario, Canada

ANN ARBOR SCIENCE
PUBLISHERS INC
P.O. BOX 1425 • ANN ARBOR, MICH. 48106

Copyright © 1976 by Ann Arbor Science Publishers, Inc.
P.O. Box 1425, Ann Arbor, Michigan 48106

Library of Congress Catalog Card Number 75-36281
ISBN 0-250-40125-8

Manufactured in the United States of America
All Rights Reserved

PREFACE

ENVIRONMENTAL BIOGEOCHEMISTRY is the chemical study of the interactions between the biological life forms and their surroundings. This broad definition has been adopted in the organization of these volumes. Present-day vernacular, however, assigns "environment" an anthropocentric meaning with the result that environmental science focuses customarily on the physical and biological processes involved in interactions between man and all systems of air, land, water, energy and life that surround him. From this anthropocentric viewpoint, environmental biogeochemistry refers to the basic and applied inquiry about chemical changes in the atmosphere, biosphere, hydrosphere and lithosphere resulting from the activities of man. The majority of the papers in these volumes conform to the narrower, anthropocentric definition.

Environmental biogeochemistry is still in the formative stage as a discipline and owes its development to the inability of the existing scientific specialties to deal holistically with the complexities of the total environmental system. Hitherto, the main contributions in the field have come from subdisciplines in geoscience, biology and chemistry. However, for environmental biogeochemistry to develop into a major integrating discipline, the heterogeneous collection of approaches must be melded into a single conceptual framework and the methods of operation standardized. This work represents perhaps the first attempt to fit biogeochemistry into environmental science and hopefully is only a precursor to other "proceedings" aimed at the cross-fertilization of ideas and techniques among scientists concerned with the interactions of biological life forms and their life support systems.

In a general way, environmental biogeochemistry uses the systems approach to evaluate the energy and mass budgets of elemental or mineral cycles. Consequently, the papers in the field tend to group themselves according to these cycles. Volume 1 contains papers on various aspects of the environmental biogeochemistry of Carbon, Nitrogen, Phosphorus, Sulfur and Selenium Cycles. The first section of Volume 2 contains papers on the transfer of various metals (either from natural sources or emitted by human activities) and their effects on biological systems. The second section of Volume 2 is made up of papers which model or emphasize mass balances in ecosystems. No apology is offered for the heterogeneity of subject matter; the topics point out the breadth and interrelatedness of the fauna and flora and their abiotic environments.

These volumes are based on papers presented at the Second International Symposium on Environmental Biogeochemistry held in Hamilton (Ontario), April 8-12, 1975. An important objective of the meeting was to provide a forum for dialogue between specialists of varying backgrounds and interests, as evidenced by the affiliations of the contributors to the volumes. The symposium was organized by the Canada Centre for Inland Waters, Burlington (Ontario) and co-sponsored by UNESCO, International Association of Geochemistry and Cosmochemistry, the Geochemical Society and the Canadian Society of Microbiologists.

Jerome O. Nriagu

ACKNOWLEDGMENTS

Many people have helped in organizing the symposium and in preparing the papers for publication. Particular thanks are given to Prof. M. Alexander, Prof. T. D. Brock, Prof. G. Eglinton, Prof. P. H. Given, Dr. R. O. Hallberg, Prof. G. W. Hodgson, Dr. W. E. Krumbein, Dr. P. A. Meyers, Prof. E. A. Paul, Dr. M. Schnitzer and Dr. J. J. Skujins who as members of the Program Committee guided us through the complicated initial stages of organizing a symposium. Special acknowledgment is due to Dr. J. J. Skujins (Utah State University, Logan) for being the driving force behind the symposia on Environmental Biogeochemistry. The actual burden of running the meeting fell on the following members of the Local Arrangement Committee: Drs. W. A. Glooschenko, A. L. W. Kemp, J. D. H. Williams, P. T. S. Wong, J. J. Miller and C. I. Dell. Particular appreciation is also extended to Mr. W. Booth and Mrs. N. Harper for their valuable assistance during the symposium. Without the understanding and support of Drs. A. R. LeFeuvre and P. G. Sly, the completion of these volumes would have been impossible.

TABLE OF CONTENTS

VOLUME 1

SECTION I   THE CARBON CYCLE

1. Variations of the Carbon Cycle at Present and in the Geological Past 3
   E. M. Galimov

2. Evaluation of ^{12}C and ^{13}C Data on Atmospheric CO_2 on the Basis of a Diffusion Model for Oceanic Mixing 13
   Angela Rebello, Klaus Wagener

3. Man's Impact on the Atmospheric Carbon Monoxide Cycle 25
   W. Seiler, H. Zankl

4. The Steady-State Concentration of Carbon Monoxide in the Troposphere 39
   Barnard Weinstock, Tai Yup Chang

5. On the Mechanisms of CO_2 and CH_4 Production in Natural Anaerobic Environments 51
   Larry M. Games, J. M. Hayes

6. Organic Geochemistry of Lake Sediments 75
   P. A. Cranwell

7. The Chemistry of Humic Substances 89
   M. Schnitzer

8. Investigation of Extracellular Electron Transport by Humic Acids .109
   J. E. Schindler, D. J. Williams, A. P. Zimmerman

9. Isolation and Characterization of Metal-Organic Complexes from Tropical Volcanic Soils 117
   S. M. Griffith, M. Schnitzer

10. Kerogen Structures in Recently-Deposited Algal Mats at Laguna Mormona, Baja California: A Model System for the Determination of Kerogen Structures in Ancient Sediments 131
    R. P. Philp, M. Calvin

11. Lipids of Recently-Deposited Algal Mats at Laguna Mormona, Baja California 149
    J. Cardoso, P. W. Brooks, G. Eglinton, R. Goodfellow, J. R. Maxwell, R. P. Philp

12. Nucleic Acid Base Contents as Indicators of Biological Activity in Sediments 175
    W. van der Velden, Alan W. Schwartz

13. Carbohydrates in Lake Ontario Sediments 185
    D. Liu

14. Soil Polysaccharides in Buried Humic Horizons Of Ashitaka Loam Formation 191
    R. Hamada, K. Yoshizaki, K. Sakagami, T. Kurobe

15. Biogeochemistry of PCB and DDT in the North Atlantic 203
    George R. Harvey, William G. Steinhauer

SECTION II BIOGEOCHEMICAL CYCLING OF NITROGEN

16. Nitrogen Cycling in Terrestrial Ecosystems . . . 225
    E. A. Paul

17. Soil Nitrogen Transformations: A Modeling Study 245
    H. E. Doner, A. D. McLaren

18. The Influence of Environmental Factors on Nitrification 259
    J. De Leval, J. Remacle

19. The Fate of Nitrogen in Soil: Losses by Denitrification and Leaching 271
    R. J. Dowdell

SECTION III BIOGEOCHEMICAL CYCLING OF SULFUR AND SELENIUM

20. A Study of Phosphorus Kinetics in a Lake
    Ecosystem 283
    D. R. S. Lean, M. N. Charlton

21. Dynamics of Phosphorus, Sulfur and Nitrogen at
    the Sediment-Water Interface 295
    *R. O. Hallberg, L. E. Bågander,
    Anna-Greta Engvall*

22. Sulfur and Carbon Isotopic Evidence for Biogeo-
    chemical Processes in the Dead Sea Ecosystem . . 309
    A. Nissenbaum, I. R. Kaplan

23. Stable Isotope Fractionation by *Clostridium
    Pasteurianum* 327
    *E. J. Laishley, R. G. L. McCready,
    R. Bryant*

24. Microbiological Contributions to the Atmospheric
    Load of Particulate Sulfate 351
    Dian R. Hitchcock

25. A Method for the Direct Determination of Sulfide
    in Water and Sediments 369
    Paul Giammatteo

26. Selenium in Biological Systems, and Pathways for
    its Volatilization in Higher Plants 389
    Barbara-Ann Gamboa Lewis

TABLE OF CONTENTS

VOLUME 2

SECTION IV BIOGEOCHEMISTRY OF METALS
IN THE ENVIRONMENT

27. Aluminum in Relation to the Environment and
    Human Health 427
    John R. J. Sorenson

28. Kinetics of Microbiological Aerobic Decomposition
    of Methylmercury 451
    Robert V. Cooley, Perry L. McCarty

29. Microbial Mobilization of Mercury in the Aquatic
    Environment 473
    *R. R. Colwell, G. S. Sayler, J. D. Nelson, Jr.,
    A. Justice*

30. Role of Mangrove Vegetation in Mercury Cycling
    in the Florida Everglades 489
    Martha Tripp, Robert C. Harriss

31. Laboratory Investigation of Mercury Transport
    through Bed Sediment Movements 499
    *Akira Kudo, Donald R. Miller, R. Ron Townsend,
    Helal Sayeed*

32. Mass Spectrometric-Isotope Dilution Determinations
    of Copper and Lead in Hudson River Water 513
    E. J. Catanzaro

33. Binding of Metal Ions by Humic Acids 519
    F. J. Stevenson

34. Adsorption of Metals by Chitin 541
    T. Yoshinari, V. Subramanian

35. Potential Effects of Metals in Precipitation on
    the Exchangeable Humus-Hydrogen in Soil and
    Surface Water 557
    Egil T. Gjessing, Merete Johannessen

36. Movement and Compartmentation of Nickel and
    Copper in an Aquatic Ecosystem 565
    *T. C. Hutchinson, A. Fedorenko, J. Fitchko,
    A. Kuja, J. VanLoon, J. Lichwa*

37. Abundance and Distribution of Some Heavy Metals
    in Recent Sediments of a Highly Polluted Limnic-
    Fluviatile Ecosystem near Mainz, West Germany . . 587
    *N. Laskowski, Th. Kost, D. Pommerenke,
    A. Schäfer, H. J. Tobschall*

38. Uranium Contents of Hydromorphic Soils and Soil
    Fractions Derived from Accumulation Sites 597
    H. W. Scharpenseel, F. Pietig, E. Kruse

39. The Role of Heterotrophic Microorganisms in the
    Deposition of Iron and Carbon in Soil Profiles . 609
    J. Berthelin, Y. Dommergues

40. Manganese as an Energy Source for Bacteria . . . 633
    Henry L. Ehrlich

SECTION V   STUDIES ON ECOLOGICAL MASS BALANCE

41. Biogeochemical Evolution of Phosphorus Limitation
    in Nutrient-Enriched Lakes of the Precambrian
    Shield 647
    D. W. Schindler

42. Ecosystem Development and the Biological Control
    of Stream Water Chemistry 665
    Peter M. Vitousek, William A. Reiners

43. Nutrient Dynamics in Running Waters: Production,
    Assimilation and Mineralization of Organic
    Matter 681
    *P. G. C. Campbell, P. Couture, L. Talbot,
    A. Caillé*

44. Patterns of Deposition of Natural and Fallout
    Radionuclides in the Sediments of Lake Michigan
    and their Relation to Limnological Processes . . 705
    David N. Edgington, John A. Robbins

45. The Effect of the Thermal Regime on the Annual
    Distribution of Selected Elements in Linsley
    Pond (North Branford, Connecticut) 731
    U. M. Cowgill

46. Computer Model for Toxicant Spills in Lake
    Ontario 743
    *D. C. L. Lam, C. K. Minns, P. V. Hodson,
    T. J. Simons, P. Wong*

47. Interactions of Beef Cattle Wastes with Soil . . 763
    Fred A. Norstadt, Lynn K. Porter

48. Is Air Chemistry Monitoring Worth its Salt? . . . 777
    M. P. Paterson

SECTION I

THE CARBON CYCLE

CHAPTER 1

VARIATIONS OF THE CARBON CYCLE AT PRESENT
AND IN THE GEOLOGICAL PAST

E. M. GALIMOV

V. I. Vernadsky Institute of Geochemistry and
Analytical Chemistry, Academy of Sciences of the
USSR, Vorobjovskoe Shosse 47a, Moskow B-334, USSR

INTRODUCTION

There is some fear that the bulk of CO_2 resulting from combustion of fuels of all kinds has led to an increase in the CO_2 concentration in the atmosphere. The increase may result in global warming and finally in the modification of the whole complex of factors controlling the climate.

Calculations based on the record of world output of the major fuels give 0.4×10^{18} g as the overall mass of technogenic CO_2 evolved since the last century. This quantity is nearly 1/6 the total CO_2 content of the earth's atmosphere. The quantity of technogenic CO_2 entering the atmosphere in 1974 alone amounted, in our estimation, to 2.1×10^{16} g, that is nearly 1% of the natural CO_2 content in the atmosphere. It is a rather substantial value.

It should be kept in mind that a continuous gas exchange between the atmosphere and the ocean exists. In order to produce noticeable variations of the CO_2 partial pressure in the atmosphere, the CO_2 flow from an external source has to be comparable in magnitude to the CO_2 content of the ocean. This amounts to 1.4×10^{20} g (Vinogradov, 1967). The question then is whether the rate of the gas exchange between the ocean and the atmosphere is sufficient to prevent the increase in CO_2 concentration in the atmosphere brought about by the technogenic CO_2.

Direct investigation of the changes of atmospheric CO_2 concentration is difficult. Several factors give rise to fluctuations in the CO_2 content of the atmosphere. These include the diurnal variations due to photosynthetic activity, the variations in the weather, the current conditions of the water regime and other peculiarities of the local carbon cycle. The quantitative analysis *per se* cannot differentiate between the noise in the natural CO_2 fluctuations and the effect due to technogenic CO_2.

For the first objective, one can use the stable carbon isotope technique. The atmospheric CO_2 differs noticeably in the $^{13}C/^{12}C$ ratio from the CO_2 derived from the combustion of the natural fuels (Figure 1.1). The carbon isotope composition of atmospheric CO_2 is known to be characterized by the average value of $\delta^{13}C = -7‰$. Combustion of fuels of all types produces CO_2 with carbon isotope composition of $-30‰$. As the accuracy of measurement of carbon isotope composition is $0.1-0.2‰$, one may estimate that about 1% variation in the atmospheric CO_2 concentration is due to the technogenic CO_2 admixture.

Inasmuch as plants assimilate the atmospheric CO_2, it may be possible to reconstruct the temporal changes in the carbon cycle by means of tree-ring dating. The sequoia-ring method was used previously by Craig (1954) to determine whether there is natural temporal alteration of

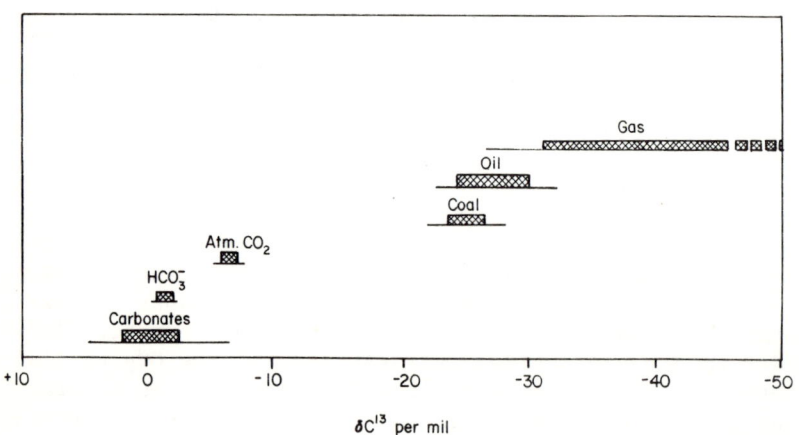

Figure 1.1 The distribution of carbon isotopes in some natural materials.

atmospheric CO_2. On the other hand, Smith and Epstein (1970) found that a systematic difference in $\delta^{13}C$ did exist between plants from the Los Angeles area (consistently 0.4 to 1.2‰ lighter) and the corresponding species from Utah or Texas. They proposed that this difference may reflect the smaller $\delta^{13}C$ values of atmospheric CO_2 in southern California compared to the less polluted rural areas.

We attempted a similar investigation. Two spruce trees in the open edge of forest on the high, steep bank of the Ocher river in the Permial district were sawed down. The trees grew on a well-aerated site, away from occasional CO_2 point sources. By using two tree samples, it was expected that some of the variations connected with individual biosynthesis peculiarities would be eliminated. In order to exclude any plant age effect, trees with different ages of 100 and 50 years were chosen.

It should be noted that the fractionation of carbon isotopes during photosynthesis depends on the environmental conditions around the plant. This factor complicates the interpretation of the data which may be obtained during the study.

METHODOLOGY

Carbon isotope compositions of the tree samples as well as all other samples discussed in this article were measured on mass spectrometer MI-1309. The procedure used in sample preparation for isotope analysis as well as the evaluation of measurement errors have been described elsewhere (Galimov, 1968). The accuracy of each analysis is ±0.2‰. The data are referenced to the standard PDB ($^{13}C/^{12}C$ = 0.0112375).

RESULTS AND DISCUSSION

The experimental results are presented in Figure 1.2. The curves are seen to be similar for the two trees (with differing ages) for the same time interval. Consequently, the observed carbon isotope variations are caused by some external influences and do not reflect individual peculiarities of biosynthesis. Either significant temporal changes in the isotopic composition in atmospheric CO_2 or the conditions of carbon fixation may be responsible for these variations. The latter may, in the first place, be controlled by the temperature variations. We have compared the observed curve of $\delta^{13}C$ variations with the changes in the average annual and monthly temperatures in the particular district but have not found any correlation. Even if the variations

6 *Environmental Biogeochemistry*

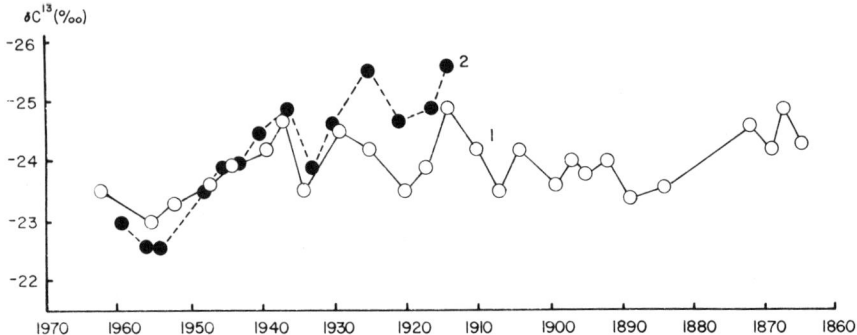

Figure 1.2 The variations of carbon isotope composition in wood rings of two spruce trees: 1) 100 years old; 2) 50 years old.

of the $\delta^{13}C$ values are due to changes in carbon isotope composition of ambient air, it is, however, clear that the changes are not caused by technogenic CO_2 as no marked increase in the ^{12}C isotope concentration is observed from the middle of the last century up to the present. Any significant influence of technogenic CO_2 should have been witnessed during several years in the life history of these trees. Two x 10^{17} g of technogenic CO_2 was calculated to have been broadcast to the atmosphere during the period between 1947 and 1967. In the absence of factors stabilizing the CO_2 concentration in the atmosphere, this should have resulted in the change of atmospheric carbon isotope composition by nearly 2‰. Accordingly, the carbon in the wood should have been enriched in the lighter isotope by 2‰ during this period. However, as can be seen from data presented, this is not the case. Systematic technogenic alteration of carbon isotope composition apparently has been masked by natural fluctuations. Anyway, the changes due to anthropogenic CO_2 cannot be differentiated from the background noise. In other words, we have not found any appreciable influence of the technogenic CO_2 on the carbon cycle.

To evaluate the effects of man's activities on the geochemical cycle of carbon, it is necessary to examine the variations in the carbon cycle during the geological past. This may also be done by means of the carbon isotope method. Carbonate samples are used in this case. The objectives are essentially the same. However, instead of wood sections, the geological sections and instead of tree rings the strata of sedimentary beds are scrutinized during the study.

This part of the problem was tackled in cooperation with
A. B. Ronov and A. A. Migdisov (1975). Elucidation of geological regularities of general significance requires a large
quantity of statistical data. More than 15,000 samples were
at our disposal. To ensure that the data presented are
representative of the average isotope composition of each
geological period, several hundreds of specimens from every
one of the geological ages were analyzed.

Figure 1.3 displays the pattern of the carbon isotope
variations in the Phanerozoic carbonates of the Russian
platform. It is immediately seen that the carbon isotope
composition correlates (positively) with the abundance of
the carbonates as well as with the occurrence of continental
conditions on the Russian platform during the different geological periods. We believe that the observed isotopic
variations reflect the influence of biogenetic CO_2. As the
biogenic carbon dioxide is isotopically light ($\delta^{13}C = -25‰$)
all the processes favoring in any way the influx of biogenetic CO_2 into the basins of sedimentation result in the
enrichment of the precipitated carbonates with the light
isotope. These processes prevail during periods of sea regression, in time of increase of clastic material in sediments, and during periods of predominant littoral and shallow
sedimentation of the carbonates.

Carbon isotope compositions of Precambrian carbonates
are plotted in Figure 1.4. The large circles correspond
to the compound samples from different localities of the
Ukrain and Baltic shields. A number of $\delta^{13}C$ values of Precambrian carbonates from our previous work (Galimov et al.,
1969) are included (small circles). The data obtained by
Perry and Tan (1972) for the most ancient rocks of African
shield are also included (open circles).

Two periods stand out clearly: 2.7 billion years and
1.8-1.9 billion years, which are characterized by unusually
heavy carbonates-carbon isotopes. It may be recalled that
the precipitaiton of carbonates occurs in isotope exchange
equilibrium with the bicarbonate of the ocean and the CO_2
of atmosphere. The isotope equilibrium constant for the
isotope exchange reaction is 1.012. This means that for
the $\delta^{13}C$ value corresponding to that of modern atmospheric
CO_2 of $-7‰$, the ultimate enrichment of carbonates in ^{13}C
isotope will yield $\delta^{13}C$ values less than $+5‰$. (There will,
of course, be exceptions under specialized circumstances
but these are of no concern when regional problems are being
examined).

The isotopic compositions of carbonates are lighter
than the limiting $+5‰$ value throughout the Phanerozoic

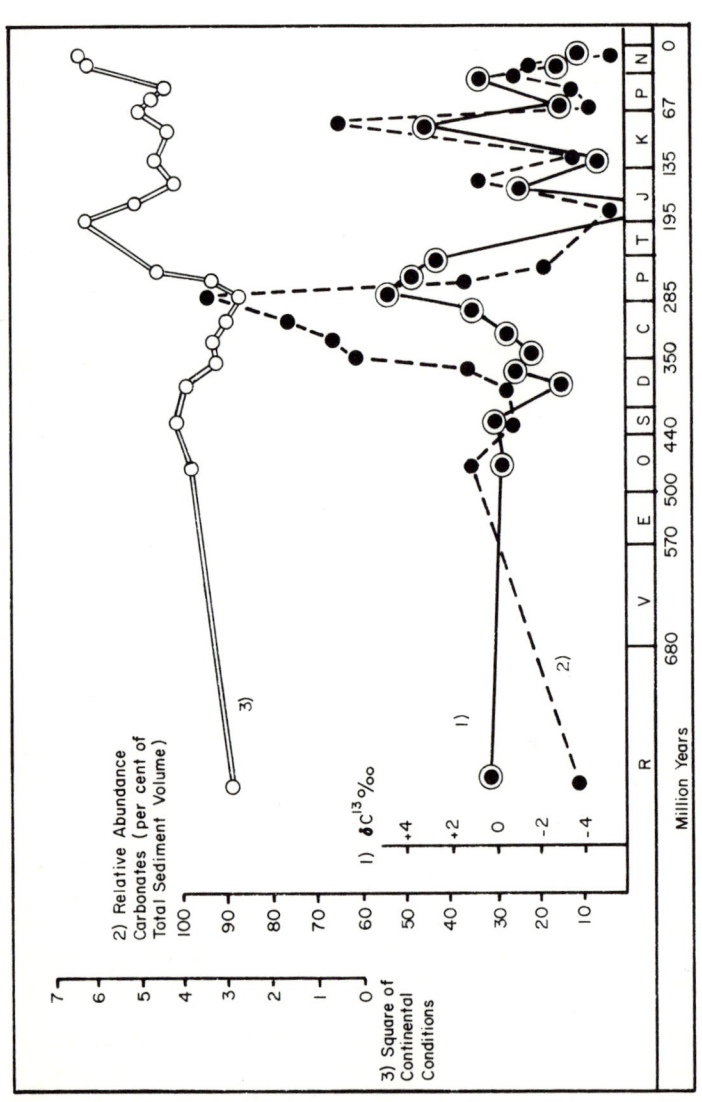

Figure 1.3 Comparison of variations of the carbon isotope composition of phanerozoic carbonates with the carbonate distribution in the deposits of each geological age and with the existence of terrestrial conditions within the Russian platform.

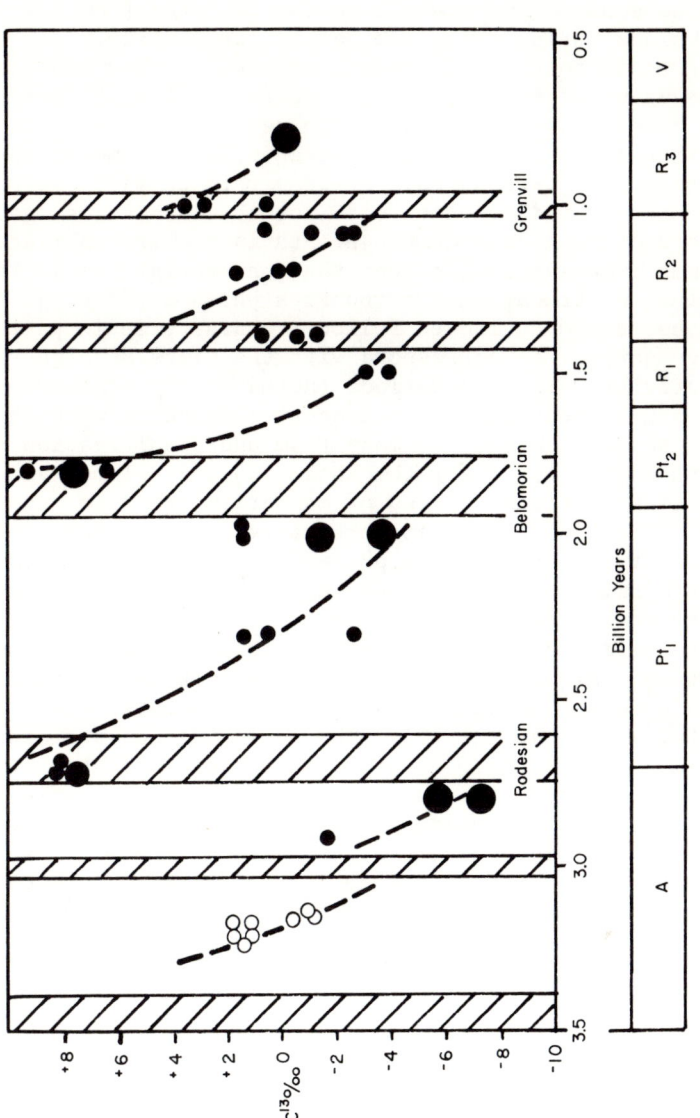

Figure 1.4 Comparison of variations of the carbon isotope composition in the Precambrian times with the history of tectonic activity.

and late Precambrian. More ancient carbonates exceed the
+5‰ limit only during the two periods mentioned above.
The implication seems to be that the isotope compositions
of the atmospheric and oceanic carbon differed from the
modern values during these two particular periods.

The enrichment of carbonates in heavy carbon isotope
over the value of +5‰ may be due to the influx of isotopically heavier CO_2 to the exchange reservoir. Such carbon dioxide may be derived for instance, by the metamorphic
decomposition of carbonates. Keeping this in mind, we have
compared the distribution of $\delta^{13}C$ for the carbonates against
the record of main geotectonic events in the Precambrian.
It becomes immediately apparent that the variations in the
$\delta^{13}C$ values of Precambrian carbonates which at first glance
seem to be random do indeed follow a characteristic pattern.
The heavy carbonates correspond with the times of high tectonic activity. The two periods (noted above) of extreme
isotope ^{13}C concentration in carbonates coincide with the
two most powerful stages of magmatism and granitization in
the Earth's history (*i.e.*, the Rhodesian and Belomorian).
Other less significant stages of magmatism are also pronounced. The ^{13}C content in carbonates progressively falls
in the intervals between the phases of tectonic activity
with the minimum values reached in the periods immediately
preceding each stage of magmatism.

More detailed discussion of the relationships presented
above as well as our data on the carbon isotope composition
of organic matter of ancient sedimentary rocks is given elsewhere (Galimov *et al.*, 1975). The following feature deserves
special mention though. In order to achieve the observed
enrichment of carbonates in ^{13}C isotope, huge quantities of
metamorphogenic CO_2 comparable to the CO_2 content in the
ocean reservoir might have been released to the earth's
atmosphere at the appropriate geological interval. Such
an influx of CO_2 potentially could have brought about catas-
trophic changes in the climate.

REFERENCES

Craig, H., 1954: Carbon-13 variations in Sequoia rings and
   the atmosphere. *Science* 119, No. 3083, 141-143.

Galimov, E.M., 1968: Geochemistry of stable carbon isotopes.
   Nedra, Moskow (in Russian).

Galimov, E. M., A. A. Migdisov and A. B. Ronov, 1975: Variations of isotope composition of carbonate and organic carbon of sedimentary rocks in the earth's history. *Geochimia*, No. 3. (in Russian).

Galimov, E. M., V. S. Prochorov and N. G. Kuznetsova, 1969: On composition of ancient atmosphere through carbon isotope composition of Precambrian carbonates. *Geochimia*, No. 9. (in Russian).

Perry, E. S. and E S. Tan, 1972: Significance of oxygen and carbon isotope variations in early Precambrian cherts and carbonate rocks of Southern Africa. *Geol. Soc. Amer. Bull.* 83.

Smith, B. N. and Epstein, 1970: Two categories of $^{13}C/^{12}C$ ratios for higher plants. *Plant Physiol.* 47, No. 380, 380-384.

CHAPTER 2

EVALUATION OF ^{12}C AND ^{13}C DATA ON ATMOSPHERIC
CO_2 ON THE BASIS OF A DIFFUSION MODEL FOR
OCEANIC MIXING

ANGELA REBELLO

　Departamento de Química
　Pontifícia Universidade Católica
　Rio de Janeiro, Brazil

KLAUS WAGENER

　Institute of Physical Chemistry
　Nuclear Research Center Julich, and
　Chair of Biophysics
　Technical University Aachen
　Federal Republic of Germany

INTRODUCTION

　　　Forecasts on varying atmospheric CO_2 levels resulting from human activity depend on proper models of the CO_2 exchange behavior of the ocean. Up to now, box models have been used to describe the ^{14}C distribution and the net uptake of CO_2 by the oceans as well (Craig, 1957; Revelle and Suess, 1957; Bolin and Eriksson, 1959; Machta, 1972). As a consequence of the first order reactions, assumed for the fluxes between boxes, the relaxation time turns out to be the same for both exchange of isotopic molecules and net uptake of CO_2. This, however, does not agree with the observations of different authors on the ^{14}C and ^{12}C perturbations caused by human activities. Therefore, a better modeling of oceanic mixing behavior is required particularly to fit phenomena on a medium time frame extending over decades.

The objective of this paper was to use mathematical models to investigate the diffusive properties of the ocean. The new models are compared against the box models which have been applied to the deep sea. Welander (1959) treated in a general way the frequency response of different models describing the exchange of matter between atmosphere and ocean, and included the case of a diffusive sea. Obviously, the mathematics have been too difficult, so that up to now nobody tackled the problem in this way.

NET UPTAKE OF CO_2 BY THE OCEANS

We used a very simple approach to manage the mathematical difficulties, and this was done in the following way:

1. Substituting the continuous change in atmospheric CO_2 (total concentration and isotopic composition as well) by discrete injections. Each injection is considered to decay by exchange with the oceans independently from all others.
2. The capacity of the oceanic exchange volume for each injection increases in time with the mean vertical penetration of the diffusion process. The penetration depth is given by the Einstein equation (1):

$$\overline{X^2} = 2 Dt \qquad (1)$$

$$\overline{X} \simeq \sqrt{2 Dt}$$

$\overline{X}$ = mean penetration depth of the diffusion process;
D = eddy diffusion coefficient for vertical mixing;
t = time

Then the decrease of the single injection i in the atmosphere is given by equation (2):

$$\Delta M_i(t) = \frac{\Delta M_i}{1 + \frac{1}{d}(2 D \Delta t)^{\frac{1}{2}}} \qquad (2)$$

ΔM_i = amount of CO_2 released into the atmosphere with the injection i;
d = the thickness of an ocean layer containing an equivalent amount of atmospheric CO_2 (about 65 meters);
Δt = $t - t_i$;

3. The real process then is the superposition of all the single processes and is described by equation (3):

$$\Delta M(t) = M(t) - M(o)$$

$$= \sum_i \frac{\Delta M_i}{1 + \frac{1}{d}(2D\Delta t)^{\frac{1}{2}}} \qquad (3)$$

In this first approach we neglected both the existence of the biosphere on land and the intermediate box of the mixed layer. The neglect of the biosphere is justified within the allowance of an error of about 10%, because of the fast recycling of the fixed carbon by respirative processes (Wagener and Forstel, 1972). The existence of the mixed layer can be neglected in this context, because of its small capacity for processes ranging over decades. Thus, in this approach, our model is reduced to a simple one-parameter model, and this parameter is the coefficient for vertical eddy mixing on a global scale. We adjusted it empirically using the measurements of Keeling (1965,1971).

Using the data for annual input of fossil fuel carbon into the atmosphere, as given by the UN Statistical Papers, a summation according to equation (3) was carried out for different years, ranging between 1880 and 2000. For practical reasons, we replaced the diffusion coefficient by the parameter α which reads as

$$\alpha = \frac{\sqrt{2D}}{d} \ [\text{year}^{-\frac{1}{2}}]$$

The result is shown in Figure 2.1. The adjustment was done by using the derivative function of the measured curve for the year 1965, the center of the Keeling measurements.

What turns out for the overall apparent eddy diffusion coefficient is 0.0185 cm^2/sec. This figure, however, is smaller than the normally accepted value for eddy diffusion in the deep sea by a factor of about 100. The explanation for this seems to be evident, on the basis of the buffering behavior of sea water for CO_2, which is usually presented in the form

$$\frac{dP}{P} = \beta \frac{dc}{c} \qquad (4)$$

16   Environmental Biogeochemistry

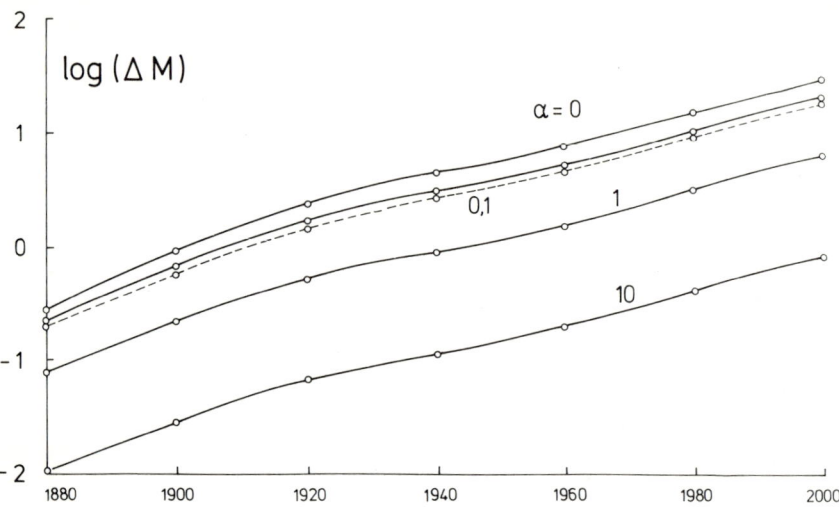

Figure 2.1 Plots of equation (3). Parameter is $\alpha = \sqrt{2\,D/d}$ in units of year$^{-\frac{1}{2}}$. $\alpha = 0$ means no uptake of CO_2 by the oceans at all. Dotted line corresponds to $\alpha = 0.166$, what fits to the actual rate of CO_2 increase in the atmosphere.

with β of the order of 10. $1/\beta$ can be considered as the capacity factor which has to be introduced into our equation (3). In case of isotopic exchange, β is equal to 1; for the net uptake of CO_2, β is equal to 10. Consequently, we obtain for the ratio, R, of the apparent diffusion coefficients:

$$R \equiv \frac{\text{Apparent Diffusion Coefficient for the Exchange of Isotopic } CO_2 \text{ Molecules}}{\text{Apparent Diffusion Coefficient for Net Uptake of } CO_2}$$

$$= \left(\frac{\alpha/\beta_{ex}}{\alpha/\beta_{net}}\right)^2$$

$$= \left(\frac{\beta_{net}}{\beta_{ex}}\right)^2 \approx 100$$

with $\beta_{ex} = 1$ and $\beta_{net} \approx 10$.

Using the figure of 0.0185 cm^2/sec for the apparent diffusion coefficient on a global scale, we made a summation of all injections, as explained earlier, over the secular time. The results are shown in Figure 2.2, from which it is evident that our results resemble quite closely the results obtained by Machta's multiparameter box model. Thus, we can realize that this very simple one-parameter model works quite well when applied to processes ranging over decades. Under these circumstances, all the factors we neglected do not seem to play any important role.

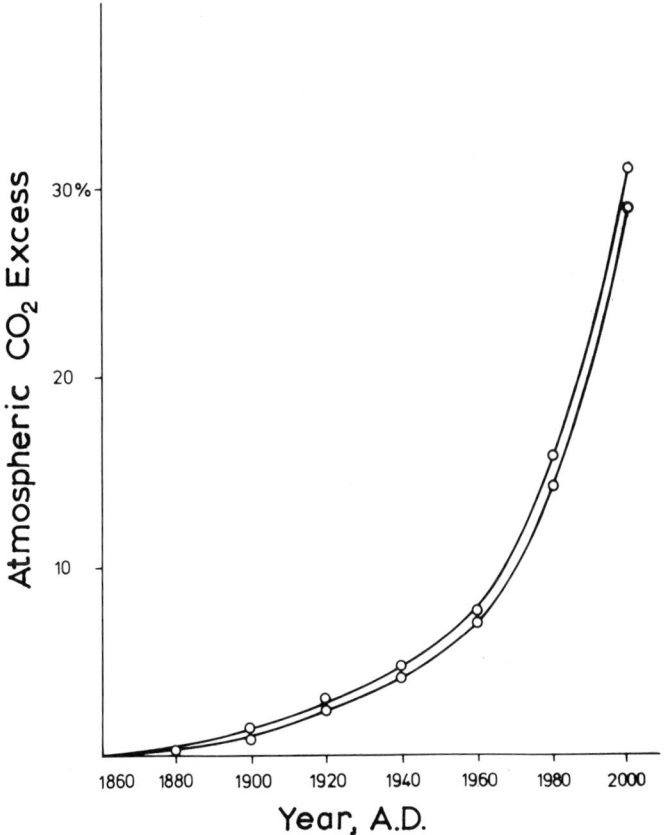

Figure 2.2   Computed atmospheric CO_2 excesses. Upper curve: after the multi-parameter box model of Machta (1972). Lower curve: after the one-parameter diffusion model of this paper.

EXCHANGE OF ISOTOPIC CO_2 MOLECULES

To apply the diffusion model to the dilution kinetics of isotopic CO_2 tracers injected into the atmosphere, we have to transform $\Delta\gamma$ values (γ = mole fraction of the carbon isotope under consideration) into ΔM values, as used in equation (2) and (3). If we release a certain amount of CO_2 (ΔM_i in mass) with a mole fraction of $^{13}C, \gamma^*$, into the atmosphere, it follows for the produced labeling after mixing with the atmospheric reservoir (M_o in mass):

$$\Delta\gamma_i = (\gamma^* - \gamma_o) \frac{\Delta M_i}{M_o + \Delta M_i} \qquad (5)$$

As a first approximation, we may neglect ΔM_i (small compared to M_o). Then equation (5) reads as

$$\Delta\gamma_i \simeq K \cdot \Delta M_i \qquad (6)$$

with

$$K = \frac{\gamma^* - \gamma_o}{M_o} = \frac{1.959 \cdot 10^{-4}}{63.54} = 3.08 \quad 10^{-6},$$

assuming a relative deviation in isotopic composition of fossil carbon from that of atmospheric carbon of $(\gamma^* - \gamma_o)/\gamma_o = -17.7\%$. Using equation (6), we can directly apply equation (3) to the variations of ^{13}C in atmospheric CO_2:

$$\Delta\gamma(t) = K \sum_i \frac{\Delta M_i}{1 + \alpha (\Delta t)^{\frac{1}{2}}} \cdot \qquad (7)$$

In this way, we applied the model to both the expected decrease in the atmospheric ^{13}C content resulting from the burning of fossil fuel, and the measured decrease of the ^{14}C activity after the end of the bomb test period. For the ^{13}C decrease during the last decades we traced back the isotope record in the cellulose of tree rings (Rebello, 1974; see next section); for ^{14}C we used the data given by Telegadas (1971). The ^{13}C record as analyzed in a Brazilian Araucaria is shown in Figure 2.3. We expected that the curve would decrease more and more with time according to the rapid increase of fossil fuel combustion. There are some speculations on the reliability of the early

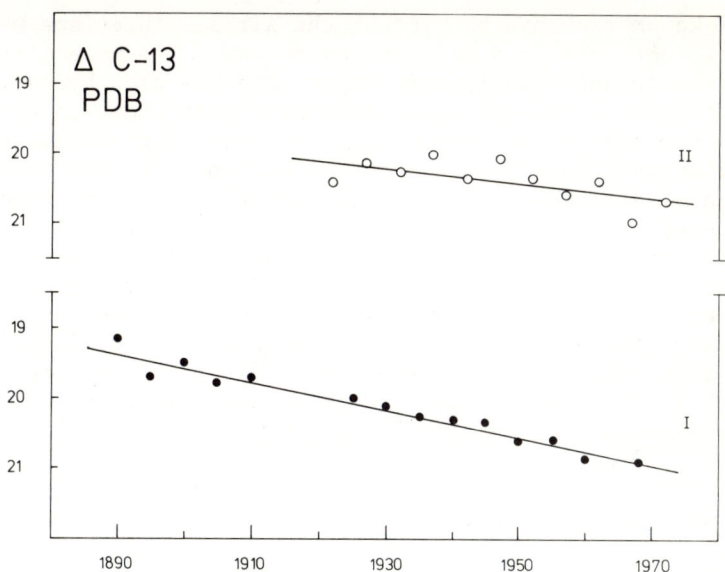

Figure 2.3   Decrease of ^{13}C content as found in the cellulose prepared from the tree rings of a Brazilian araucaria (Rebello, 1974). I: cross section 0.65 m above ground level; II: cross section 12.60m above ground level. Note that the centers of both curves are the same relative to PDB standard.

part of the curve, considering the possibly different behavior of a young tree with respect to isotope fractionation. For our evaluation, we used the time span between 1930 and 1970. Inserting these data into our model, we obtain an apparent diffusion coefficient for isotopic exchange, D_{ex}, of 0.1 cm^2/sec. The same figure turns out from the decrease of the ^{14}C excess between 1963 and 1969 when the data of Telegadas are used.

The coincidence of the two results obtained from ^{13}C and ^{14}C, respectively, may just be accidental. The ^{13}C exchange process was derived for a period of 40 years. However, this record may have been influenced by some unknown factors, such as climatic effects on the isotope fractionation of the tree, or the additional production of CO_2 from the cultivation of natural land. In any case, the evaluated figure of D_{ex} = 0.1 cm^2/sec is smaller than expected by a factor of 10. This, most probably, is the effect of the well-known exchange barrier at the air/sea interface which we did not take into account, up to now. The molecular passage through the laminar layer, and the slow hydration reaction of dissolved CO_2, is obviously the rate-controlling step for short-term variations. However, for processes ranging over decades the vertical transport within the ocean's water mass becomes rate-controlling. This seems to be strongly indicated by the results of Figure 2.2.

An improvement of the diffusion model in this simplest form by taking into account a first-order reaction at the interface before the diffusion transport is under preparation. At least the diffusion model is the first one giving different time constants for the behavior of trace components (such as ^{13}C and ^{14}C) and the principal component, ^{12}C, respectively.

SELECTION OF TREE AS ISOTOPE RECORDER AND PREPARATION OF CELLULOSE

The *Araucaria (Brasiliensis) Angustifolia* is a coniferous plant which has its natural habitat mainly in the southern highlands of Brazil. The selection of an Araucaria was imposed by the fact that it grows in regions where the seasonal temperature modifications are sufficiently pronounced to allow the formation of annual tree rings. As is well known, tropical wood plants in general do not show regular growing layers.

A free-standing specimen was collected far from the pollutant urban centers, where the air masses could not be considered as homogeneous, about 1,600 m above the sea level and 200 km from the coast. Dating techniques were used to estimate its age to be 85 ± 3 years. The dated wood samples were obtained by grouping five rings, taken from three different directions, and this for two different cross sections of the tree.

Up to now, most of the investigations on ^{13}C variations in atmospheric CO_2 (by analyzing the isotopic record in tree rings) were performed in natural wood. On account of the following facts, however, it turned out to be necessary to use pure cellulose:

1. The wood is permeated by mobile resins of unknown age;
2. the second principal component of wood, the lignin, has an isotopic composition different from that of cellulose (Wiesberg, 1974; Rebello, 1974), as a result of differences in the biochemical pathways. It must also be

taken into account that the ratio of lignin to cellulose is not constant for a sequence of rings. In addition, it has been supposed that there is a time shift between the formation of cellulose and the deposition of lignin for a single ring.

In order to obtain pure cellulose from natural wood, some well-known chemical procedures (Ott, 1965; Nikitin, 1966) were combined in such a way as to minimize degradation and consequent isotopic fractionation. Temperature, chemical concentrations and time were carefully determined and controlled for all the reaction pathways. In Figure 2.4 a schematic representation of the separation process shows the

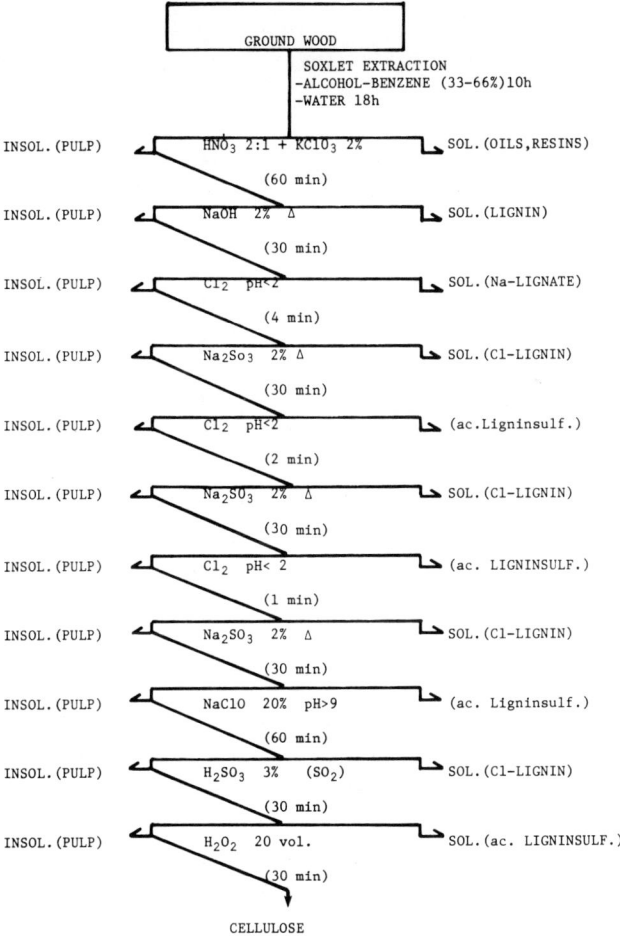

Figure 2.4 Chemical procedure for the preparation of pure cellulose from natural wood.

proper concentration, temperatures (Δ = ebullition), pH and time required for each reaction. Qualitative tests were performed with each of the cellulose samples in order to check for purity and possible damage. To accomplish these tests Reactive-C for pulps and cupramonium solutions were used.

The cellulose samples were quantitatively transformed into CO_2, and the isotopic measurements were performed with a precision of ±0.01‰ on a Micromass 602-Spectrometer.

REFERENCES

Bolin, B. and Eriksson, E., 1959: Changes in the carbon dioxide content of the atmosphere and sea due to fossil fuel combustion. Rossby Memorial Volume, Esselte, Stockholm, pp. 130-142.

Craig, H., 1957: The natural distribution of radiocarbon and the exchange time of carbon dioxide between atmosphere and sea. *Tellus* IX, 1-17.

Machta, L., 1972: The role of the oceans and biosphere in the carbon dioxide cycle. In: D. Dyrssen and D. Jagner, eds.: *The Changing Chemistry of the Oceans.* Stockholm, pp. 121-145.

Nikitin, N. I., 1966: *The Chemistry of Cellulose and Wood.* Israel Program for Scientific Translations, Jerusalem.

Ott, E., and H. Spurlin, 1965: In: *Cellulose, High Polymers,* Vol. V, Interscience, New York.

Pales, J. C. and Keeling, C. C., 1965: The concentration of atmospheric carbon dioxide in Hawaii. *J. Geophys. Res.* 70; Cf. also: *Man's Impact on the Global Environment.* MIT Press, Cambridge (1971), p. 48.

Rebello, A., 1974: The ^{13}C record in tree rings and its correlation with the increase of atmospheric CO_2 resulting from fossil fuel combustion. A critical investigation on isotope records in biological material. Thesis, Pontificia Universidade Catolica, Rio de Janeiro.

Revelle, R. and H. E. Suess, 1957: Carbon dioxide exchange between atmosphere and ocean, and the question of an increase of atmospheric CO_2 during the past decades. *Tellus* IX, 18-27.

Telegadas, K., 1971: U. S. Atomic Energy Commission Report, HASL-243.

Wagener, K. and H. Forstel, 1972: Relaxation phenomena in the biological carbon cycle under conditions of variable atmospheric CO_2 content. *Z. Naturforsch.* 27a, 821-815.

Wiesberg, L., 1974: Die ^{13}C-Abnahme in Holz von Baumjahresringen, eine Untersuchung zur anthropogenen Beeinflussung des CO_2-Haushaltes der Atmosphäre. *Thesis,* Aachen.

Welander, P., 1959: On the frequency response of some different models describing the transient exchange of matter between atmosphere and sea. *Tellus* XI, 348-354.

CHAPTER 3

MAN'S IMPACT ON THE ATMOSPHERIC
CARBON MONOXIDE CYCLE

W. SEILER AND H. ZANKL

Max-Planck-Institute for Chemistry
Mainz, West Germany

INTRODUCTION

Besides the main gases oxygen, nitrogen, argon and carbon dioxide, the atmosphere contains a great number of trace gases with mixing ratios ranging from a few ppm (parts per million) down to a few ppt (parts per trillion). In spite of the low mixing ratios, some of the trace gases are of great importance for the earth's environment and are even fundamental for some forms of life on earth. Nearly all trace gases undergo a natural cycle in the atmosphere: they are produced by natural sources and are destroyed by natural sinks, mostly within the troposphere. Sources and sinks are provided by various biological, microbiological, chemical and physical processes. In recent time, these natural cycles have been disturbed by man's activity at an increasing rate, *e.g.*, by disposal of pollutants into the air and the water, or by fertilization with a corresponding change of the microbiological ecosystem in the soil. These disturbances result in a change of the mixing ratios and the distribution of several trace gases in the atmosphere. The consequences are changes of the global environment. Climatic changes and changes of the solar radiation reaching the earth's surface may be cited as examples. Both are also very important factors for the survival of man on earth. Environmental changes will become greater with increasing industrialization so that there exists a severe potential for global environmental catastrophes, as have been predicted already by several well-known scientists.

It is therefore of great importance to predetermine, by calculations, the quantity and the quality of possible changes of the global environment so that the effects of the different pollutants may be ascertained. For such calculations, we must have a good knowledge of both the natural cycles and the anthropogenic production rates for the individual trace gases involved. At present, our knowledge about the natural cycles and the contribution by man's activity is clearly unsatisfactory. The main reasons are a) the complexity of the natural cycles and sinks; b) the lack of instruments with a sufficient sensitivity for measurements in clean air (low concentration); and c) the great variety of anthropogenic processes leading to the formation of the individual gaseous pollutant.

Carbon monoxide may be used as an example to illustrate the problems discussed above. First we give a survey of the major sources and sinks of the atmospheric CO-cycle where it will turn out that man's impact on the CO-cycle is not negligible. Subsequently the problems of estimation of the amount of the anthropogenic sources are discussed in detail.

MAIN SOURCES AND SINKS OF ATMOSPHERIC CO

Since the detection of CO in the atmosphere by Migeotte and Neven in 1949 and prior to 1969, CO was thought to be produced exclusively by man. None of the sinks was known so that CO was considered to be a major problem of cumulative global air pollution. Today, we have identified several natural sources and sinks (Table 3.1) which will be discussed briefly in this section.

Anthropogenic CO-production

Carbon monoxide is produced by a variety of technological processes, mainly by incomplete combustion of carbonous fossil fuel in automobiles, industry, power plants and home heating. The published data about the total worldwide anthropogenic CO-production, based on the known emission factors and the consumption rate of fossil fuel, vary between 2×10^{14} g/yr (Bates and Witherspoon, 1952; Robinson and Robbins, 1969) and 4×10^{14} g/yr (Jaffe, 1972). More recently Seiler (1974) reported a total anthropogenic CO-production of 6.4×10^{14} g/yr for 1973 out of which about 70% is generated by automobiles.

The Carbon Cycle 27

Table 3.1
Summary of the Estimated Annual Production
and Distribution Rates of the Individual
CO sources and sinks

	Total	Northern Hemisphere	Southern Hemisphere
Sources x 10^{14} g/yr:			
Anthropogenic	6.4	5.4	1.0
Oceans	1.0	0.4	0.6
Rain water		negligible	
Bush- and forest-fires and Burning of agricultural waste	0.6	0.4	0.2
Oxidation of hydrocarbons	0.6	0.4	0.2
Total CO production	8.6	6.6	2.0
Oxidation of methane	15–40		
Sinks x 10^{14} g/yr:			
Uptake by soil	4.5	3.0	1.5
Flux into the stratosphere	1.1	0.9	0.2
Total CO consumption	5.6	3.9	1.7
Oxidation by OH	20–50		

CO-production by Natural Sources

Microbiological activity in the surface layers of the oceans and lakes contributes about 1 x 10^{14} g/yr. Some CO is also formed by bush and forest fires as well as by burning of agricultural waste (0.6 x 10^{14} g/yr). A similar amount may be generated by the photochemical oxidation of higher hydrocarbons in the troposphere, *e.g.*, terpenes (Robinson and Robbins, 1969). At present, the influence of plants on the CO-cycle is uncertain. On the one hand Wilks (1959), Sigel *et al.* (1962), and others reported a CO-production by green plants. On the other hand, Bidwell and Fraser (1972), using $C^{14}O$ as a radioactive tracer, could show a consumption of CO by higher plants. Our own unpublished measurements made under atmospheric conditions with respect to the mixing ratios of CO, CO_2, H_2O and O_2 as well as the temperature indicate a production of CO by plants which seems to be in the range of 0.2 x 10^{14} g/yr. A much higher

CO-production rate (20–40 x 10^{14}g/yr) is assumed by several authors (Weinstock and Nicki, 1972), Wofsy et al., 1972; Levy, 1972) to occur by the photochemical oxidation of methane in the troposphere as a result of reaction with OH radicals.

Sinks of CO

The CO sources are balanced mainly by three sinks. The first is the consumption of CO by microorganisms at the soil surface (Liebl, 1971; Inman et al., 1971), which is estimated by Seiler (1974) to be 4.5 x 10^{14}g/yr. Carbon monoxide (1.1 x 10^{14}g/yr) is also removed from the troposphere by a flux of CO through the tropopause into the stratosphere (Seiler and Warneck, 1972) where CO is destroyed photochemically. Most of the atmospheric CO (20–50 x 10^{14}g/yr) is suggested by Weinstock and Nicki (1972), Wofsy et al. (1972) and Levy (1972) to be oxidized by OH in the troposphere.

The production and destruction rates given in Table 3.1 seem to indicate that the CO-cycle is mainly controlled by the photochemical processes. However, we have to take into account that all these data are only estimated values and consequently are beset with great errors. The greatest uncertainty concerns the estimation of the photochemical CO production and destruction, which are both based exclusively on model calculation. At present, we know neither the averaged OH concentration in the troposphere nor the exact photochemical mechanism for this type of CO-production. The starting point is the oxidation of methane by OH with formaldehyde acting as an intermediate product. The first calculations of the OH-production rates in the troposphere as a function of latitude and altitude averaged over the seasons were recently published by Warneck (1975). Using these data, one calculates a photochemical CO-Production and destruction of about 4 x 10^{14}g/yr, significantly lower than the values given in Table 3.1. The discrepancy between the estimated photochemical CO-production and destruction rates demonstrates the degree of uncertainty which can only be reduced by appropriate measurements of the OH concentrations in the troposphere.

Although the estimations of the anthropogenic CO-production were based on measured emission factors and known consumption rates of fossil fuel, the accuracy fo the CO-production rate from technological processes as given in Table 3.1 is not better than ±50%. The main reasons for this uncertainty are as follows:

1. At present we do not know all the individual technological CO-sources. It may be safely assumed that industrial CO-sources exist in addition to those already considered whose emission cannot be estimated because of the lack of published data on production, type of equipment of emission factors. Some of these CO-producing processes are ammonia- and methanol-producing industries, organic chemical manufacture and others.

2. Most of the emission factors of the individual CO-sources are dependent, to a high degree, on the operating conditions of the equipment. Consequently the emission factors for a single CO-source vary over a wide range, for example by one order of magnitude for a single automobile at a velocity of 15 km/hr and 80 km/hr. Furthermore, the emission factors for automobiles vary by at least a factor of two. Similar differences may be expected also for other CO-producing processes. In addition, the emission factors vary from country to country depending on the degree of pollution control. For some countries they are even unknown, $e.g.$, for the Eastern European countries, where a considerable amount of fossil fuel is consumed. For these reasons it is nearly impossible to give CO-emission factors representative for the individual technological CO-sources on a global basis.

3. Data on the consumption rate of fossil fuel by different technological processes are only available for some western countries. Little or no data are available for developing countries or eastern countries like Russia or China. For these countries the consumption rates have to be estimated.

Due to these factors the estimation of the total amount of CO produced by technological processes is and will remain uncertain to a large degree.

ESTIMATION OF THE RATE OF ANTHROPOGENIC
CO-PRODUCTION IN A LARGER CITY: MUNICH

In order to check the published data, we approached the problem of estimating anthropogenic CO-production rate with a completely different method. We measured the CO-distribution over Munich, a European city with a population of one million, and used these measurements to calculate the total anthropogenic CO-production in the area covered. The production rates obtained in this way do not depend on our knowledge of emission factors of the individual technological CO-sources, the consumption rates of fossil fuels etc., but include all the technological CO-sources in Munich. Measurements were made using continuous registration instruments (Seiler and Junge, 1970) which were installed in a two-propeller aircraft, type Beachcraft 65. The lower detection limit of the instruments is about 1 ppb, much less than the CO-mixing ratios expected to range from 0.2 ppm up to a few ppm. The accuracy is better than ±5% at a CO-mixing ratio of 0.1 ppm. Sampling of outside air was performed at the nose of the aircraft so that all contamination by the aircraft exhaust could be excluded. Several checks, were made to establish that the sampling occurred indeed without contamination. In addition to the registrations of CO, simultaneous measurements were carried out for several other trace gases and for other meteorological parameters such as windspeed, wind direction, temperature, humidity and pressure.

Figure 3.1 shows the grid for the flight routes over Munich, with an extension of about 50 km in North-South and East-West directions. The grid covers more than three times the developed area of Munich. The measurements were started at an altitude level of 300m above ground which was the lowest level allowed for air traffic. The flight levels were increased by steps of 100 m up to altitudes of 600, 700 and 800 m depending of the weather situation.

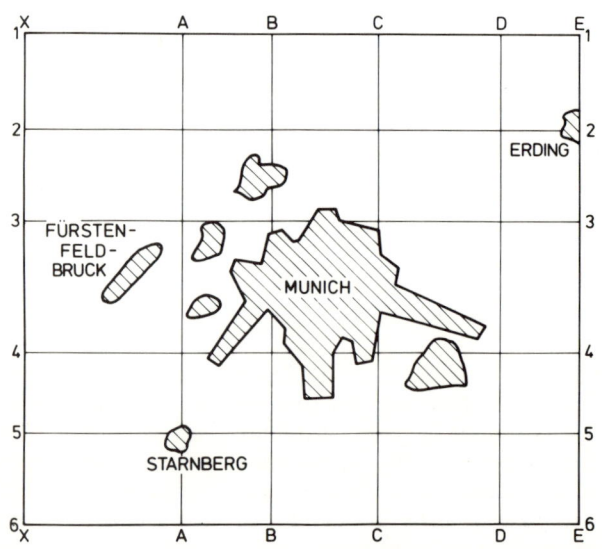

Figure 3.1 Flight grid over Munich, Germany (FRG).

Most important for our purpose were flights at the
weatherside and leeside of Munich, carried out near the
outer boundaries of the grid (for example, for westerly
winds between E1 and E6 and X1 and X6, respectively). Here
the measurements were started at altitudes of about 100 m
and extended in steps of 100 m up to 800 m. The measurements
could be made in a relatively short time of 2 hours so that
we can assume reasonably constant conditions for each set
of flights with regard to wind direction and wind speed.

The CO-measurements over Munich were carried out for
different weather situations during the period from December 1974 to April 1975. During these 5 months, 10 vertical
profiles each at the leeside and the weatherside of Munich
were obtained. Up to now three of these have been analyzed,
corresponding to completely different weather situations.
Figure 3.2 shows the vertical CO-distribution at the leeside,
obtained on January 29, 1975, during a low pressure system
over Europe with strong westerly winds (270°) and wind
speeds up to 50 km/hr at an altitude of 600 m. The air
temperature at ground was +4°C, which is relatively high
compared with average temperatures of the season. The solid
lines in Figure 3.2 represent the measured CO-mixing ratios
whereas the dashed lines below 100 m and above 600 m height
represent extrapolated data. The highest CO-mixing ratios
were found at an altitude of about 400 m above ground with
values of about 0.55 ppm. The CO-mixing ratios decrease
in horizontal and vertical direction to 0.22 ppm, a value
identical with the CO-mixing ratio at the weatherside where
CO is uniformly distributed. Figure 3.3 shows the CO-
distribution at the leeside of Munich on February 27, 1975.
On that day a high pressure system with corresponding lower
windspeeds of about 20 km/hr existed over Central Europe.
In this case the maximum CO-mixing ratio is obtained at a
lower altitude (250 m) caused by a strong temperature inversion at about 800 m height. Similar to the previous
figure, the CO-mixing ratio decreases to about 0.30 ppm in
horizontal and vertical directions, a value which is slightly
higher than those at the weather side. Notable is the sharp
decrease of the CO-mixing ratio at the temperature inversion
layer from 0.35 ppm to about 0.20 ppm. The latter value
is typical for clean air masses in the middle latitudes of
the northern hemisphere. The strong gradient demonstrates
that temperature inversions prevent a rapid mixing of air
masses in vertical direction and consequently cause an
increase of the concentration of pollutants below the inversion layer during high pressure weather conditions. On the

32   Environmental Biogeochemistry

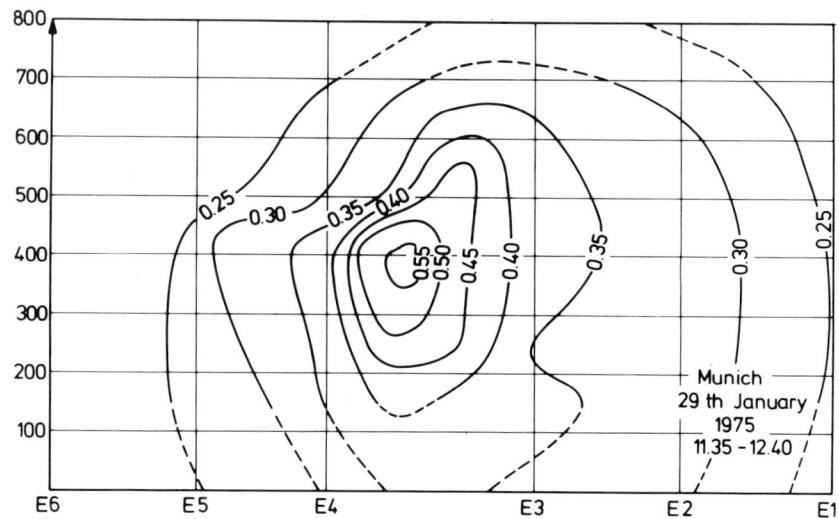

Figure 3.2   Vertical and horizontal CO-distribution at
             leeside of Munich (January 29,1975).

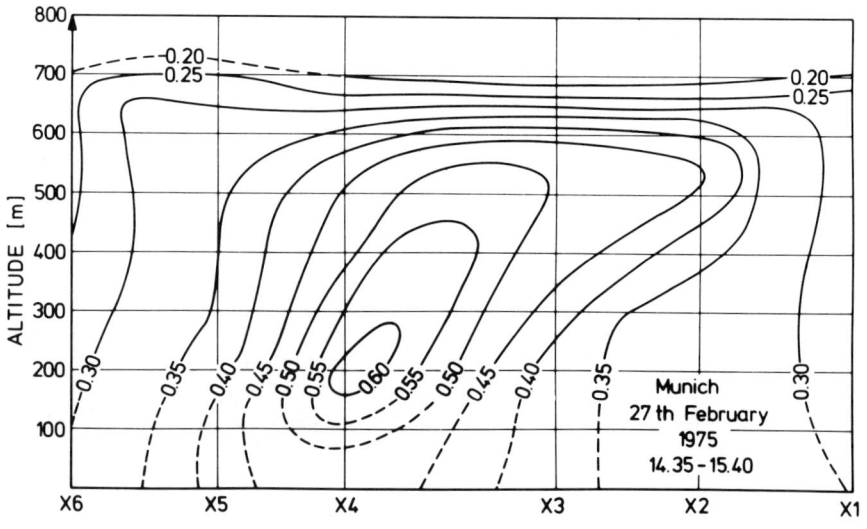

Figure 3.3   Vertical and horizontal CO-distribution at
             leeside of Munich (February 27, 1975).

basis of the measured CO-distributions and the horizontal
wind speeds at the leeside and the weatherside, the total
transport of CO through the vertical planes was calculated.
Since natural CO-sources and sinks, for example the uptake
by soil, are comparatively minor in importance, the difference of the CO-fluxes at the weatherside and the leeside
must be identical with the total anthropogenic CO-production
within the area of Munich during the time of observations.

With the data of Figures 3.2 and 3.3, the total anthropogenic CO-production rate is calculated to be 1.8×10^8 g/hr
and 1.0×10^8 g/hr, respectively. These data were obtained
on the 29th of January at noontime and on the 27th of February at 3 pm, respectively. Due to strong daily variations
of the CO-emission rates by changes in the traffic pattern
etc., the given values are only valid for the time of measurements (around noon). They do not represent average
values for one day. Such averages would be needed for
the calculation of the daily CO-production as a function
of time was not obtained. At present, the best indicators
for this dependency are the continuous registrations of the
CO-mixing ratio at the ground in different regions of the
city where the absolute CO-values are directly influenced
by the emission rates of the individual CO-sources. Figure
3.4 shows the average of three CO-registrations at three
different stations as a function of time during the 29th
of January, 1975. The CO-mixing ratios show high daily

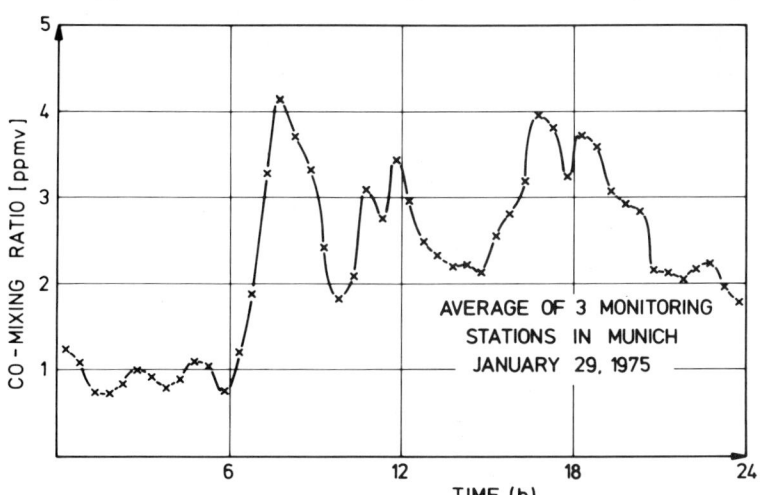

Figure 3.4 Average of CO-registrations monitored at three
different stations in the city of Munich as a
function of daytime for January 29, 1975.

variations with maxima in the early morning and late afternoon caused by the higher traffic density (rush hour) and relatively low values during the nighttime. Similar daily variations are also reported for cities in the USA and other countries in Western Europe. As Figure 3.4 shows, the noontime CO-mixing ratios have values which are typical for about 14 hours per day. This behavior has been observed also in all other similar investigations. If we assume the same correlation for the CO-emission rates with time for the *total* daily CO-production in Munich, the resulting average value will be about 20×10^8 g which represents a minimum value. If traffic and industrial processing account for more than 94% of the total CO derived from technical sources (Jaffe, 1968), the value given above is representative for the whole year. However, if coal and oil burning for home heating contribute about 30% of the total anthropogenic CO as assumed by Baum (1972), the CO-emission rates should show seasonal variations with higher values in winter and correspondingly lower values in summer. For these alternative cases, the CO-emission from technological sources will result in 0.7 and 0.5×10^{12} g/yr, respectively, for the city of Munich. These values may be extrapolated to 0.4 and 0.3×10^{14} g/yr, respectively, for the entire Federal Republic of Germany.

From the known emission factors and the consumption of fossil fuel, the total anthropogenic CO-production in the FRG has been estimated as 0.2×10^{14} g/yr (NATO Report N.10, 1973). For 1975 this amount must have increased by about 10% due to a corresponding increase of the consumption of fossil fuel. In view of the uncertainties concerning the absolute values of the emission factors as discussed before and the assumptions made for the daily and seasonal variations of the CO-emission rates, the total anthropogenic CO-production rates obtained by these two completely different methods are in good agreement. The methods indicate that the anthropogenic CO-production of about 6×10^{14} g/yr as assumed by Seiler for 1974 is certainly not overestimated. Compared with the known natural CO-sources, as listed in Table 3.1, the CO-production by technological sources is again shown to be very important and may even dominate, if the photochemical CO-production should turn out to be lower than was assumed by Seiler and Schmidt (1974).

The great influence of anthropogenic CO-production on the natural CO-cycle can best be demonstrated by the latitudinal CO-distribution in the troposphere (Figure 3.5). If the anthropogenic CO-production played only a minor role, one should expect higher CO-mixing ratios in the southern

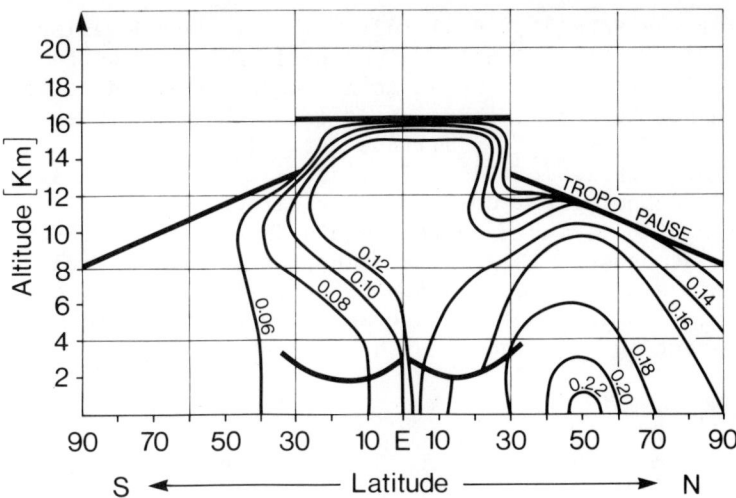

Figure 3.5 General CO-distribution in marine air in the troposphere and lower stratosphere.

hemisphere than are actually found due to the higher CO-production rate by the oceanic CO-source in the southern hemisphere, the higher CO-destruction rate at the soil surface in the northern hemisphere, both resulting from the given land-sea area distribution between the two hemispheres. However, although only CO-data in unpolluted air masses were used in Figure 3.5, the observed latitudinal CO-distribution shows the highest CO-mixing ratios in the northern hemisphere at middle latitudes in surface air. Here, the mixing ratios decrease with altitude from a mean value of 0.22 ppm at the surface to about 0.15 ppm just below the tropopause and then drop rapidly to a constant CO-value of 0.05 ppm in the lower stratosphere. In contrast to the northern hemisphere, the CO-mixing ratios increase with altitude in the southern hemisphere from the equator to about 20°S, due to the interhemispheric exchange in the upper troposphere (10 km height) and a corresponding flux of CO-rich air from the northern hemisphere into the southern hemisphere where the CO is mixed into the lower troposphere. South of 40°S the CO-mixing ratio is uniformly distributed in horizontal and vertical directions.

Most important is the interhemispheric difference of the CO-distribution, with average CO-values of 0.15 ppm in the northern and 0.06 ppm in the southern hemisphere. Calculations of the latitudinal stationary CO-distribution using a one-dimensional interhemispheric exchange model

(Seiler and Schmidt, 1974) indicate that the CO-maximum at middle latitudes and the interhemispheric CO-difference can be explained only by an anthropogenic CO-production in the northern hemisphere of at least the same order of magnitude as that of all the natural CO-sources taken together. These considerations as well as the estimation of the technological CO-sources by different methods as discussed above indicate that man's impact on the CO-cycle in the atmosphere is of great importance. The overall implications of this conclusion for the global environment as far as CO is concerned are as yet unknown and a further assessment is required.

ACKNOWLEDGMENTS

We are very much obliged to Prof. Junge and Prof. Warneck for discussions. We are indebted to Mr. Geins for temporarily operating the instruments and to the DFVLR, Oberpfaffenhofen, in performing the flights.

The work was performed as part of the program of the "Sonderforschungsbereich 73, Atmospheric Trace Components," which receives partial funds through the Deutsche Forschungsgemeinschaft.

REFERENCES

Bates, D. R., and A. E. Witherspoon, 1952: The photochemistry of some constituents of the earth's atmosphere (CO_2, CO, CH_4, N_2O). *Mon. Notic. Roy. Astron. Soc.* **112**, 101-124.

Baum, F., 1972: CO-Emissionen aus Hausbrand-Feuerstätten. *Staub* **32**, 54-59.

Bidwell, R. G. S., and D. E. Fraser, 1972: Carbon monoxide uptake and metabolism by leaves. *Can. J. Bot.* **50**, 1435-1439.

Inman, R. E., and R. B. Ingersoll, 1971: Soil: a natural sink for carbon monoxide. *Science* **172**, 1229-1231.

Jaffe, L. S., 1972: Carbon monoxide in the biosphere: sources, distributions and concentrations. Presented at the Meeting on Sources, Sinks and Concentrations of Carbon Monoxide and Methane in the Earth's Environment, Petersburg, U. S. A.

Levy, H., 1972: Photochemistry of the lower troposphere. *Planet. Space Sci.* **20**, 919-939.

Liebl, K. H., 1971: The soil as a sink and source of atmospheric CO. Master's degree thesis, University and Max-Planck-Institute for Chemistry, Mainz.

Migeotte, M. V., and L. Neven, 1952: Recent progress in the observation of solar infrared spectrum at Jungfraujoch. *Mem. Soc. Roy. Sci., Liege* 12, 165-178.

NATO/CCMS-Report No. 10, 1972: Air quality criteria for carbon monoxide. B-1110, Bruxelles, Belgium.

Robinson, E., and R. C. Robbins, 1969: Sources, abundance, and fate of gaseous atmospheric pollutants supplement. SRI-Project-PR 6755.

Seiler, W., 1974: The cycle of atmospheric CO. *Tellus* 26, 116-135.

Seiler, W., and C. Junge, 1970: Carbon monoxide in the atmosphere. *J. Geophys. Res.* 77, 3204-3214.

Seiler, W., and U. Schmidt, 1974: New aspects on CO and H_2- cycles in the atmosphere. *International Conference on Structure, Composition and General Circulation of the Upper and Lower Atmospheres and Possible Anthropogenic Perturbations*, Volume 1, pp. 192-222.

Siegel, S. M., G. Renwick and L. A. Rosen, 1962: Formation of carbon monoxide during seed germination and seedling growth. *Science* 137, 683-684.

Warneck, P., 1975: OH production rates in the troposphere. *Planet. Space Sci.*, in press.

Weinstock, B., and H. Nicki, 1972: Carbon monoxide balance in nature. *Science* 176, 290-292.

Wilks, S. S., 1959: Carbon monoxide in green plants. *Science* 129, 964-966.

Wofsy, S., J. C. McConnell and M. B. McElroy, 1972: Atmospheric CH_4, CO and CO_2. *J. Geophys. Res.* 77, 4477-4495.

CHAPTER 4

THE STEADY-STATE CONCENTRATION OF CARBON
MONOXIDE IN THE TROPOSPHERE

BERNARD WEINSTOCK
TAI YUP CHANG

Scientific Research Staff, Ford Motor
Company, Dearborn, Michigan 48121

INTRODUCTION

It has been concluded that the residence time of CO in the troposphere is controlled in a major way by hydroxyl radicals (OH). The principal source of CO is predicted to be the oxidation of methane (CH_4) by OH and the principal sink to be the oxidation of CO by OH to form carbon dioxide. This possibility was first brought forth from ^{14}CO considerations (Weinstock, 1969), which predicted a short CO residence time. It was further supported by steady-state analysis for stable CO and ^{14}CO (Weinstock and Niki, 1972). Independently, photochemical models of the troposphere (Levy, 1971; McConnell et al., 1971) predicted OH concentrations of sufficient magnitude to maintain the CO balance and first suggested ahat the oxidation of methane provided the major source of CO (McConnell et al., 1971). An experimental verification of this picture was provided from stable isotope measurements of atmospheric CO (Stevens et al., 1972). The correctness of the OH prediction has now been directly confirmed by measurements of the ambient OH concentration (Wang et al., 1975). The measurements, although limited and preliminary, nevertheless agree with the predictions of the earlier analyses.
    Extensive measurements of ambient CO concentrations in remote areas show a marked difference between the northern

and southern hemispheres (Seiler, 1974; Wilkniss *et al.*, 1973). The concentrations of CO observed in the northern hemisphere are similar to those used in the earlier steady-state analysis (Weinstock and Niki, 1972), but are decidedly lower in the southern hemisphere. This was taken into account by Weinstock and Chang(1974) in their steady-state analysis of the CO concentrations in the northern hemisphere. Since the ^{14}CO data were for the northern hemisphere, the previous conclusions (Weinstock and Niki, 1972) remained essentially unchanged. The CO stable isotope data of Stevens *et al.* 1972) also were for the northern hemisphere as well as the OH measurements of Wang *et al.* (1975). The previous conclusions about the role of OH thus appeared to be confirmed for the northern hemisphere.

A problem that remains is to explain the lower concentrations of CO that are observed in the southern hemisphere. From the photochemical models of Levy (1972) and of McConnell *et al.* (1971), no distinction between the two hemispheres is predicted for the steady-state CO concentration. A revision of these photochemical models will be proposed here to explain the lower concentration of CO observed in the southern hemisphere.

STEADY-STATE MODEL

Before discussing the photochemical model, the steady-state model will be briefly reviewed and the calculation updated on the basis of more recent estimates. The steady-state equations for stable CO and radioactive CO are:

$$d(CO)/dt = P_1 + P_2 - k(CO) \qquad (1)$$

$$d(^{14}CO)/dt = NP_1 + P_3 - k(^{14}CO) \qquad (2)$$

In these equations, $d(CO)/dt$ and $d(^{14}CO)/dt$ are the rate of change with time of the total number of moles of stable CO and of radioactive ^{14}CO in the troposphere, respectively. The term P_1 is the sum of the production rates of CO from all processes that produce CO from "living" carbon, *i.e.*, carbon in equilibrium with ^{14}C in nature. The term P_2 is the sum of the production rates of CO from all processes that produce CO from fossil carbon, *i.e.*, carbon containing a negligible amount of ^{14}CO. The term P_3 is the production rate of ^{14}CO in the troposphere by cosmic-ray neutrons. The term NP_1 is the production rate of ^{14}CO from "living" carbon, N being the mole fraction of ^{14}C in "living" carbon. The terms (CO) and (^{14}CO) are the total amount of stable

CO and of radioactive ^{14}CO in the troposphere, respectively. The term k is a rate constant for the removal of CO, which is assumed to be the same for stable CO and for ^{14}CO in this approximation. The difference in k for CO and ^{14}CO was derived by Weinstock and Chang (1974) and found to be less than the uncertainty of the calculations in this model, and therefore the assumption that k is the same for CO and ^{14}CO will be used. Because of the difference in the value of (CO) for the northern hemisphere and southern hemisphere, the calculations for the two hemispheres must be treated separately.

Northern Hemisphere

The previous calculations for the northern hemisphere (Weinstock and Chang, 1974) will be updated to include new estimates for the sources and sinks of CO (Table 4.1). The values used to solve equations (1) and (2) for P_1 and k are listed below (the previous values used are given in parentheses).

$P_2 = 1.93 \times 10^{13}$ (1.4×10^{13}) mole·yr^{-1}

$P_3 = 145$ mole·yr^{-1}

$\underline{N} = 1.17 \times 10^{-12}$

$(CO) = 1.05 \times 10^{13}$ (1.0×10^{13}) mole

$(^{14}CO) = 22.5$ mole

The value for (CO) corresponds to a mixing ratio of 0.15 ppm (Seiler, 1974) compared with 0.14 ppm used previously. The solutions are:

$P_1 = 1.07 \times 10^{14}$ (1.0×10^{14}) mole·yr^{-1},

$k = 12.0$ yr^{-1}.

The product, k(CO), gives the total CO removal rate or sink, which is 1.26×10^{14} mole·yr^{-1}. If 0.20×10^{14} mole·yr^{-1}, the estimated total non-OH sink (Table 4.1), is subtracted from this, the difference, 1.06×10^{14} mole·yr^{-1}, is the sink arising from the OH-CO reaction. Using 1.35×10^{-13} cm^3 · molecule^{-1}·sec^{-1} (Stuhl and Niki, 1972) for the OH-CO bimolecular rate constant, the average concentration of OH deduced is 2.36×10^6 molecule·cm^{-3} or 4.7×10^6 molecule·cm^{-3} during daylight. This agrees with the previous value of 2.8×10^6 Molecule·cm^{-3} (Weinstock and Chang, 1974), for which correction for non-OH sinks was not made. The direct OH measurements of Wang *et al.* (1975) have now confirmed the correctness of this prediction.

It is now of interest to see if the value of $P_1 = 1.07 \times 10^{14}$ mole·yr^{-1} obtained from the solution of the steady-state equations can be explained. If we subtract the non-OH sources of living CO, 0.22×10^{14} mole·yr^{-1} (Table 4.1), from this value, the difference is 0.85×10^{14} mole·yr^{-1}, which is to be accounted for by the OH + CH$_4$ reaction. If all of the CH$_4$ is converted to CO by that reaction, the product of $k = 9.2 \times 10^{-15}$ cm^3·molecule^{-1}·sec^{-1} (Greiner, 1970), OH = 2.36×10^6 molecule·cm^{-3}, and (CH$_4$) = 1.4 ppm (Ehhalt, 1974) gives 0.67×10^{14} mole·yr^{-1} as the CO production rate from CH$_4$. This is 20% less than the value deduced from P_1.

In summary, it is concluded that the balance of CO in the northern hemisphere is well understood. The role of OH as the major removal mechanism for CO in the northern hemisphere must be accepted because of the direct observation of OH of sufficient concentration to provide the OH sink. The value of $P_1 = 1.07 \times 10^{14}$ mole·yr^{-1}, deduced from the steady-state equations, is to be compared with a production rate of 0.67 mole·yr^{-1} from OH-CH$_4$ reaction and of 0.22×10^{14} mole·yr^{-1} from other CO sources of living carbon, or a total of 0.89×10^{14}. The discrepancy of about 20% is will within the 50% uncertainty assigned to the steady-state model (Weinstock and Niki, 1972).

Southern Hemisphere

A solution of equations (1) and (2) cannot be made at this time for the southern hemisphere because of a lack of (^{14}CO) measurements there. An estimate of P_1 for the southern hemisphere can be made from equation (1), if it is assumed that the concentration of OH is the same in both hemispheres. This is a reasonable assumption because, on the basis of the photochemical model to be discussed, the OH concentration is found to be relatively insensitive to a variation of the pertinent global atmospheric concentrations. For this calculation, the following values are used:

$P_2 = 0.36 \times 10^{13}$ mole·yr^{-1} (Seiler, 1974)

(CO) = 0.06 ppm = 0.42×10^{13} mole (Seiler, 1974)

Table 4.1
Sources and Sinks of CO (Exclusive of OH Source and Sink)

Sources	Northern Hemisphere (10^{14} g·yr^{-1})	(10^{13} mole·yr^{-1})	Southern Hemisphere (10^{14} g·yr^{-1})	(10^{13} mole·yr^{-1})
Fossil Fuel[a]	5.4	1.93	1.0	0.36
Ocean[a]	0.4	0.14	0.6	0.21
Fires and Agricultural Waste[a]	0.4	0.14	0.2	0.07
Oxidation of HC[a]	0.4	0.14	0.2	0.07
Degradation of Chlorophyll[b]	5.0	1.79	1.7	0.61
Flux: N.H. to S.H.[c]			1.6	0.57
TOTAL	11.6	4.14	5.3	1.89
Sinks				
Soil[a]	3.0	1.07	1.5	0.54
Flux into Stratosphere[a]	0.9	0.32	0.2	0.07
Flux: N.H. to S.H.[c]	1.6	0.57		
TOTAL	5.5	1.96	1.7	0.61

[a]Seiler (1974).
[b]The value in the northern hemisphere is taken from Stevens et al. (1972). The value in the southern hemisphere is assumed to be one-third of the value in the northern hemisphere.
[c]Newell et al. (1974).

$$(OH) = 2.36 \times 10^6 \text{ molecule} \cdot cm^{-3}$$

The removal rate by OH + CO is then 4.24×10^{13} mole·yr^{-1} to which must be added 0.61×10^{13} mole·yr^{-1}, the non-OH sink (Table 4.1) to give 4.85×10^{13} mole·yr^{-1}, the total CO sink, k(CO). P_1 is then k(CO) - P_2 = 4.49×10^{13} mole·yr^{-1}. The non-CH$_4$ production of "living" CO is 1.53×10^{13} mole·yr^{-1}, so that 2.96×10^{13} mole·yr^{-1} must be accounted for by the OH + CH$_4$ reaction.

The CH$_4$ concentration in the southern hemisphere is about the same as that in the northern hemisphere (Wilkness et al., 1973) and the OH concentration has been also taken to be the same. The calculated production of CO by CH$_4$ + OH will then be the same in the southern hemisphere as that in the northern hemisphere, 6.7×10^{13} mole·yr^{-1}, provided all of the CH$_4$ is converted into CO as had been assumed for the northern hemisphere. This value is more than a factor of 2 greater than that predicted above, and the major purpose of this paper is to propose an explanation of this difference.

One possible explanation is that the assumption of the OH concentration being the same in the two hemispheres is a poor one. The calculation was repeated with the OH concentration in the southern hemipshere taken equal to one half that in the northern hemisphere. Poorer agreement was found. The OH + CH$_4$ rate calculated from P_1 was 0.84×10^{13} mole·yr^{-1} compared with 3.35×10^{13} mole·yr^{-1} from the rate equation. There was a small improvement if the OH concentration in the southern hemisphere was taken to be twice that in the northern hemisphere: 7.2×10^{13} mole·yr^{-1} compared with 13.4×10^{13} mole·yr^{-1}. In order to get more reasonable agreement in this model, the OH concentration in the southern hemisphere would have to be very much larger. This possibility appears to be unreasonable and a different explanation is proposed.

The above discrepancy would be resolved, if not all of the CH$_4$ is converted to CO in the southern hemisphere, contrary to what has been assumed for the northern hemisphere. For the case when the OH concentration is the same in the two hemispheres, 44% of the CH$_4$ would be converted to CO to achieve agreement. In the next section, a photochemical model will be proposed to demonstrate this possibility.

There is an interesting prediction that also results from this analysis. Using the value of $P_1 = 4.49 \times 10^{13}$ mole·yr^{-1} for the southern hemisphere, the mole fraction of ^{14}CO is derived to be 4.2×10^{-12}. This is about two times greater than for the northern hemisphere and about four

times greater than "living" carbon. A determination of
the specific activity of CO in the southern hemisphere
and a redetermination in the northern hemisphere would be
of value to improve the quantitative understanding of this
problem.

TROPOSPHERIC PHOTOCHEMICAL MODEL

From the steady-state analyses discussed in the previous sections, it was concluded that there is a significant difference between the northern and southern hemisphere with respect to the oxidation of methane to form CO. For the northern hemisphere, the analysis was consistent with virtually complete conversion of CH_4 to CO. For the southern hemisphere, the analysis required that only 44% of the CH_4 results in CO production. It is proposed that the origin of this different behavior is the difference in concentration of the trace reactants that participate in the photochemistry that drives the CH_4 oxidation. In particular, we have focused on the probable difference in concentration of the nitrogen oxides between the two hemispheres as the cause of this different behavior.

There are very few measurements of NO and NO_2 (NO_x) in remote areas. However, the nitrate deposition rate is found to be much lower in the southern hemisphere than in the northern hemisphere, indicating a corresponding difference in NO_x concentrations (Eriksson, 1952; Robinson and Robbins, 1970). It is not possible to make a reliable estimate of the difference in NO_x concentrations for the two hemispheres from Eriksson's data at this time. Robinson and Robbins' analysis would suggest about a factor of 3 difference, but it could be much greater than that.

The photochemical model of Levy (1972) was essentially used in this analysis and the equations are listed in the Appendix. In order to provide a mechanism for the partial conversion of CH_4 to CO, a rainout step for methyl hydroperoxide, CH_3OOH, was added [equation (36)]. In addition, equations (34) and (35), which are rainout steps for nitric acid and hydrogen peroxide, respectively, were added.

In this model, the rate at which methane carbon is removed from the atmosphere without first producing CO is given by $k_{36}(CH_3OOH)$, *i.e.*, the rate of rainout of methyl hydroperoxide. The rate at which methane carbon is converted to CO is given by:

$$[k_{27}(OH) + k_6 + k_7](HCHO)$$

The ratio, R, of these two rates is the ratio of the amount
of CH_4 that does not produce CO to the amount that does:

$$R = \frac{k_{36}(CH_3OOH)}{[k_{27}(OH)+k_6+k_7] \, (HCHO)}.$$

In the steady-state approximation, this is equivalent to:

$$R = \frac{k_{36}k_{29}(HO_2)}{k_{23}(NO) \, [k_{31}(OH)+k_{36}+k_9]+k_9k_{29}(HO_2)}$$

Qualitatively, this is the desired result. The equation
predicts that the lower concentration of NO expected in the
southern hemisphere will result in a smaller fraction of
CH_4 converted to CO as a product.

Preliminary numerical steady-state solutions for the
proposed mechanism show that the term, $k_9k_{29}(HO_2)$, is much
smaller than the left-hand term in the demominator of R.
Also, (OH) does not change appreciably for a large variation
in (NO). The ratio, R, is not inversely proportional to
(NO), however, because (HO_2) increases or decreases in the
same sense as (NO). This change in (HO_2) is relatively insensitive to the change in (NO), so that the qualitative
conclusion is still the same, *i.e.*, a decrease in (NO) will
result in an increase in R.

It is not possible to make a reliable quantitative conclusion about the dependence of R on (NO) from any photochemical model at this time because of important uncertainties in many critical rate constants. In addition, the (NO)
is not known for either hemisphere. A numerical evaluation
has been made nevertheless and is presented here as an illustration of the qualitative behavior. For the most part, the
same rate constants used by Levy (1972) were used, except
for a number of reactions involving nitrogen compounds; for
$k_{12}, k_{14}-k_{18}$, k_{23}, and k_{25}, the values of Niki *et al.* (1972)
and of Wu *et al.* (1973) were used. In addition, assignments
were made for k_{34}, k_{35}, and k_{36}, the three rainout steps,
which were not included in Levy's model.

For this illustration, (NO_x) was taken to be 3×10^{-3} ppm
for the northern hemisphere and 6×10^{-4} ppm for the southern
hemisphere. The values of R computed were 0.04 and 0.15,
respectively. Thus for a decrease in (NO_x) of a factor of
5, R increased by a factor of 4. The qualitative behavior
is correct, but quantitative agreement with the steady-state
analysis is poor. The 4% nonconversion of CH_4 to CO for
the northern hemisphere will make the 20% discrepancy for

the CO balance derived from the steady-state equations 4% greater. But this is still well within the 50% uncertainty of the analysis. For the southern hemisphere, R = 0.15 corresponds to 13% nonconversion of CH_4 to CO, compared to 56% predicted from the steady-state analysis. This discrepancy is clearly more serious. It would be improved if the (NO_x) were much lower in the southern hemisphere than that used in the illustration.

In order to make a critical test of this model that has been offered to explain the difference in (CO) between the two hemispheres, more extensive data are needed. Measurements of a number of atmospheric species in both hemispheres are needed. Of most importance would be (NO_x) and (HO_2). The former could be done with some improvement in current technology. The latter would require a new instrumental development, which probably could be achieved in a few years. In addition, measurements of (OH), (HCHO), and (CH_3OOH) would be of great interest as well as measurements of a number of critical rate constants and rainout coefficients. Furthermore, as mentioned earlier, measurements of the specific activity of CO in the southern hemisphere and remeasurement of it in the northern hemisphere would also be important.

APPENDIX

Photochemical Model

(1) $NO_2 + h\nu \xrightarrow{O_2} NO + O_3$

(2) $HNO_3 + h\nu \rightarrow OH + NO_2$

(3) $HNO_2 + h\nu \rightarrow HO + NO$

(4) $NO_3 + h\nu \xrightarrow{O_2} 2NO_2 + O_3$

(5) $O_3 + h\nu \rightarrow O(^1D) + O_2$

(6) $HCHO + h\nu \xrightarrow{O_2} 2CO + 2HO_2$

(7) $HCHO + h\nu \rightarrow CO + H_2$

(8) $HOOH + h\nu \rightarrow OH + OH$

(9) $CH_3OOH + h\nu \rightarrow CH_3O + OH$

(10) $NO + O_3 \rightarrow NO_2 + O_2$

(11) $NO_2 + O_3 \rightarrow NO_3 + O_2$

(12) $NO_3 + NO \rightarrow NO_2 + NO_2$

(13) $NO_3 + NO_2 \rightarrow N_2O_5$

(14) $N_2O_5 \rightarrow NO_2 + NO_3$

(15) $N_2O_5 + H_2O \rightarrow HNO_3 + HNO_3$

(16) $OH + NO_2 \rightarrow HNO_3$

(17) $OH + NO \rightarrow HNO_2$

(18) $NO + NO_2 + H_2O \rightarrow HNO_2 + HNO_2$

(19) $O(^1D) + M \xrightarrow{O_2} O_3 + M$

(20) $O(^1D) + H_2O \rightarrow OH + OH$

(21) $OH + CO \xrightarrow{O_2} HO_2 + CO_2$

(22) $OH + CH_4 \xrightarrow{O_2} CH_3O_2 + H_2O$

(23) $CH_3O_2 + NO \rightarrow CH_3O + NO_2$

(24) $CH_3O + O_2 \rightarrow HCHO + HO_2$

(25) $HO_2 + NO \rightarrow OH + NO_2$

(26) $OH + HOOH \rightarrow HO_2 + H_2O$

(27) $OH + HCHO \xrightarrow{O_2} HO_2 + CO + H_2O$

(28) $HO_2 + HO_2 \rightarrow HOOH + O_2$

(29) $HO_2 + CH_3O_2 \rightarrow CH_3OOH + O_2$

(30) $OH + HO_2 \rightarrow H_2O + O_2$

(31) $OH + CH_3OOH \rightarrow CH_3O_2 + H_2O$

(32) $OH + H_2 \rightarrow HO_2 + H_2O$

(33) $CH_3O_2 + CH_3O_2 \rightarrow CH_3O + CH_3O + O_2$

(34) $HNO_3 \rightarrow$ rainout

(35) $HOOH \rightarrow$ rainout

(36) $CH_3OOH \rightarrow$ rainout

REFERENCES

Ehhalt, D. H., 1974: The atmospheric cycle of methane. *Tellus* 26, 58-70.

Eriksson, E., 1952: Composition of atmospheric precipitation: I nitrogen compound. *Tellus* 4, 215-232.

Greiner, N. R., 1970: Hydroxyl radical kinetics by kinetic spectroscopy. VI. Reactions with alkanes in the range 300-500°K. *J. Chem. Phys.* 53, 1070-1076.

Levy II, H., 1971: Normal atmosphere: large radical and formaldehyde concentrations predicted. *Science* 173, 141-143

Levy II, H., 1972: Photochemistry of the lower troposphere. *Planet Space Sci.* 20, 919-935.

McConnell, J. C., M. B. McElroy and S. C. Wofsy, 1971: Natural sources of atmospheric CO. *Nature* 233, 187-188.

Newell, R. E., G. J. Boer, and J. W. Kidson, 1974: An estimate of the interhemispheric transfer of carbon monoxide from tropical general circulation data. *Tellus* 26, 103-107.

Niki, H., E. E. Daby and B. Weinstock, 1972: Mechanisms of smog reactions. *Adv. Chem. Ser.* 113, 16-57.

Robinson, E. and R. C. Robbins, 1970: Gaseous nitrogen compound pollutants from urban and natural sources. *J. Air Poll. Control Assoc.* 20, 303-306.

Seiler, W., 1974: The cycle of atmospheric CO. *Tellus* 26, 116-135.

Stevens, C. M., L. Krout, D. Walling, A. Venters, A. Engelkemeir and L. E. Ross, 1972: The isotopic composition of atmospheric carbon monoxide. *Earth and Planet, Sci. Letters* 16, 147-165.

Stuhl, F. and H. Niki, 1972: Pulsed vacuum-uv photochemical study of reactions of OH with H_2, D_2 and CO using a resonance-flourescent detection method. *J. Chem. Phys.* 57, 3671-3677.

Wang, C. C., L. I. Davis Jr., C. H. Wu, S. M. Japar, H. Niki and B. Weinstock, 1975: Hydroxyl radical concentrations measured in ambient air. *Science* 189, 797-800.

Weinstock, B., 1969: The residence time of carbon monoxide in the atmosphere. *Science* 166, 224-225.

Weinstock, B. and H. Niki, 1972: Carbon monoxide balance in nature. *Science* 176, 290-292.

Weinstock, B. and T. Y. Chang, 1974: The global balance of carbon monoxide. *Tellus* 26, 108-115.

Wilkniss, P. E., R. A. Lamontagne, R. E. Larson, J. W. Swinnerton, C. R. Dickson and T. Thompson, 1973: Atmospheric trace gases in the southern hemisphere. *Nature* 245, 45-47.

Wu, C. H., E. D. Morris, and H. Niki, 1973: The reaction of nitrogen dioxide with ozone. *J. Phys. Chem.* 77, 2507-2511.

CHAPTER 5

ON THE MECHANISMS OF CO_2 AND CH_4 PRODUCTION
IN NATURAL ANAEROBIC ENVIRONMENTS

LARRY M. GAMES
J. M. HAYES

Departments of Chemistry and Geology, and
Water Resources Research Center, Indiana
University, Bloomington, Indiana 47401

Methane and CO_2 produced in the anaerobic decay of organic matter have been investigated in a large variety of environments, including landfills (Games and Hayes, 1974), sewage sludges (Nissenbaum et al., 1972; Jeris and McCarty, 1965), mixed bacterial cultures (Rosenfeld and Silverman, 1959), lake and marine sediments (Koyama et al., 1973; Nissenbaum et al., 1972), paddy soils (Takai, 1970), and glacial drifts (Wasserburg et al., 1963). The microorganisms responsible for the conversion of organic matter to CH_4 are collectively termed methane-producing bacteria (Doelle, 1969), and are generally considered to be the final member of the food chain existing in an anaerobic ecology (Pine, 1971). This food chain is exceedingly complex, although its general outline can be shown in relatively simple terms, as in Figure 5.1.

The exact mode of production of the methane has been the subject of considerable interest, partly as a result of the need to understand the anaerobic process of organic waste decomposition (Wolfe, 1971). In addition, an understanding of the decomposition of organic matter in natural environments hinges partly on the full resolution of the process of methane production.

Because the carbon isotopic contents of the CO_2 and CH_4

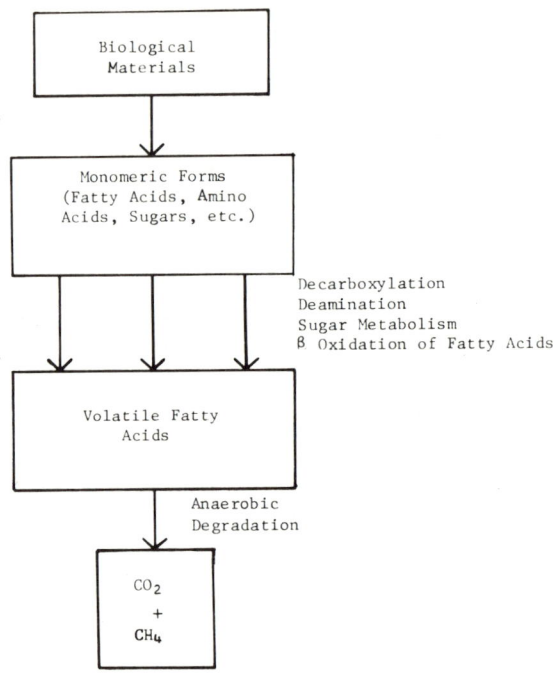

Figure 5.1 Generalized degradation of organic material in an anaerobic environment.

can be partly controlled by the biochemical mechanism of gas formation, the isotope ratios may provide information useful in understanding anaerobic degradation. Table 5.1 summarizes all of the previously known fractionations for biologically-produced CH_4 and CO_2. The isotope ratios are expressed in terms of $\delta^{13}C_{PDB}$ (Craig, 1953), where $\delta^{13}C_{PDB}(s) = [(R_s/R_{PDB}) - 1]10^3$, $R = {}^{13}C/{}^{12}C$, and s and PDB denote the sample and the PDB standard, respectively. The δ notation is thus a *differential* measurement of the ^{13}C abundance. Samples with isotope ratios greater than that of the standard will have $\delta > 0$, and are described as "heavy" relative to the standard. The units for δ are parts per thousand, noted as ‰, and termed "per mil." The isotopic fractionation between two materials, A and B, is described in terms of the fractionation factor, $\alpha(A/B)$, defined as follows:

$$\alpha(A/B) = \frac{R_A}{R_B} = \left(\frac{\delta^{13}C_{PDB}(A) - \delta^{13}C_{PDB}(B)}{\delta^{13}C_{PDB}(B) + 1000}\right) + 1$$

Table 5.1

Previously Reported Isotopic Fractionation Between CO_2, CH_4, and Organic Matter

	$\delta^{13}C_{PDB}$(‰)			$T(°C)$	$\alpha(CO_2/CH_4)$	$\dfrac{R_{CO_2}}{R_{org.}}$	$\dfrac{R_{org.}}{R_{CH_4}}$
	CO_2	CH_4	Organic Matter				
Lake Mud Oana and Deevey, 1960	−5	−77	−30	7	1.078	1.026	1.051
Marine Sediment Interstitial Water Nissenbaum et al., 1972	+16.7	−55	−20	10	1.076	1.037	1.037
Glacial Drift Gas Wasserburg et al., 1963	−7.9	−75	---	---	1.072	---	---
Sewage Sludge Nissenbaum et al., 1972	+4.1	−47.1	−22.6	35[a]	1.054	1.027	1.026
Landfill Gases Games and Hayes, 1974	+20	−50	−25	17	1.074	1.046	1.026
Lake Kivu Deuser et al., 1973	−2	−45	---	26[b]	1.045	---	---

[a] Normal sewage sludge tank, actual temperature not reported.

[b] Highest temperature measured in deep part of lake.

The calculation of a fractionation factor is sometimes regarded as appropriate only when an isotopic equilibrium has been established between A and B. Here we imply only that A and B are related in some way, and do not assume any degree of equilibration.

There appear to be three separate types of fractionations (see Table 5.1). Two groups can be recognized on the basis of $\alpha(CO_2/CH_4)$, and the group with $\alpha \sim 1.07$ can be further subdivided on the basis of the relation between the gases and the organic matter. The CO_2 in the landfill and in marine sediments is isotopically enriched relative to the organic matter to a greater extent than the CO_2 in the lake mud sediment, while CH_4 is depleted in ^{13}C in these two to a lesser degree. Figure 5.2 shows these relationships graphically and emphasizes the differences that have been observed.

Many hypotheses have been advanced to explain the variations in isotopic content of CO_2 and CH_4 in the differing environments. After obtaining the results for the landfill noted in Table 5.1, four experiments that would test the various theories to some extent were designed. These experiments were designed to 1) determine if the CO_2 and CH_4 isotopic compositions were similar in identical environments located in different places, 2) measure the change in isotopic composition of these two gases as a function of

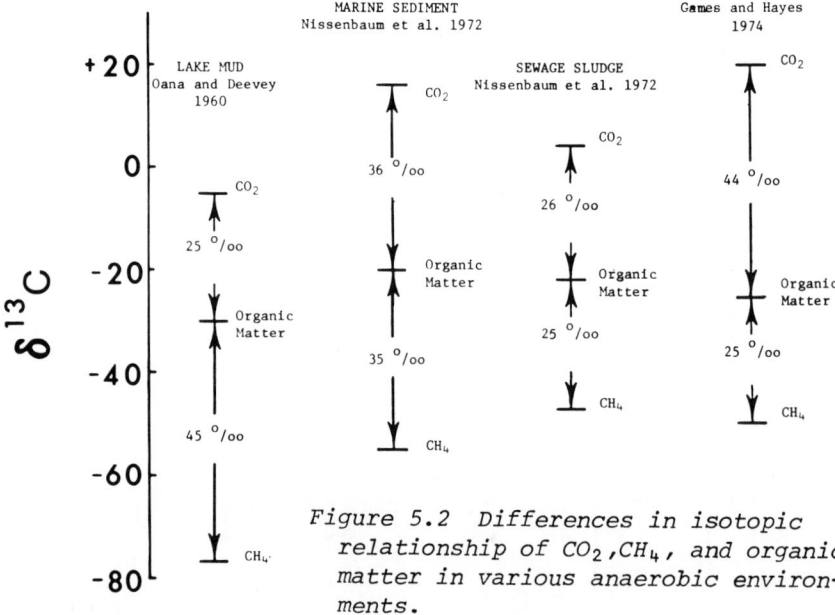

Figure 5.2 Differences in isotopic relationship of CO_2, CH_4, and organic matter in various anaerobic environments.

the extent of nutrient utilization, and 3) determine if the isotopic fractionation caused by various methane-producing bacteria was markedly different.

All of the work that will be described involves the analysis of the amount and isotopic content of biologically produced CO_2 and CH_4 in various anaerobic environments, and the results will be used to consider each question listed above. The four investigations were:

1) *Landfill leachate gases.* Because of the unusual isotopic composition of the CO_2 and the CH_4 in one landfill (Games and Hayes, 1974), the leachate from two other landfills was analyzed to determine whether the initial observation might have been unique. These landfills differed somewhat in the type of refuse material deposited in them, and also varied in the amount of material and length of time since they had been actively used.

2) *Sewage sludge gases.* The gases being released from a sewage sludge tank have been previously examined by Nissenbaum et al. (1972). To determine if the fractionation factor observed in that case was a common one, a second sewage sludge reactor was investigated in this work. The 400-liter tank utilized was of experimental design, located in the laboratory of J. T. Pfeffer in the Sanitary Engineering Department at the University of Illinois (Pfeffer and Liebman, 1974). It was operated at 60°C, much above the normal temperature of operation of sludge tanks. Although Nissenbaum et al. (1972) did not report the temperature of the sludge system they investigated, a temperature of 35°C is normal (Smith and Mah, 1966). The Pfeffer reactor was cycled daily, producing gas at the rate of 30 l. per hour from a sludge throughput of $\sim$20 kg/day.

3) *Isotopic fractionation by pure cultures of methane-producing bacteria.* The isotopic fractionation brought about by pure cultures of methane-producing bacteria has yet to be reported in the literature. Most workers who have analyzed the CO_2 and CH_4 in natural environments or mixed cultures have considered the isotope ratios of these two gases entirely in terms of the methane-producing bacteria This approach ignores other metabolic processes and individual bacteria that contribute to CO_2 production and utilization in an anaerobic environment. The possibility remains that the observed isotopic relationships are to a large extent controlled by these other processes. The measurement of the range of fractionation caused by various methane-producing bacteria in pure culture should allow some assessment of this possibility.

4) *Laboratory refuse reactor study.* An apparatus was constructed in the laboratory which duplicated landfill

conditions, but permitted the control of temperature, degree of saturation of the wastes, the amount of material entering from outside sources, and the degree of decomposition of the waste material. The reactor was constructed specifically to monitor the relationship of the isotope composition of the two gases as a function of the extent of nutrient utilization. An initial charge of relatively undecomposed garbage was allowed to decompose in an anaerobic environment in which no nutrients were continuously added.

EXPERIMENTAL

Gas Analysis

Methods of analysis of the CO_2 and CH_4 will be described in more detail elsewhere (Games and Hayes, 1976). Methane was removed from water by stripping with prepurified oxygen and then combusted to CO_2 by passage over platinum at 1000°C. Carbon dioxide was removed from water by acidification followed by cryogenic pumping, and all CO_2 values reported are the sum of carbonate, bicarbonate, and free CO_2.

The amounts of gas were measured as CO_2 utilizing standard manometric procedures (Sanderson, 1948). Isotopic analyses were performed using a mass spectrometric system like that described by McKinney et al. (1950). The 95% confidence limits for $\delta^{13}C$ values reported are ±0.2‰.

Laboratory Refuse Reactor

Approximately 6 liters of unconsolidated waste material was collected from the upper, dry layers of a soil-covered landfill. This dry material was presumably not yet fully degraded. Care was taken to minimize the exposure of the refuse to the air, and it was returned to the laboratory in a closed light-sealed flask. The material was placed into a 6-liter flask in a nitrogen atmosphere and flooded with 2.5 liters of water (CO_2 and O_2 removed) to which 250 mg of $NH_4H_2PO_4$ had been added. The flask was then sealed and maintained in darkness.

An approximate 300-ml gas volume was maintained above the flooded refuse. As the pressure increased the gases escaped through a mercury bubbler which maintained the pressure at 80 torr above ambient pressure. Gas sampling was made through a port located between the bubbler and the refuse. Since gases were constantly escaping, a gas sample at any time reflected the gases being produced during a short period (less than a day) up to and including the time of analysis.

Bacterial Growth

Pure cultures of *Methanosarcina barkeri*, *Methanobacterium* strain M.o.H., and *Methanobacterium thermoautotrophicum* were grown in the laboratory of Prof. R. S. Wolfe, University of Illinois, with the aid of R. Gunsalus. Details of the media and culture conditions have been published elsewhere (Zeikus and Wolfe, 1972; Ferry et al., 1974; Bryant et al., 1968). The first two species listed were grown at 40°C and the last was grown at 65°C.

All three bacteria were cultured on a $CO_2:H_2$ (20:80 v/v) substrate and carried out the transformation:

$$CO_2 + 4 H_2 \longrightarrow CH_4 + 2 H_2O$$

Measurements of the amount and isotope ratio of the methane produced were made during the log phase of growth for the *M. barkeri* and *M.* strain M.o.H., and at various stages of growth for the *M. thermoautotrophicum*. Substrate CO_2 and H_2 were provided continuously to the *M. barkeri* and *M.* strain M.o.H., but only an initial amount was added to the *M. thermoautotrophicum*.

RESULTS AND DISCUSSION

Pure Cultures

The results for *Methanosarcina barkeri* and *Methanobacterium* strain M.o.H. are shown in Table 5.2 and, because of differences in the methods of growth and time of harvesting, are best considered separately from the *Methanobacterium thermoautotrophicum* data. Duplicate cultures were grown and harvested at the same time for both bacteria. The amounts of methane being produced were reproducible for the duplicate cultures, as were the isotope ratios.

There appears to be a large difference in isotopic composition of the cell mass of the two bacteria, with the cell mass of the *M.* strain M.o.H. being even heavier than the CO_2 utilized in its growth. This value represents the average of the duplicate analyses with a spread of less than 1‰. It seems inconceivable that this effect could occur if the metabolism of the cell involved the conversion of CO_2 to cell mass and methane, but further experiments to substantiate this anomalous effect have not been conducted. The cell mass of the *M. barkeri* is intermediate in isotopic content between the CO_2 and CH_4, as expected.

Table 5.2
Carbon Isotope Ratios of Gases in Pure Cultures
of Two Methane-Producing Bacteria

	$\delta^{13}C_{PDB}$ (‰)	
	Methanosarcina barkeri	Methanobacterium strain M.o.H.
CO_2(feed)[a]	-42.3	-42.0
CO_2(out)	-41.6 , -41.3[b]	-40.0
CH_4(out)	-82.8 , -82.4	-95.7 , -95.5
Total Cell Mass	-55.6 , -52.2	-32.7
$\alpha(CO_2/CH_4)$	1.044	1.059

[a] The terms "feed" and "out" refer to gases entering and leaving the culture vessels.

[b] Where duplicate entries appear, they represent complete duplicate experiments, including the growth of separate cultures.

The results for the M. thermoautotrophicum cultures are more complicated, because the CO_2 substrate is not continually replenished during growth. The isotopic difference between the CO_2 and CH_4 is only a measure of the fractionation factor $\alpha(CO_2/CH_4)$ when an insignificant portion of the CO_2 has been utilized. During depletion of the CO_2, its isotope ratio will continually increase because the lighter isotope ^{12}C is being preferentially incorporated into the CH_4. The isotope ratio of the remaining CO_2 is then a measure of the extent of its depletion. In the same manner, the amount and isotope ratio of methane produced will be a measure of the depletion of the substrate. Table 5.3 shows the results for the CO_2, CH_4, and cell mass isotope ratio measurements for harvesting at different extents of CO_2 depletion. Unlike the results for the other two bacteria, which are measurements of the isotope ratios and concentrations at a single point in time, these analyses are for the pooled CH_4 produced and the total CO_2 remaining. The results are arranged in order of increasing ^{13}C content of the CO_2 present, which coincides with an increased amount of CH_4 present in the culture flask. No measurements of the amounts of CO_2 or CH_4 are available for the 48-hr incubation cultures, and they are placed in the progression solely on the basis of the isotope ratio results for the CO_2, CH_4, and cell mass.

Table 5.3
Isotope Ratios and Concentrations of Methane and CO_2 in Growing Cultures of *Methanobacterium thermoautotrophicum*

	CO_2		CH_4		Total Cell Mass $\delta^{13}C_{PDB}(‰)$
	mg C/l of gas	$\delta^{13}C_{PDB}(‰)$	mg C/l of gas	$\delta^{13}C_{PDB}(‰)$	
CO_2 (feed)	---	-41.4	---	---	---
5½-hr incubation	9.1	-39.8	.56	-64.5	---
13-hr incubation	5.0	-37.0	2.03	-65.1	-49.1
24-hr incubation	6.7	-27.9	4.68	-65.0	-53.1
~48-hr incubation	---	+3.9	---	-55.3	-38.2
~48-hr incubation	---	+9.8	---	-50.5	-34.6
7-day incubation	--- (no CO_2 present)	---	8.7	-43.0	-31.2

60 *Environmental Biogeochemistry*

The calculation of a fractionation factor in this case is more difficult, as it requires the extrapolation of the isotope ratio of CH_4 to zero methane production. Ordinarily this would be done graphically (Raaen et al., 1968) but inspection of Table 5.3 shows that the isotopic content of the CH_4 is not changing significantly in the early stages of methane production. Thus, adopting $\delta^{13}C_{PDB}(CH_4) = -64.9‰$ (the average of the results for the first three measurements in Table 5.3) for the CH_4 and the isotope ratio of the feed CO_2 $[\delta^{13}C_{PDB}(CO_2) = -41.4‰]$ for the CO_2, the calculated fractionation factor is $\alpha(CO_2/CH_4) = 1.025$. This is much smaller than the fractionation factors calculated for the other two bacteria, partially because of the 25°C higher temperature of incubation. The fractionation factor for *M. thermoautotrophicum* at 40°C is probably about 1.031, an estimate based on the temperature dependence of the equilibrium CO_2/CH_4 fractionation (Bottinga, 1969), and on the observed temperature dependence of $\alpha(CO_2/CH_4)$ in sewage sludge reactors (Tables 5.1 and 5.4).

Laboratory Refuse Reactor

Figures 5.3 and 5.4 show the concentration and isotope ratio of CO_2 and CH_4 emanating from the laboratory refuse reactor during the time period it was monitored. The close relationship between the concentration of the two gases can be seen quite clearly in Figure 5.3. The CO_2 concentration falls off slightly, but begins rising again on about day 80. The rate of increase in the isotope ratio of CO_2 in Figure 5.4 becomes faster at the same time that the concentration increase takes place, about day 80. The isotope ratio results in Figure 5.4 vary more smoothly than the gas production rates, but calculation of a fractionation factor is difficult because no steady-state relationship between the isotope content of the two gases was reached until about 100 days had passed. At that point, the fractionation factor was the same as that of the landfill itself $[\alpha(CO_2/CH_4) = 1.074]$. The change of the isotope ratios of the gases with time will be discussed in more detail later.

Comparison with Previous Results

The fractionation factors observed in previous investigations (Table 5.1) fall into two groups. All systems except the sewage sludge and the Lake Kivu gases have fractionation factors in the range 1.070-1.078, with the two excluded values forming a poorly defined "group" at lower

Table 5.4
Summary of Fractionation Effects Observed in this Work

	$\delta^{13}C_{PDB}$ (‰)					$\dfrac{R_{CO_2}}{R_{org.}}$	$\dfrac{R_{org.}}{R_{CH_4}}$
	CO_2	CH_4	Organic Matter	T (°C)	(CO_2/CH_4)		
Methanosarcina barkeri	−42.3	−82.6	----	40	1.044	----	----
Methanobacterium strain M.o.H.	−42.0	−95.6	----	40	1.059	----	----
Methanobacterium thermoautotrophicum	−41.4	−64.9	----	65	1.025	----	----
Landfill A	+16.6	−52.1	−25[a]	20[b]	1.072	1.043	1.028
Landfill B	+16.1	−48.5	−25[a]	20[b]	1.068	1.042	1.025
Sewage Sludge Tank	−5.5	−49.1	−25[a]	60	1.046	1.020	1.026
Laboratory Reactor 30–60 days	−5 to −10	−40 to −47	----	25	1.035–1.048	----	----
Laboratory Reactor 85–120 days	+10 to +17	−56 to −58	----	25	1.070–1.075	----	----

[a] Assumed on basis of other landfills and sewage sludges.
[b] Assumed on basis of previous data.

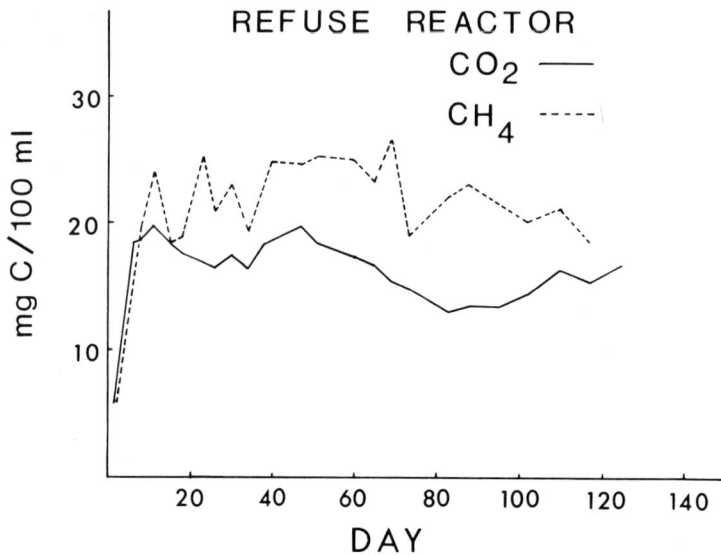

Figure 5.3 Change of concentration of CO_2 and CH_4 as a function of time of incubation of refuse in laboratory reactor.

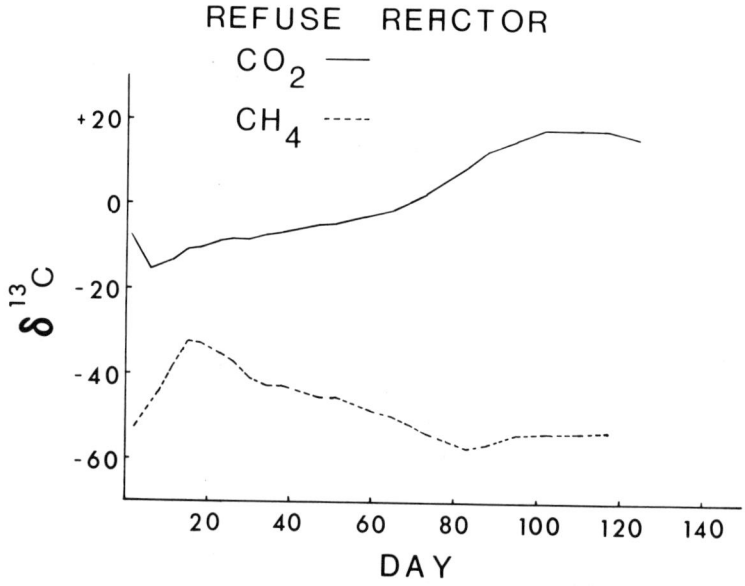

Figure 5.4 Change in isotope ratios of CO_2 and CH_4 as a function of time of incubation of refuse in the laboratory reactor.

values of $\alpha(CO_2/CH_4)$. The new results contribute heavily to the lower group, with both mixed cultures (sewage sludge, 30-60 day period in laboratory reactor) and pure cultures falling outside the 1.070-1.078 range. Where similar environments have been examined both in the previous reports and in the present work, good agreement is obtained. For example, the new landfill observations *do* fall in the high fractionation group, increasing the range only to 1.068-1.078. Similarly, the old and new sewage sludge observations are in good agreement, particularly when the temperature difference is taken into account. This uniformity within any given environment is probably not universal, however. For example, Presley and Kaplan (1968) found interstitial waters in other marine sediments that did not contain dissolved CO_2 which was isotopically enriched similar to the CO_2 reported by Nissenbaum *et al.* (1972), although no CH_4 was found in such cases.

Metabolism of Methane-Producing Bacteria

In order to interpret the data, some knowledge of the metabolism of methane-producing bacteria is required. Studies of the methane-producing bacteria, which are strictly anaerobic, have been hampered by the difficulty encountered in growing them in pure culture (Doelle, 1969; Wolfe, 1971), since exposure to even small amounts of oxygen leads to rapid death. Despite these problems, some work has been done on the metabolic pathways leading to methane production. The direct transformations that have been unequivocally shown by pure culture studies to occur are limited to (Wolfe, 1971):

$$*CH_3 - CO_2H \longrightarrow CO_2 + *CH_4,$$

$$CO_2 + 4 H_2 \longrightarrow 2 H_2O + CH_4$$

$$4 CH_3OH \longrightarrow 3 CH_4 + CO_2 + 2 H_2O,$$

$$4 HCO_2H \longrightarrow CH_4 + 3 CO_2 + 2 H_2O$$

$$2 H_2O + 4 CO \longrightarrow CH_4 + 3 CO_2$$

Unless further work on the metabolism of these organisms uncovers still more metabolic pathways, these constitute the range of biological relationships between CO_2 and CH_4 that lead directly to the production of methane in an anaerobic environment. One further metabolic pathway may be of importance in considering the isotopic relationship between the CH_4 and CO_2, and this is the conversion of CO_2 to acetate, which has been demonstrated in a sludge digester (Chynoweth and Mah, 1971) and in metabolic studies of anaerobic bacteria (Doelle, 1969).

Various studies have been made of the relative importance of the possible metabolic pathways of methane production. Smith and Mah (1966) and Jeris and McCarty (1965) have shown in carbon-14 labeling experiments that the distribution of isotopic species can be explained if the production of methane in mixed culture sewage sludges is distributed roughly as follows:

(1) $\quad *CH_3 - CO_2H \longrightarrow *CH_4 + CO_2 \qquad 70\%$

(2) $\quad *CO_2 + 4 H_2 \longrightarrow *CH_4 + 2 H_2O$

(3) $\quad \begin{cases} CH_3 - *CO_2H \longrightarrow CH_4 + *CO_2 \\ *CO_2 + 4 H_2 \longrightarrow *CH_4 + 2 H_2O \end{cases} \Bigg\} 30\%$

Jeris and McCarty (1965) also showed that the relative contributions of these mechanisms are apparently unaffected by the type of organic compound (fat, protein, or carbohydrate) being degraded. These processes have also been shown to be the most important ones in some lake sediments and paddy soils, and the relative proportions are about the same (Koyama et al., 1973; Takai, 1970).

INTERPRETATION OF THE OBSERVED ISOTOPIC FRACTIONATIONS

As can be seen from these considerations, the CO_2 and CH_4 present in an anaerobic environment are intimately related in a metabolic sense, and from this we can hypothesize that the isotopic content of the two components should be related in some straightforward manner. To whatever extent it can be understood, this relationship should be of considerable use in determining the degree of similarity or difference between various anaerobic environments. To work toward a more realistic understanding of this problem we can consider a series of hypotheses possibly explaining the observed isotopic compositions.

1. The realtively heavy CO_2 observed in the marine sediment and landfill systems represents a residue remaining after the initial preferential use of $^{12}CO_2$. The relatively heavy CH_4 in these systems is produced by microbiological reduction of the heavy CO_2.

This hypothesis, which is due to Nissenbaum et al. (1972), would predict that as the CO_2 was further depleted it should become progressively heavier. In fact, in marine sediments, the ^{13}C abundance in the CO_2 both *increases* and *decreases* as the amount of CO_2 present increases, depending on the particular depth and type of sediment studied (Nissenbaum *et al.*, 1972; Claypool and Kaplan, 1975). This hypothesis also calls for the isotopic composition of the CH_4 to depend on that of the CO_2. However, in the laboratory refuse reactor, which presumably duplicates the landfill system, the ^{13}C contents of the evolved CO_2 and CH_4 are divergent rather than covariant (see Figure 5.4).

An example of a system which does clearly show the effect of $^{12}CO_2$ depletion is provided by the *M. thermoautotrophicum* pure culture. The evolution of the isotopic compositions of the CH_4 and CO_2 in this culture (which was grown in a closed system) is summarized in Table 5.3. The ^{13}C contents of the CH_4 and CO_2 are covariant, and, because there is apparently little increase in the cell mass at the expense of the CO_2, the final isotopic composition of the CH_4 duplicates the initial isotopic composition of the CO_2. The laboratory refuse reactor was an open system with both CO_2 and CH_4 being withdrawn as they were produced, and is thus not directly comparable. It is clear, however, that the isotopic compositions of the CO_2 and CH_4 would still diverge, even if the products were allowed to accumulate, and it appears unlikely that the isotopic fractionations in the landfill and marine sediment systems are satisfactorily explained by this hypothesis alone.

2. The observed different isotopic fractionations are due to the activities of different methane-producing bacteria in the various systems.

Three pure cultures of methane-producing bacteria have been examined in this work and are noted as the first entries in Table 5.4. The fractionations are generally less than those observed in nature, perhaps because each of the microorganisms studied here produces CH_4 only by CO_2 reduction, while other pathways--especially acetate dissimilation--are important in the natural systems.

The isotopic fractionations observed in pure cultures differ among themselves by more than a factor of two. While temperature differences might provide some of the explanation, it is especially significant that: 1) two organisms grown at the same temperature (*Methanosarcina barkeri* and *Methanobacterium* strain M.o.H.) show fractionations differing by 28‰; 2) correction of the third pure culture fractionation factor for the effect of temperature is unlikely to bring it into agreement with either of the other values; and 3) this diversity exists even though these investigations have entirely omitted acetate-dissimilating species. The morphologies of the three species are very different (Wolfe, 1971; Zeikus and Wolfe, 1972), and some differences in the kinetics of CO_2 uptake might exist and be able to affect the isotopic content of the intracellular CO_2 (Park and Esptein, 1960). In addition, the range of possible substrates for methane production by the *Methanosarcina barkeri* is broader than that for the other two bacteria (Wolfe, 1971; Zeikus and Wolfe, 1972), indicating differences in the enzyme systems present that might affect the isotope fractionation. Regardless of the reasons, the differences do exist, and, in support of the second hypothesis, it can be observed that the types of bacteria present may have an effect on the overall $\alpha(CO_2/CH_4)$ observed in complex environments.

The isotopic compositions of methane and CO_2 reported in an African rift lake, Lake Kivu, are pertinent to this discussion (see Table 5.1). Deuser *et al.* (1973) and Degens *et al.* (1973) report that the concentrations of methane in the deep part of the lake are much too high to be accounted for by decomposition of the organic matter present, and suggest that the methane is being produced from inorganic CO_2 and H_2 by methane bacteria (Bryant *et al.*, 1968). The fractionation factor observed in Lake Kivu is 1.045, almost the same as that observed for CO_2 reduction in the pure culture of *Methanosarcina barkeri*. Little comfort can be taken from this agreement, however, when it is noted that an equal fractionation factor prevails in sewage sludge reactors (see Tables 5.1 and 5.4) in which the predominant mode of methane production is known to be acetate dissimilation (Jeris and McCarty, 1965).

3. The observed differences in the isotopic compositions of the CO_2 and CH_4 are due to differences in the starting organic material. Specifically, the same microbial ecology will handle different inputs using different metabolic pathways, and these changes are responsible for the observed isotopic contrasts.

This hypothesis is difficult to test on the basis of the available evidence, but a number of relevant observations can be made. In sewage sludges, Jeris and McCarty (1965) have observed that acetate dissimilation is the predominant source of CH_4, independent of changes in the input organic matter. While isotopic compositions are not available from those experiments, it can be concluded--in opposition to this hypothesis--that in sewage sludges the last link in the food chain, at least, is largely unaffected by differences in the starting organic materials.

In the landfills and in the sewage sludge, many of the
same organic compounds should be present, especially in the
Pfeffer sewage reactor, which partially uses landfill refuse as organic feed material. In comparisons between
these systems, the differences in $\alpha(CO_2/CH_4)$ do not correlate with the chemical differences in organic inputs.
On the basis of these considerations, it appears unlikely
that the observed isotopic fractionations are satisfactorily explained by this hypothesis alone.

Quite apart from the isotopic fractionation between
the CO_2 and the CH_4, we can consider the isotopic composition of the gases relative to that of the starting organic
material. Substantial differences are observed in this
regard, and while the isotopic composition of the organic
matter always lies somewhere between that of the CO_2 and
that of the CH_4, there is no clear-cut pattern. These
differences are, no doubt, related to the effect suggested
in this hypothesis. This arises because the "organic matter"
isotopic compositions represent measurements of all the
combustible reduced carbon present in each environment. The
microorganisms may or may not feed on all this material,
and the metabolic pathways leading from the organic matter
to the substrates most directly related to CH_4 and CO_2
production can be complex and can differ substantially
depending on the chemical nature of the input material.

4. The isotopic fractionation between the CO_2 and CH_4
reflects a balance between differing production
mechanisms, and their isotopic composition relative
to the organic matter reflects variations in the pathways linking the organic matter and the substrates
most directly involved in gas production.

This "hypothesis" might be noted as "all of the above,"
inasmuch as it combines elements of many of the foregoing
possibilities. A consideration of the sources and fates
of acetate is particularly relevant.

Acetate dissimilation has been shown to be the most
important methane production mechanism in sewage sludges
(Jeris and McCarty, 1965; Smith and Mah, 1966) and has
been suggested as an important contributor to CH_4 production
in lake sediments (Koyama et al., 1973) and paddy soils
(Takai, 1970). By analogy to the lake sediment, acetate
dissimilation is also possibly an important mode of CH_4
production in marine sediments. Unfortunately, the isotope
effects occurring on the fermentation of acetate are not
known, since only one species that carries out this transformation has been isolated in pure culture, and it does so

only very slowly (Pine, 1971; Stadtman and Barker, 1951).
   The metabolism of acetate in living systems is much more complex than that of CO_2 or CH_4, since it is a possible precursor of almost every biologically produced molecule and is also a product of the catabolism of a large variety of bacterial substrates. The involvement of acetyl CoA in the catabolism and anabolism of carbohydrates, fats and proteins has been well-documented (Mahler and Cordes, 1971). In addition, its production from CO_2 and H_2 has been shown to occur in sewage sludges (Chynoweth and Mah, 1971). Finally, Smith and Mah (1966) have shown that at least 95% of the total acetate present in sewage sludge is in an extracellular form. Thus, it is accessible to the whole spectrum of organisms present, and its isotopic composition can be controlled by the whole bacterial ecology.
   The situation is further complicated by the fact that even when the dominant source of CH_4 is the dissimlation of acetate, the direct reduction of CO_2 to CH_4 is also an important pathway. Since CO_2 is both produced during acetate dissimilation and consumed by transformation to methane, its isotopic composition will be the result of many interrelated metabolic events. Carbon dioxide is also produced during the fermentation of other larger molecules, and as a result, although the isotopic compositions of CO_2 and CH_4 will be interrelated, their precise relationship will be dependent upon a large number of factors which are in turn dependent on the microbial ecology.
   A critical factor controlling the microbial ecology will be the availability of nutrients or substrates required by specific organisms or metabolic pathways. An example of such a nutrient and its possible relation to CH_4 and CO_2 production is provided by H_2. A limited availability of H_2 may be responsible for limiting the activity of CO_2-reducing species in systems where acetate dissimilation is the predominant mode of CH_4 production. Other anaerobic organisms, many of which are probably more energetically efficient than the methane bacteria, can also utilize H_2. Where other conditions allow their growth, such organisms might succesfully compete for limited supplies of H_2. On the other hand, when H_2 is abundant, CO_2 reduction might become the predominant form of CH_4 formation. For example, Nissenbaum et al. (1972) postulated that this would occur in marine sediments after sulfate-reducing bacteria (which would compete for the H_2), had fully depleted all available sulfate. It is difficult to judge the correctness of this postulate, which is certainly logical. However, if it is correct, an alternative explanation must be found for the similar

CO_2/CH_4 isotopic fractionation observed in the landfill systems, in which supplies of sulfate have not been depleted.

A second example of a possible relationship between nutrient utilization, the microbial ecology and, consequently, the isotopic composition of the CO_2 and CH_4 produced is provided by a comparison among the systems considered here. The sewage sludge systems represent environments in which substrates or nutrients of any kind are unlikely to be limiting. Complex organic matter is continually supplied and the microbial ecologies are maintained in uncnanging condition. Although the two sewage sludge systems represent widely separated observations, the CO_2/CH_4 isotopic fractionation factors are nearly equal, and are characteristically low in comparison to the fractionation factors observed in all other mixed culture systems. The landfill, lake mud, marine sediment, and glacial drift gases have all been derived from systems which are similar in that a large initial input of organic matter has been successively depleted with little, if any, recent replenishment. This similarity in histories may or may not be the cause of the observed similarity in CO_2/CH_4 isotopic fractionation factors, but there is certainly a correlation. Interestingly, the gap between these two groups of natural systems is bridged by the results obtained in the laboratory refuse reactor study.

To the extent that the laboratory refuse reactor represents a natural ecology, the observed evolution of the isotopic compositions of the CO_2 and CH_4 may reflect the course of events in some natural systems. At early stages, the isotopic fractionation approximates that observed in sewage sludge systems provided with abundant nutrients. As time passes, the isotopic fractionation increases to values representative of the systems in which substantial depletion of important nutrients has presumably occurred.

5. The system tends to thermodynamic equilibrium, as far as isotopic composition is concerned.

This is an observation rather than an explanation. Nissenbaum et al. (1972) considered the temperatures too low to allow equilibrium of CO_2 and CH_4, and, in terms of nonbiological systems, this must be correct. Nevertheless, the most prominent cluster of biological systems lies around $\alpha(CO_2/CH_4) = 1.071$, the thermodynamic value at 25°C (Bottinga, 1969), and it seems likely that thermodynamic considerations play a role in this (Galimov, 1973). We are left to ponder the course of biological events which leads to this correlation.

SUMMARY AND CONCLUSIONS

The isotopic compositions of CO_2 and CH_4 are metabolically interrelated, but this relationship is complicated by involvement of acetate. Because methane can arise from acetate or CO_2, and CO_2 itself can be produced from acetate and other molecules, no simple relationship can be discerned in complex environments.

The various environments considered show different fractionation factors for CO_2 and CH_4 and also for each as they relate to the organic matter. These differences are apparently a result of changes in the microbial ecology in response to changes in the type and availability of nutrients. This can be concluded from the results of the laboratory reactor operated as a closed system, insofar as this experiment approximates the changing conditions observed when organic matter is utilized in the natural environment.

ACKNOWLEDGMENTS

We are grateful to Robert Gunsalus and to Professor R. S. Wolfe of the University of Illinois for helpful discussions and for cultures grown according to our needs, to the National Institutes of Health for a biochemical traineeship awarded to L. M. Games, to D. J. DesMarais, S. A. Studley and D. A. Schoeller for mass spectroscopic assistance, and to the National Aeronautics and Space Administration for support of our laboratory under Grant No. NGR 15-003-118.

REFERENCES

Bottinga, Y., 1969: Calculated fractionation factors for carbon and hydrogen isotope exchange in the system calcite-carbon dioxide-graphite-methane-hydrogen-water vapor. *Geochim. Cosmochim Acta* 33, 49-64.

Bryant, M. P., B. C. McBride and R. S. Wolfe, 1968: Hydrogen-oxidizing methane bacteria - I. Cultivation and methanogenesis. *J. Bacteriology* 95, 1118-1123.

Bryant, M. P., E. A. Wolin, M. J. Wolin and R. S. Wolfe, 1967: *Methanobacillus omelianskii*, a symbiotic association of two species of bacteria. *Arch. Mikrobiol.* 59, 20-31.

Chynoweth, D. P. and R. A. Mah, 1971: Volatile acid formation in sludge digestion. In: Gould, R. F. (ed.), *Adv. in Chem. Ser.* 105, 41-54, American Chemical Society, Washington, D. C.

Claypool, G. E. and I. R. Kaplan, 1975: The origin and distribution of methane in marine sediments. In: *Natural Gases in Marine Sediments.* I. R. Kaplan, ed. (New York: Plenum), pp. 99-139.

Craig, H., 1953: The geochemistry of the stable carbon isotopes. *Geochim. Cosmochim. Acta* 3, 53-92.

Degens, E. T., R. P. von Herzen, H. K. Wong, W. G. Deuser and H. W. Jannasch, 1973: Lake Kivu: structure, chemistry, and biology of an East African rift lake. *Geologische Rundschau* 62, 245-277.

Deuser, W. G., E. T. Degens, G. R. Harvey and M. Rubin, 1973: Methane in Lake Kivu: new data on its origin. *Science* 181, 51-54.

Doelle, H. W., 1969: *Bacteria Metabolism,* (New York: Academic Press).

Ferry, J. T., P. H. Smith, and R. S. Wolfe, 1974: *Methanospirillum,* a new genus of methanogenic bacteria, and characterization of *Methanospirillum hungatii*, sp. nov. *Int. J. Systematic Bacteriology* 24, 465-467

Galimov, E. M., 1973: Izotopy ugleruda v neftegazovoy geologii. Nedra Press, Moscow (NASA technical translation TT F-682).

Games, L. M. and J. M. Hayes, 1974: Carbon in ground water at the Columbus, Indiana landfill. In: Waldrip, D. B. and R. V. Ruhe (eds) *Solid Waste Disposal by Land Burial in Southern Indiana* (Bloomington, Indiana: Indiana University Water Resources Research Center) pp. 81-110.

Games, L. M. and J. M. Hayes, 1976: Isotopic and quantitative analysis of the major carbon fractions in natural water samples. *Anal. Chem.,* in press.

Jeris, J. S. and P. L. McCarty, 1965: The biochemistry of methane fermentation using C^{14} tracers. *J. Water Pollution Control Fed.* 37, 178-192.

Koyama, T., M. Nikaido, T. Tomino and H. Hayakawa, 1973: Decomposition of organic matter in lake sediments. In: *Proceedings of a Symposium on Hydrogeochemistry and Biogeochemistry, Tokoyo* (Washington, D. C. : The Clarke Co.), pp. 512-535.

Mahler, H. R. and E. H. Cordes, 1971: *Biological Chemistry* (2nd ed.), (New York: Harper and Row)

McKinney, C. R., J. M. McCrea, S. Epstein, H. A. Allen and H. C. Urey, 1950: Improvements in mass spectrometers for the measurement of small differences in isotope abundance ratios. *Rev. Sci. Instrum.* 21, 724-730.

Nissenbaum, A., B. J. Presley, and I. R. Kaplan, 1972: Early diagenesis in a reducing fjord, Saanich-Inlet, British Columbia - I. Chemical and isotopic changes in major components of interstitial water. *Geochim. Cosmochim. Acta* 36, 1007-1027.

Oana, S. and E. S. Deevey, 1960: Carbon 13 in lake waters, and its possible bearing on paleolimnology. *Amer. J. Sci.* 258-A, 253-272.

Park, R. and S. Epstein, 1960: Carbon isotope fractionation during photosynthesis. *Geochim. Cosmochim. Acta* 21, 110-126.

Pfeffer, J. T. and J. C. Liebman, 1974: Biological conversion of organic refuse to methane. University of Illinois at Urbana-Champaign, Department of Civil Engineering, Report No. UILU-ENG_74-2019.

Pine, M. J., 1971: The methane fermentations. In: Gould, R. F., (ed.), *Adv. in Chem. Ser.* 105, , 1-10, American Chemical Society, Washington, D. C.

Presley, B. J. and I. R. Kaplan, 1968: Changes in dissolved sulfate, calcium, and carbonate from interstitial water of near-shore sediments. *Geochim. Cosmochim. Acta* 32, 1037-1048.

Raaen, V. F., G. A. Ropp and H. P. Raaen, 1968: *Carbon-14*. (New York: McGraw-Hill).

Rosenfeld, W. D. and S. R. Silverman, 1959) Carbon isotope fractionation in bacterial production of methane. *Science* 130, 1658-1659.

Sanderson, R. T., 1948: *Vacuum Manipulation of Volatile Compounds*. (New York: John Wiley and Sons, Inc.).

Smith, P. H. and R. A. Mah, 1966: Kinetics of acetate metabolism during sludge digestion. *Appl. Microbiol.* 14, 368-371.

Stadtman, T. C. and H. A. Barker, 1951: Studies on the methane fermentation IX. the origin of methane in the acetate and methanol fermentations by *Methanosarcina*. *J. Bacteriology* 61, 67-80.

Takai, Y., 1970: The mechanism of methane fermentation in flooded paddy soil. *Soil Sci. and Plant Nutrition* **16**, 238-244.

Wasserburg, G. S., E. Mazor and R. E. Zartman, 1963: Isotopic and chemical composition of some terrestrial natural gases. In: Geiss, J. and E. D. Goldberg (eds.), *Earth Science and Meteoritics*. (Amsterdam: North-Holland), pp. 219-240.

Wolfe, R. S., 1971: Microbial formation of methane. *Adv. in Microbial Physiology* **6**, 107-146.

Zeikus, J. G. and R. S. Wolfe, 1972: *Methanobacterium thermoautotrophicus* sp. n., an anaerobic autotrophic, extreme thermophile. *J. Bacteriology* **109**, 707-713.

CHAPTER 6

ORGANIC GEOCHEMISTRY OF LAKE SEDIMENTS

P. A. CRANWELL

Freshwater Biological Association
The Ferry House, Ambleside,
Cumbria LA22 OLP, England

INTRODUCTION

Organic matter in the lacustrine environment is derived from primary production within the aquatic ecosystem (autochthonous sources) and also from terrestrial biota (allochthonous sources), by transport of leached and eroded material into the lake (Likens, 1972). The primary product of photosynthesis passes through a food chain of varying complexity before mineralization occurs, principally mediated by non-photosynthetic bacteria and fungi (Stanier *et al.*, 1971). Organic material which survives degradation in the water column and in a shallow zone at the sediment surface is incorporated into the sediment and undergoes a slower process of diagenesis leading to equilibration of the sedimentary organic matter.

The stability of the cycle of matter in the aquatic environment has been affected by local man-made environmental changes such as nutrient enrichment, arising from the discharge of sewage effluents and the increased use of fertilizers. Such enrichment leads to an increase in algal productivity, while the subsequent decomposition of the algal material may deplete the dissolved oxygen in the lower part of the water column. Comparative studies of lakes varying in the degree of enrichment have shown a higher rate of deposition in the eutrophic (productive) compared with the more oligotrophic (unproductive) lakes, as determined by ^{137}Cs dating of the sediment (Pennington et al., 1973) or by reference to dated pollen horizons (Kemp et al., 1974). Analyses of seston (particulate matter in the water column) and surface sediments indicate a lower proportional loss of carbon during the transformation from seston to sediment in eutrophic compared with oligotrophic lakes (Pennington, 1974).

In recent sediments (those less than 10^4 yr old) diagenesis is minimal so that the presence of a compound which occurs in the biosphere only in a specific group of organisms may be used to infer the presence of that group of organisms in the environment of deposition. Examples of this approach in sediment studies include certain carotenoid pigments (Brown, 1969), isomeric methylheptadecanes (Eglinton et al., 1974) and branched/cyclic acids (Cranwell, 1973a).

Variations in source material and in conditions influencing the decomposition rate of sedimentary organic matter, and the use of a chemotaxonomic approach for interpretation, together enable the sediment profile to be used as a chemical record of past changes complementary to the record provided by the morphological microfossils in sediments (Frey, 1974). The results not only have paleolimnological significance, indicating the rate and extent of past changes, but also may aid our understanding of contemporary man-made changes. The organic chemistry of the freshwater environment has been reviewed (Cranwell, 1975). This paper is concerned with sedimentary constituents which are indicators of source material and with distributions within a class which vary with trophic status.

DISCUSSION

Carbohydrates

The carbohydrates of aquatic source organisms were reviewed by Swain (1970), who estimated that in a productive lake 1-20% of these carbohydrates were preserved in the sediment as hydrolyzable sugars, which reached levels of 28 mg/g dry sediment. A marked increase in abundance of ribose (1) among the hydrolyzable sugars obtained from the 600 cm horizon of a core from Hall Lake, Minnesota was attributed to a high level of bacterial activity, because ribose is an important constituent of bacteria. The abundance of fucose (2) in hydrolysates of Lake Suwa surficial sediments and the similarity in composition between these hydrolysates and those obtained from the native

diatom species were cited as evidence that in Lake Suwa
phytoplanktonic algae were principal sources of sedimentary
carbohydrates (Handa and Mizuno, 1973).

A correlation of high sugar content with high productivity was found in surface sediments from 14 lakes, mainly in Sweden (Fleischer, 1972). Glucose (3) was the major free sugar in eutrophic lakes while oligotrophic lakes contained only traces of free sugars. The higher free sugar content in the former lakes was attributed to sorption on clays and diatom frustules. Galactose (4) was the most abundant hydrolyzable sugar in oligotrophic lakes, in quantities up to 470 mg/l fresh sediment, whereas in productive lakes glucose, galactose, arabinose (5) and xylose (6) were all abundant in the hydrolysate, with individual concentrations in the range 300-600 mg/l sediment. In both types of lake glucose was the dominant hydrolyzable sugar in the seston, indicating a more rapid turnover rate in oligotrophic lakes where a change in distribution pattern occurs during sedimentation.

Organic Nitrogen Compounds

Amino Acids

The concentration of hydrolyzable amino acids in sediments from a variety of glacial and nonglacial lakes in the USA ranged from 0.1-41 parts per 10^4 parts of wet sediment (Swain, 1970). The high values were found in productive

lakes, but variation in moisture and inorganic content
of the sediment may obscure a correlation between trophic
status and percentage of sedimentary organic C and N present as bound amino acid. In the Dead Sea, however, the
difference in amino acid content between shallow, oxidizing
sediments (1.5-3.2% of the organic carbon) and the deep,
reducing sediments (7.6-11.7%) was not a reflection of the
organic carbon content, but due to different degrees of
preservation (Nissenbaum et al., 1972). In Lake Ontario
90% of nitrogen in the surface mud is organic nitrogen
(2,400 µg/g dry sediment) of which insoluble combined amino
acids and amino sugars accounted for 49-55% of the total
(Kemp and Mudrochova, 1973). The presence of αε-diaminopimelic
acid (7), a nonprotein amino acid found in bacteria, among
the combined amino acids from the interstitial waters indicates
that a part of the nonprotein amino acids is formed by
microbial alteration though some of these acids may be
artifacts of the hydrolysis.

(7) (8) (9)

Amino Sugars

The high amino sugar content of the surface sediment
of Lake Ontario, and the degree of preservation in the core
together indicate a greater resistance to degradation than
the amino acids. The significant concentrations of hexosamine [glucosamine (8) and galactosamine (9)] found in
surface muds from a number of Wisconsin lakes differing
in trophic status has been used as evidence for turnover
of organic matter by microorganisms, since hexosamines are
a major component of microbial cell walls (Keeney et al.,
1970). A trend towards increasing amino acid nitrogen and
decreasing hexosamine nitrogen content of the sediments
with increasing fertility was interpreted as an indication
of a greater net turnover of nitrogen in oligotrophic
than in eutrophic lake sediments.

The Carbon Cycle 79

Purines and Pyrimidines

The purine derivatives adenine (10), guanine (11), and hypoxanthine (12), and the pyrimidines cytosine (13), thymine (14), and uracil (15) have been identified in sediments from Lake Erie (Van der Velden and Schwartz, 1974). These compounds represent 1-2% of the total nitrogen in the surface sediments and it was suggested that the steep increase in concentration in the most recent sediments could be related to cultural eutrophication. However, rapid alteration of these compounds within the sediment had not been eliminated as a possible explanation. These bases are derived from nucleic acids and thus represent a possible source of genetic information. Uracil, which is a component of ribonucleic acid (RNA) but not of deoxyribonucleic acid (DNA), is much less abundant in the sediment than in zooplankton from the lakes, suggesting preferential degradation of RNA before incorporation in the sediment.

Pigments

Evidence from several sources (reviewed by Swain, 1970) indicated that degradation products of chlorophyll (16) preserved in sediments provided a more sensitive index of lake productivity than organic carbon content. The surface sediments from a series of lakes of increasing trophic status showed a three-fold increase in carbon (7-19% of dry weight) but a thirty-fold increase in chlorophyll derivatives (0.2-6.9 units/g sediment). In more recent studies (Gorham et al., 1974) a close correlation between algal standing crop,

(16)

epilimnetic chlorophyll, and sedimentary pigments in 19 English lakes supported the use of sedimentary pigments as empirical indices of present lake fertility. Comparative studies of pigment diversity in sediments and in aquatic and terrestrial source material, both living and dead, showed that the greater diversity in eutrophic sediments compared with oligotrophic sediments was algal in origin (Sanger and Gorham, 1970). This pigment diversity was not a sensitive index of lake productivity but showed the greater importance of autochthonous sources in productive lakes.

Chlorophyll diagenesis occurs both in the water column and at the oxidized zone of the sediment. Two unrelated processes have been observed: rupture of the tetrapyrrole ring, which required light and oxygen, and derivative formation giving colored, identifiable products (Daley and Brown, 1973; Daley, 1973). It was suggested that the phaeophorbide concentration in sediments was a measure of herbivore grazing pressure and that the allomerized phorbin content was a measure of anoxia in the water column, and therefore an indirect index of lake trophic development in the post-glacial period.

The use of individual carotenoids as selective indicators of a contribution to the sediment of particular source organisms has been reviewed (Brown, 1969).

Hydrocarbons

The C_{23}-C_{35} n-alkanes found in recent sediments have been considered to represent a higher plant contribution (Eglinton and Hamilton, 1967) on account of the marked odd/even predominance. The chain-length distribution pattern of n-alkanes from the sediments of oligotrophic glacial lakes showed a clear correlation with the predominant plant cover within the drainage basin, as deduced from the pollen content of the same horizons (Cranwell, 1973b). Sediments derived from acidic peat showed C_{31} as the dominant alkane, while in sediments derived from more base-rich forest soils C_{27} and C_{29} were most abundant. Contemporary soils of these types showed fully consistent alkane distribution patterns. In regions where flat topography and/or low rainfall do not favor erosional input of organic matter, different distributions may occur. In reducing sediments from the Dead Sea (Nissenbaum et al., 1972) and in sediment from Lake Kivu (Degens et al., 1973) predominant alkanes were C_{15} and C_{17}, which are abundant in algae and photosynthetic bacteria.

(17)

(18)

(19) CH$_2$OH

(20)

Among branched/cyclic alkanes the presence in sediments of 7- and 8-methylheptadecane has been interpreted as a contribution from blue-green algae which are known to afford these isomers (Eglinton et al., 1974). The isoprenoid alkanes pristane (17) and phytane (18) have been found in sediments of all ages (Maxwell et al., 1971) and are thought to be derived from phytol (19), the diterpenoid side chain of chlorophyll.

Cyclic alkanes are potential biological markers because of the structural specificity and stability of the carbon skeleton. Isomeric pentacyclic triterpanes with a hopane-type skeleton (20) and 27-32 carbon atoms are widespread

in sediments (Eglinton et al., 1974) and include compounds
not yet found in living organisms, e.g., members of the
17αH, 21βH-hopane series. Their origin is uncertain but
it has been postulated that their presence demonstrates
the occurrence of active microbial processes during sedimentation (Van Dorsselaer et al., 1974).

Fatty Acids

Sedimentary fatty acids include both saturated and
unsaturated components. The former can be subdivided into
normal and branched/cyclic components, as in the case of the
alkanes. Eglinton (1973) suggested that differences in
distribution of n-alkanoic acids in sediments might reflect
differences in the environment. Eutrophic English lakes
show a bimodal distribution of n-alkanoic acids, with maxima
at C_{16} and at C_{22}, C_{24}, or C_{26}, while oligotrophic lakes
only show the second maximum (Cranwell, 1974; Eglinton et al.,
1974). The acids of lower molecular weight have been
attributed to autochthonous input while those in the C_{22}-
C_{32} region are abundant in soils (Morrison, 1969). Most
of the n-alkanoic acids of autochthonous origin in these
lakes are derived from algae, which contain saturated
and unsaturated acids mainly in the C_{14}-C_{20} range, though
radio-labeling studies have shown that the latter are rapidly converted in sediments into shorter chain length
saturated acids (Eglinton, 1973). These results suggest a
mainly terrestrial origin for n-alkanoic acids in oligotrophic lakes formed by glacial action, with an additional
component derived from phytoplankton in productive lakes of
similar origin.

The branched/cyclic alkanoic acids include iso- and
anteiso-acids in the C_{13}-C_{18} region which are believed to
be of microbial origin and also 10-methyl branched C_{16} and
C_{18} acids and the cyclopropanoid acids dihydrosterculic (21a) and
lactobacillic (21b) for which a microbial origin has been
suggested (Cranwell, 1973a). Also present in the lower
molecular weight region are the isoprenoid acids pristanic
(22) and phytanic (23), shown by ^{14}C labeling to be derived
from phytol (Brooks and Maxwell, 1974). Among components

$$CH_3(CH_2)_xCH\overset{CH_2}{\overset{\diagup\diagdown}{\text{—}}}CH(CH_2)_yCO_2H$$

(21a) x=y=7
(21b) x=5, y=9

(22)

(23)

(24)   (25)

of higher molecular weight, triterpanoic acids with hopane type skeleton and 31 or 32 carbon atoms have been found in a wide variety of contemporary and ancient lacustrine sediments (Eglinton et al., 1974; Van Dorsselaer et al., 1974), including several horizons of a post-glacial sediment core from a Scottish loch (Cranwell, unpublished observations). In these five sediment horizons, corresponding with different stages of the vegetation succession and soil development within the catchment, a 17βH,21βH-C_{32} hopanoic acid homologue (24) was the most abundant constituent of the branched/cyclic alkanoic acid fraction. The origin of these hopane-derived acids, as with the hopanes themselves, is uncertain, but a tetrahydroxy C_{35} hopane homologue (25) recently isolated from a bacterium (Förster et al., 1973) is a possible precursor. A higher abundance of branched/cyclic alkanoic acids, expressed as a percentage by weight of the total sedimentary monocarboxylic acids, was found in sediments from productive lakes compared with unproductive lakes. The observation that surface sediments of the former lakes contain a higher bacterial population than those from unproductive lakes may explain the greater abundance of branched/cyclic acids in productive lakes, since many of the constituents occur in bacteria and aquatic fungi (Cranwell, 1973a) or, in the case of the hopane acids, may be derived from microbial precursors.

The presence, in sediments, of αω-dicarboxylic acids and α-, β-, and ω-hydroxyacids has been reviewed (Cranwell, 1975). The parallel distribution in chain length between the diacids and ω-hydroxyacids was believed to indicate that the former are derived from the latter, which occur in cutin (Eglinton and Hamilton, 1967), while the α- and β-hydroxyacids are believed to be intermediates in the degradation of fatty acids.

Ketones

A homologous series of methyl n-alkyl ketones with 19-33 carbon atoms and strong odd-carbon predominance has been obtained from several sediments (Cranwell, 1973c). These ketones are also found in soils, and may be formed by mi-

crobial attack on alkanes (Morrison, 1969). 6,10,14-
Trimethylpentadecan-2-one has also been obtained from a
sediment and shown to be derived from phytol by incubation
of the sediment with ^{14}C-labeled phytol (Brooks and Maxwell, 1974).

Alcohols

A homologous series of n-alkanols, mainly in the C_{18}-C_{30} range, with strong even-carbon predominance as found in higher plant waxes (Eglinton and Hamilton, 1967), has been found in lacustrine sediments (Cranwell, 1973c). The isoprenoid alcohols phytol, dihydrophytol and phytol isomer were found in the surface sediments of a productive lake, while ^{14}C-labeled phytol was converted into dihydrophytol and a phytol ester in this sediment (Brooks and Maxwell, 1974).

(26)   (27)

Sterols

The distribution of sterols in eukaryotic microorganisms has been reviewed (Goodwin, 1973) and a chemotaxonomic approach to the classification of algae, using the sterol content, has been reported (Patterson, 1971). In general, sitosterol (26 R=C_2H_5) and stigmasterol (27), the main sterols in leafy material of higher plants, are not abundant in eukaryotic microorganisms, so that sterol composition could be a potentially useful indicator of sedimentary source materials. In addition to the Δ^5-unsaturated C_{27}, C_{28}, and C_{29} sterols abundant in the biosphere, the related 5α- and 5β-stanols also occur in lacustrine sediments (Henderson et al., 1972; Gaskell and Eglinton, 1974, 1975), the 5α-isomer being more abundant. Gaskell and Eglinton interpreted the parallel distribution of these Δ^5-sterols and related stanols as evidence for the operation of a reduction process in the sediment, rather than the input of saturated and unsaturated sterols. The degree of reduction increased with depth, and evidence for the reduction process was obtained by incubation of 4-^{14}C-cholesterol in the sediment. The predominance of cholesterol (26,R=H) in the surface sediment, compared with the dominance of sitosterol in the horizon immediately below, was believed to indicate a shift towards phytoplankton in the balance of sediment input. In agreement with this conclusion, cholesterol was the main component of the contemporary phytoplankton.

Additional evidence that the sedimentary sterol distribution reflects the balance of input of allochthonous and autochthonous material has been obtained by Cranwell (1973c and unpublished material). He found a low abundance of cholesterol and 5α-cholestanol in soils, in which sitosterol was dominant, but an increased quantity in oligotrophic lake sediments. The highest combined abundance of cholesterol and 5α-cholestanol occurred in sediments from modern eutrophic lakes and in older post-glacial horizons containing both chemical (*e.g.*, bimodal n-alkanoic acid distribution) and fossil (diatom frustules) evidence of a large autochthonous component.

SUMMARY

Changes in the relative contribution of terrestrial and aquatic source materials to sediments can be discerned by analysis of the organic constituents preserved therein. Most studies have concerned lakes which have a high proportion of terrestrial input, on account of climatic and topographic factors. In paleolimnological studies chemical analysis provides useful evidence, about the source and nature of sedimentary organic matter, which complements other techniques.

REFERENCES

Brooks, P. W. and J. R. Maxwell, 1974: Early stage fate of phytol in a recently-deposited lacustrine sediment. In: Tissot, B. and F. Bienner (eds.), *Advances in Organic Geochemistry 1973* (Paris: Editions Technip), pp. 977-991.

Brown, S. R., 1969: Paleolimnological evidence from fossil pigments. *Mitt. Int. Ver. Theor. Angew. Limnol.* 17, 95-103.

Cranwell, P. A., 1973a: Branched-chain and cyclopropanoid acids in a recent sediment. *Chem. Geol.* 11, 307-313.

Cranwell, P. A., 1973b: Chain-length distribution of n-alkanes from lake sediments in relation to post-glacial environmental change. *Freshwater Biol.* 3, 259-265.

Cranwell, P. A., 1973c: Organic compounds in lake sediments. *Proc. Challenger Soc.* **4**, 231-232.

Cranwell, P. A., 1974: Monocarboxylic acids in lake sediments: indicators, derived from terrestrial and aquatic biota, of paleoenvironmental trophic levels. *Chem. Geol.* **14**, 1-14.

Cranwell, P. A., 1975: Environmental organic chemistry of rivers and lakes, both water and sediment. In: *Environmental Chemistry*, Vol. 1 (London: The Chemical Society) pp. 22-54.

Daley, R. J., 1973: Experimental characterization of lacustrine chlorophyll diagenesis II. *Arch. Hydrobiol.* **72**, 409-439.

Daley, R. J. and S. R. Brown, 1973: Experimental characterization of lacustrine chlorophyll diagenesis I. *Arch. Hydrobiol.* **72**, 277-304.

Degens, E. T., R. P. von Herzen, H. K. Wong, W. G. Deuser and H. W. Jannasch, 1973: Lake Kivu: Structure, chemistry and biology of an East African rift lake. *Geol. Rundschau.* **62**, 245-277.

Eglinton, G., 1973: Chemical fossils: a combined organic geochemical and environmental approach. *Pure Appl. Chem.* **34**, 611-632.

Eglinton, G. and R. J. Hamilton, 1967: Leaf epicuticular waxes. *Science*(Wash. DC) **156**, 1322-1334.

Eglinton, G., J. R. Maxwell, and R. P. Philp, 1974: Organic geochemistry of sediments from contemporary aquatic environments. In: Tissot, B. and F. Bienner (eds.) *Advances in Organic Geochemistry 1973*, (Paris: Editions Technip), pp. 941-961.

Fleischer, S., 1972: Sugars in the sediments of Lake Trummen and reference lakes. *Arch. Hydrobiol.* **70**, 392-412.

Förster, H. J. , K. Biemann, W. G. Haigh, N. H. Tattrie and J. R. Colvin, 1973: The structure of novel C_{35} pentacyclic terpenes from *Acetobacter xylinum*. *Biochem. J.* **135**, 133-143.

Frey, D. G., 1974: Paleolimnology. *Mitt. Int. Ver. Theor. Angew. Limnol.* **20**, 95-123.

Gaskell, S. J. and G. Eglinton, 1974: Short-term diagenesis of sterols. In: Tissot, B. and F. Bienner (eds.) *Advances in Organic Geochemistry 1973* (Paris: Editions Technip), pp. 963-976.

Gaskell, S. J. and G. Eglinton, 1975: Rapid hydrogenation of sterols in a contemporary lacustrine sediment. *Nature* (London) 254, 209-211.

Goodwin, T. W., 1973: Comparative biochemistry of sterols in eukaryotic microorganisms. In: Erwin, J. A. (ed.) *Lipids and Biomembranes of Eukaryotic Microorganisms* (New York: Academic Press), pp. 1-41.

Gorham, E., J. W. G. Lund, J. E. Sanger and W. E. Dean, 1974: Some relationships between algal standing crop, water chemistry, and sediment chemistry in the English lakes. *Limnol. Oceanogr.* 19, 601-617.

Handa, N. and K. Mizuno, 1973: Carbohydrates from lake sediments. *Geochem. J.* 7, 215-230.

Henderson. W., W. E. Reed and G. Steel, 1972: The origin and incorporation of organic molecules in sediments as elucidated by studies of the sedimentary sequence from a residual Pleistocene lake. In: Von Gaertner, H. R. and H. Wehner (eds.), *Advances in Organic Geochemistry 1971* (Oxford: Pergamon), pp. 335-352.

Keeney, D. R., J. G. Konrad and G. Chesters, 1970: Nitrogen distribution in some Wisconsin lake sediments. *J. Water Poll. Control Fed.* 42, 411-417.

Kemp, A. L. W., T. W. Anderson, R. L. Thomas and A. Mudrochova, 1974: Sedimentation rates and recent sediment history of Lakes Ontario, Erie and Huron. *J. Sediment. Petrol.* 44, 207-218.

Kemp, A. L. W. and A. Mudrochova, 1973: The distribution and nature of amino acids and other nitrogen-containing compounds in Lake Ontario surface sediments. *Geochim. Cosmochim. Acta* 37, 2191-2206.

Likens, G. E., 1972: Eutrophication and aquatic ecosystems. In: *Nutrients and Eutrophication,* Special Symposium, American Society for Limnology and Oceanogrophy, Lawrence, pp. 3-13.

Maxwell, J. R., C. T. Pillinger and G. Eglinton, 1971: Organic geochemistry. *Q. Rev. Chem. Soc.* (London) 25, 571-628.

Morrison, R. I., 1969: Soil lipids. In: Eglinton, G. and M. T. J. Murphy (eds.), *Organic Geochemistry* (Berlin: Springer-Verlag), pp. 558-575.

Nissenbaum, A., M. J. Baedecker and I. R. Kaplan, 1972: Organic geochemistry of Dead Sea sediments. *Geochim. Cosmochim. Acta* 36, 709-727.

Patterson, G. W., 1971: The distribution of sterols in algae. *Lipids* 6, 120-127.

Pennington, W., 1974: Seston and sediment formation in five Lake District lakes. *J. Ecol.* 62, 215-251.

Pennington, W., R. S. Cambray and E. M. Fisher, 1973: Observations on lake sediments using fallout ^{137}Cs as a tracer. *Nature* (London) 242, 324-326.

Sanger, J. E. and E. Gorham, 1970: The diversity of pigments in lake sediments and its ecological significance. *Limnol. Oceanogr.* 15, 59-69.

Stanier, R. Y., M. Doudoroff and E. A. Adelberg, 1971: Microorganisms as geochemical agents. In: *General Microbiology* (London: Macmillan), pp. 686-705.

Swain, F. M., 1970: *Non-marine Organic Geochemistry*. (Cambridge: University Press).

Van Dorssalaer, A., A. Ensminger, C. Spyckerelle, M. Dastillung, O. Sieskind, P. Arpino, P. Albrecht, G. Ourisson, P. W. Brooks, S. J. Gaskell, B. J. Kimble, R. P. Philp, J. R. Maxwell and G. Eglinton, 1974: Degraded and extended hopane derivatives (C_{27} to C_{35}) as ubiquitous geochemical markers. *Tetrahedron Lett.* 1349-1352.

Van der Velden, W. and A. W. Schwartz, 1974: Purines and pyrimidines in sediments from Lake Erie. *Science* (Washington D. C.) 185, 691-693.

CHAPTER 7

THE CHEMISTRY OF HUMIC SUBSTANCES

M. SCHNITZER

Soil Research Institute, Agriculture
Canada, Ottawa, Ontario K1A 0C6 Canada

INTRODUCTION

Humic substances, the major organic components of soils and sediments, arise from the chemical and biological degradation of plant and animal residues and from synthetic activities of microorganisms. The products so formed tend to associate into complex polymeric structures that are more stable than are the starting materials. Important characteristics of humic substances are their ability to form water-soluble and water-insoluble complexes with metal ions and hydrous oxides and to interact with clay minerals and hydrophobic organic compounds such as alkanes, fatty acids, dialkyl phthalates, pesticides, etc. Of special concern is the formation of water-soluble complexes of fulvic acids with toxic metals and organics which may increase the concentrations of these constituents in soil solutions and in natural waters to levels that are far in excess of their normal solubilities.

The purpose of this paper is to present: a) some general background information on the chemistry of the principal humic materials that are found in soils and waters, and b) to describe some recent advances in the chemistry of these substances.

Definition and Fractionation
of Humic Substances

Soils and sediments contain a large variety of organic materials that can be grouped into humic and nonhumic substances. The latter include those whose physical and chemical characteristics are still recognizable, that is, carbohydrates, proteins, peptides, amino acids, fats, waxes and low-molecular-weight organic acids. Most of these compounds are attacked relatively readily by microorganisms and have usually only a short life-span in soils and sediments.

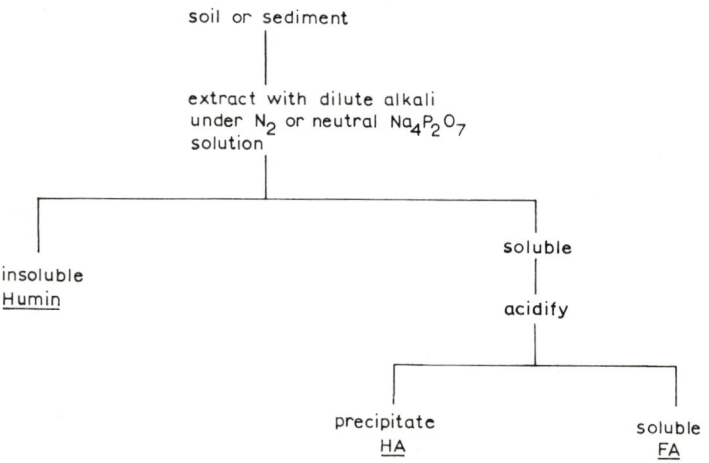

Figure 7.1 *Extraction and fractionation of humic materials.*

By contrast, humic substances exhibit no longer specific physical and chemical characteristics (such as melting and boiling points, refractive index, exact elementary composition, definite IR spectrum, etc.) normally associated with well-defined organic compounds. Humic substances are dark-colored, acidic, predominantly aromatic, hydrophilic, chemically complex polyelectrolytes that range in molecular weights from a few hundred to several thousand, and are usually partitioned into three main fractions. These are: a) humic acid (HA), which is soluble in dilute alkali but is precipitated by acidification of the alkaline extract; b) fulvic acid (FA), which is that humic fraction which remains in solution when the alkaline extract is acidified; that is, it is soluble in both dilute alkali and dilute acid; and c) humin, which is that humic fraction that cannot be

extracted from the soil or sediment by dilute base or acid.
From analytical data published in the literature (Schnitzer
and Khan, 1972) it becomes apparent that structurally the
three humic fractions are similar, but that they differ
in molecular weight, ultimate analysis and functional group
content, with FA having a lower molecular weight, containing
more O but less C and N, but having a higher content of
oxygen-containing functional groups (CO_2H, OH, C=O) per
unit weight than the other two humic fractions. Figure 7.1
summarizes the fractionation procedures that are used by
most soil scientists. Detailed descriptions of methods of
extraction and purification of HA's and FA's have been
published in a recent monograph (Schnitzer and Khan, 1972).

Table 7.1
Elementary Analysis of HA's and FA's

Element	% dry, ash free wt	
	HA	FA
C	50-60	40-50
H	4-6	4-6
N	2-6	<1-3
S	0-2	0-2
O	30-35	44-50

Some Analytical Characteristics
of Humic Substances

As shown in Table 7.1, the major elements in humic substances are C and O. The C content of HA's ranges from 50 to 60%, and that of O from 30 to 35%. The C content of FA's is lower than that of HA's and varies from 40 to 50%, while the O content ranges from 44 to 50%. Thus, HA's contain more C and N but less O than do FA's. The elementary analysis of humins resembles that of HA's.

Average values for oxygen-containing functional groups in 12 different humic preparations are listed in Table 7.2. An inspection of the data shows that FA's contain, per unit weight, considerably greater amounts of CO_2H and alcoholic OH groups but fewer quinonoid C=O groups than do the other two humic fractions. Humins appear to be enriched in ketonic and quinonoid C=O groups, but are otherwise similar to HA's. Compared to lignin, the methoxyl-content of all humic materials is low.

Table 7.2
Oxygen-Containing Functional Groups
in Humic Substances (meq/g)

Type of Material	CO_2H	Phenolic OH	Alcoholic OH	Ketonic C=O	Quinonoid C=O	Methoxyl
HA	4.4	3.3	1.9	1.2	1.0	0.3
Humin	3.1	2.2	-	3.1	2.0	0.4
FA	8.1	3.9	4.0	1.4	0.6	0.3

Methods Available for the Characterization of Humic Substances

The methods that are most frequently used for the characterization of humic substances can be divided into nondegradative and degradative ones. Nondegradative methods (Table 7.3) include chemical, spectrophotometric, spectrometric, X-ray, electrometric, electron-microscopic and radiochemical procedures. Degradative methods (Table 7.4) include oxidations in alkaline and acidic media, reduction, hydrolysis, thermal and biological degradations. With complex materials such as HA's and FA's, degradation is often a useful approach to obtaining information on their chemical structures. The expectation here is to produce simpler compounds that can be identified and whose chemical structures can be related to those of the starting materials. The method of choice should not be too drastic or lead to the formation of unwanted by-products and/or artifacts. Recent advances in the development of efficient gas chromatographic-mass spectrometric-computer systems, that make possible the separation and the qualitative and quantitative identification of micro-amounts of organic compounds in complex mixtures, have greatly enhanced the efficacy of chemical and possibly also of biological degradation as structural tools. Applications to humic substances of most of the methods listed in Tables 7.3 and 7.4 have been described in numerous papers and are summarized in two books (Kononova, 1966; and Schnitzer and Khan, 1972). I shall now describe some of our more recent work on the chemical degradation of HA's and FA's and on the isolation of alkanes and fatty acids from humic substances.

Table 7.3
Nondegradative Methods Used on HA's and FA's

(1)	Functional group analysis
(2)	Spectrophotometry in the UV and visible
(3)	IR
(4)	NMR (proton and C_{13})
(5)	ESR
(6)	X-Ray
(7)	Electron microscope, electron diffraction
(8)	Electrometric titrations
(9)	Molecular weight measurements
(10)	Radioisotopes

Table 7.4
Degradation Methods Employed on Untreated
and Methylated HA's and FA's

(1)	$KMnO_4$ oxidation
(2)	CuO-NaOH oxidation
(3)	CuO-NaOH + $KMnO_4$ oxidation
(4)	CuO-NaOH + $KMnO_4$ + H_2O oxidation
(5)	Peracetic acid oxidation at 40° and 80°C
(6)	HNO_3 oxidation
(7)	Na-amalgam reduction
(8)	Zn-dust distillation
(9)	Hydrolysis with 2 N NaOH at 170°C
(10)	Hydrolysis with 6 N HCl
(11)	Hydrolysis with H_2O
(12)	Thermal analysis: DTG and DTA
(13)	Biological degradation (?)

The Chemical Degradation
of Humic Substances

The general procedure used in our laboratory for the degradation of humic substances and for the separation and identification of degradation products is shown in Figure 7.2. Following degradation, the products are extracted into organic solvents, methylated and then separated by column and thin-layer chromatography. Portions of the fractions are then further separated by preparative gas chromatography into well-defined compounds which are identified by comparing their mass and micro-IR spectra with those of authentic specimens. Other portions of each fraction are injected directly into a GLC-MS-computer system. Preliminary identification of gas chromatographic peaks is

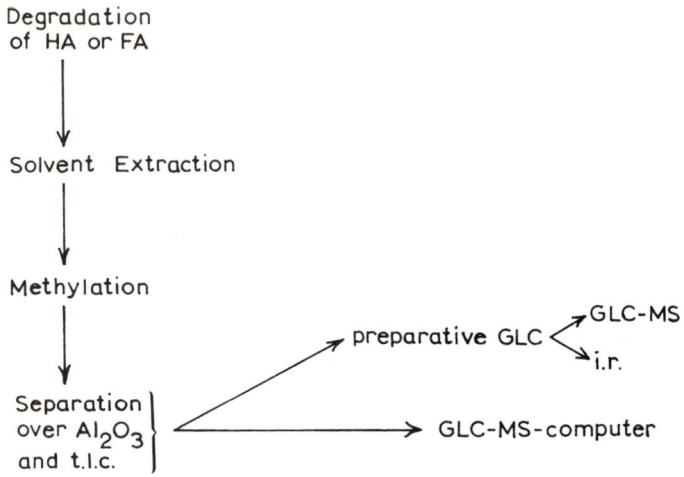

Figure 7.2 General procedure used in our laboratory for the degradation of humic substances and for the separation and identification of degradation products.

first made by recording "mass chromatograms" for fragments characteristic of specific compounds or groups of compounds expected to occur in the mixture by searching through the stored spectral data for specific m/e ratios (Skinner and Schnitzer, 1975). The identity of the compound in each peak is then confirmed by: a) running its mass spectrum; b) eluting it from the gas chromatograph and recording its micro-IR spectrum; c) matching its mass and micro-IR spectrum with the known compound to which it corresponds; and d) co-chromatography (on the gas chromatograph) of known and unknown (Neyroud and Schnitzer, 1974).

Major Degradation Products

Major compounds produced by the oxidation of methylated and unmethylated humic substances from widely differing environments in alkaline as well as in acidic solutions are aliphatic carboxylic, benzenecarboxylic and phenolic acids. In addition, smaller amounts of n-alkanes, substituted furans and dialkyl phthalates can also be identified. The most abundant aliphatic degradation products consist of n-fatty acids, especially the n-C_{16} and n-C_{18} acids, and di- and tri-carboxylic acids (Figure 7.3). Prominent benzenepolycarboxylic acids are the tri-, tetra-, penta- and hexa-forms (Figure 7.4). Major phenolic acids isolated

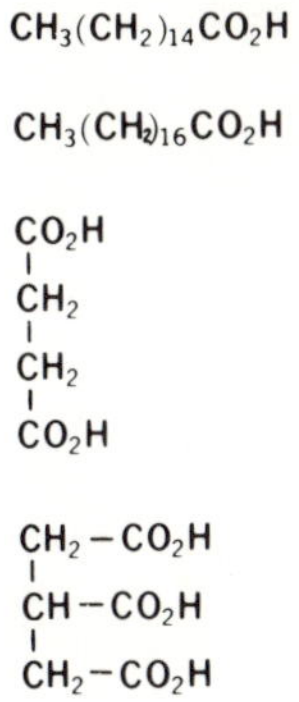

Figure 7.3 Major aliphatic degradation products.

$CH_3(CH_2)_{14}CO_2H$

$CH_3(CH_2)_{16}CO_2H$

$\begin{array}{l} CO_2H \\ | \\ CH_2 \\ | \\ CH_2 \\ | \\ CO_2H \end{array}$

$\begin{array}{l} CH_2-CO_2H \\ | \\ CH-CO_2H \\ | \\ CH_2-CO_2H \end{array}$

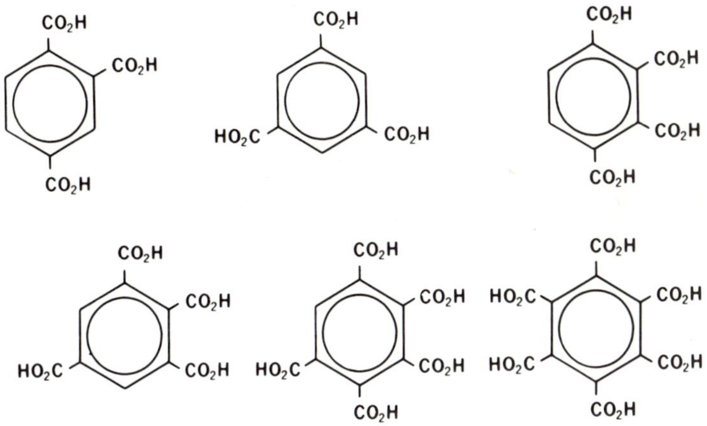

Figure 7.4 Major benzenecarboxylic degradation products.

include those with between 1 and 3 OH groups and between 1 and 5 CO_2H groups on the aromatic ring (Figure 7.5). The main chemical structures in six tropical volcanic HA's and FA's as revealed by the $KMnO_4$ oxidation of unmethylated as well as methylated materials (Griffith and Schnitzer, 1975) are shown in Figure 7.6. In this investigation 52 degradation products were identified. In structures 31, 38, 39, 48 and 51 the aromatic ring is substituted by C-atoms only. Structures 36, 43, and 50 are produced by the oxidation of

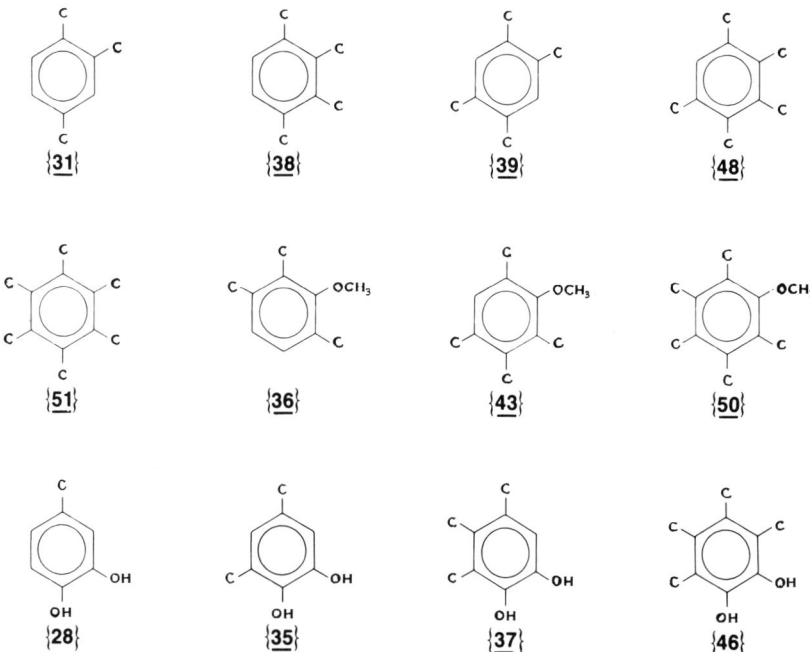

Figure 7.5 Major phenolic degradation products.

Figure 7.6 Major chemical structures in tropical volcanic HA's and FA's. C stands for a CO_2H group or a group forming CO_2H on $KMnO_4$ oxidation.

both unmethylated and methylated HA's and FA's. Because these structures resist destruction by electrophilic alkaline $KMnO_4$, it is very probable that the OCH_3-groups on the rings of these structures occur in that form in the initial humic materials, rather than as OH groups. By contrast, the OH-groups in 28, 35, 37, and 46 must have occurred as

OH or, partly as OCH_3 and as OH, in the initial HA's and
FA's, since these structures were found among oxidation
products from methylated humic materials only and were
apparently destroyed in unmethylated HA's and FA's by $KMnO_4$.
The structures listed in Tables 7.3 to 7.6 may be considered
to constitute the major humic "building blocks." It is
noteworthy that these are similar in HA's, FA's and humins,
regardless of whether the humic materials are extracted
from surface or subsurface soils. Even the climatic conditions under which humic materials are formed appear, with
one exception, to have little effect on interrelationships
between the major "building blocks" (Table 7.5).

Table 7.5
Weight Ratios of Benzenecarboxylic to Phenolic
Compounds among $KMnO_4$ Oxidation Products of
Methylated HA's and FA's Extracted from Soils
Formed under Widely Differing Climates

	Benzenecarboxylic Phenolic	
Origin of HA and FA	HA	FA
Arctic soil	1.4	-
Acid soils, cool, temperate zones	1.1-1.6	1.5
Near-neutral soils, cool temperate zones	1.8-2.2	1.8-2.1
Near-neutral, subtropical soils	5.3-6.1	1.2-2.8
Acid, tropical, volcanic soils	1.2-2.4	1.4-3.5

Table 7.6 lists the maximum amounts of principal structures that can be produced by the oxidative degradation of
1.0 g of HA and 1.0 g of FA and the methods that are most
effective for this purpose. The HA that we used for these
experiments originated from the surface horizon of a Black
Chernozemic soil in Central Alberta. The FA was extracted
from the Bh horizon of a poorly-drained Podzol in Prince
Edward Island (Neyroud and Schnitzer, 1974). These two
humic preparations were selected because they were among the
best developed naturally occurring HA's and FA's that we
were able to find and also because we have assembled a
considerable amount of information on thses materials.

Thus, alkaline cupric oxide (CuO-NaOH) oxidation is
an efficient method for releasing phenolic structures from
HA's and FA's but is relatively ineffective for degrading
aromatic structures bonded by C-C bonds and which are poor

Table 7.6
Maximum Yields of Major Degradation Products
(mg/g of humic material)

	HA	FA	Method	Reference
Aliphatic	120.0	111.1	CuO-NaOH + $KMnO_4$[a]	Neyroud and Schnitzer (1974a)
Phenolic	101.3	150.8	CuO-NaOH + $KMnO_4$[a]	Neyroud and Schnitzer (1974a)
Benzenecarboxylic	160.1	114.9	CH_3CO_2OH, $KMnO_4$[a]	Schnitzer and Skinner (1974); Khan and Schnitzer (1971)
Total	381.4	376.8		
Aromaticity (%)	69	71		

[a] Of methylated HA and FA

in oxygen. The latter types of structures are more readily degraded by more drastic oxidation with peracetic acid or alkaline $KMnO_4$. The data in Table 7.6 show that 12.0% of the HA-oxidation products are aliphatic, 10.1% phenolic and 16.0% benzenecarboxylic. Similarly, 11.1% of the FA-oxidation products are aliphatic, 15.1% phenolic, and 11.5% benzenecarboxylic. If we assume that losses during the lengthy separation and purification procedures are of the order of 50%, then aliphatic structures may constitute 24.0% of the HA-structure and 22.2% of the FA-structure, phenolics 20.2 and 30.2%, and benzenecarboxylic structures 32.0 and 23.0% totaling 76.2 and 75.4%, respectively. From the data in Table 7.6 it is possible to approximate the aromaticity of the HA and FA by expressing yields of phenolic + benzenecarboxylic acids as percentages of total yields. The aromaticity of both humic materials is about 70%. The major difference between the two humic materials is that the FA is richer in phenolic structures than is the HA.

The Isolation of Alkanes and
Fatty Acids from Humic Substances

Earlier work in our laboratory (Ogner and Schnitzer, 1970, Schnitzer and Ogner, 1970) had shown that HA's and FA's contained small amounts of n-alkanes and n-fatty acids. To obtain more definite information on the qualitative and quantitative distribution of these compounds in humic substances, we proceeded to assess effects of the following pretreatments on the extractability by organic solvents of alkanes and fatty acids from HA and FA (Schnitzer and Neyroud, 1975): a) no pretreatment; b) ultrasonic dispersion; c) hydrolysis with H_2O at 170°C in an autoclave; d) a combination of ultrasonic dispersion and H_2O-hydrolysis; e) methylation; and f) saponification with 2 N NaOH at 170°C. The HA and the FA were the same preparations as those employed for the oxidative degradations listed in Table 7.6.

As shown in Table 7.7, H_2O-hydrolysis at 170°C and a combination of ultrasonic dispersion + H_2O-hydrolysis are the most efficient pretreatments for the extraction of n-alkanes from the HA and FA. Maximum amounts of n-alkanes that can be extracted are 3.7 mg/g of HA and 17.0 mg/g of FA. The alkanes extracted from the two humic preparations vary from n-C_{12} to n-C_{38}, the majority being in the n-C_{18} to n-C_{38} range, and with an odd to even C ratio of 1.0. The distribution of n-alkanes and the odd to even C ratio are similar to those of microbial hydrocarbons (Jones, 1969).

Table 7.7
Extraction of Alkanes and Fatty Acids from HA and
FA after Different Pretreatments

Type of Pretreatments	HA		FA	
	Alkanes (mg/g)	Fatty acids (mg/g)	Alkanes (mg/g)	Fatty acids (mg/g)
None	0.4	1.4	0.4	1.2
Ultrasonic dispersion	1.3	4.5	3.0	7.6
H_2O-hydrolysis at 170°C for 3 hr	3.7	4.3	2.6	7.4
Ultrasonic dispersion + H_2O-hydrolysis	2.7	4.6	17.0	24.4
Methylation	3.4	4.7	1.6	1.0
2 N NaOH hydrolysis at 170°C for 3 hr	0.1	98.2	0.1	91.7

The most effective pretreatment for the release of fatty acids from the HA and FA is saponification, which makes it possible to extract between 90 and 100 mg of fatty acids per g of each humic material. The other pretreatments are considerably less effective in this respect. The fatty acids range from $n-C_{12}$ to $n-C_{38}$, with most being in the C_{14} to C_{22} range. Fatty acids extracted from the HA and FA have even to odd C ratios of 9.6 and 2.6, respectively; $n-C_{16}$ and $n-C_{18}$ constitute between 60 and 80% of the fatty acids extracted from the two humic preparations, which suggests a microbiological origin (Breger, 1960).

The data in Table 7.7 can be interpreted in the following manner: the relatively mild pretreatments (ultrasonic dispersion, H_2O-hydrolysis, methylation) facilitate the extraction of "loosely-held" fatty acids, whereas the harsher saponification liberates total fatty acids. Thus, in the case of the HA, about 5% of the fatty acids is "loosely held," while 95% is "tightly bound." In the case of the FA, about 25% of the fatty acids isolated is "loosely held," whereas 75% is "tightly bound." Two points are of special interest: a) the two humic materials investigated come from widely differing pedological and geochemical environments, geographically 3000 miles apart, yet they contain almost identical amounts of fatty acids per unit weight; and b) yields of fatty acids reported here are considerably higher than any reported so far for humic substances.

Along with fatty acids, saponification also releases a number of phenolic acids from the HA and FA (Table 7.8). The data show a phenolic to fatty acid molar ratio of 0.76 for HA and of 1.90 for FA. This suggests that in the humic substances fatty acids react with phenolic OH groups to form esters of the following type:

$$R_1-\underset{R_6\ R_5}{\overset{R_2\ R_3}{\bigcirc}}-O-\overset{O}{\underset{\|}{C}}-(CH_2)_n-CH_3$$

where R_1 (Table 7.8) = CO_2H or $COCH_3$ or OH; R_2 = H or OH or CO_2H; R_3 = H or OH or OCH_3 or CO_2H; R_4 = OH esterified to fatty acid; R_5 = H or OH or OCH_3; R_6 = H or CO_2CH_3 and n = mainly 14 and 16 for the HA and 14, 15, 16, and 18 for the FA. Esters could form via any free OH group on the

Table 7.8
Phenolics and n-Fatty Acids Released by
Hydrolysis of HA and FA with 2 N NaOH at
170°C for 3 hr
(mmol/g of humic material)

Phenolics	HA	FA
4-OH-benzenecarboxylic acid	0.03	0.08
3,4-diOH-acetophenone	0.02	0.01
3,5-diOH-benzenecarboxylic acid	0.04	0.21
3,4-diOH-benzenecarboxylic acid	0.05	0.16
3,4,5-triOH-benzenecarboxylic acid	0.10	0.07
3,4,5-triOH-acetophenone	0.04	0.02
4-OH-1,2-benzenedicarboxylic acid	0.01	0.03
2-OH-1,3,5-benzenetricarboxylic acid	0	0.03
	0.29	0.61
Fatty acids		
n-C_{14}, n-C_{15}, n-C_{17}, n-C_{19}, n-C_{20}, n-C_{22}	0.07	0.13
n-C_{16}	0.23	0.14
n-C_{18}	0.06	0.05
	0.36	0.32

aromatic ring, that is, at R_1, R_2, R_3 and R_5 in addition to R_4, so that in humic substances a considerable number of different phenol-fatty acid esters is a distinct possibility.

Thus, while small proportions of the fatty acids are "loosely-held," possibly physically adsorbed, on the large humic surfaces and in internal voids, most of the fatty acids form esters with phenolic humic "building blocks." By this type of reaction mechanism humic substances may fix, stabilize and preserve over long periods of time relatively large amounts of hydrophobic organic compounds, provided that these contain at least one CO_2H group. These observations suggest that the role of humic substances in transport, dispersion and sedimentation in soils and waters of hydrophobic organic compounds, including toxic pollutants and petroleum source materials, may have been underestimated.

The Chemical Structure of Humic Substances

From the data presented in the preceding paragraphs it appears that substantial proportions of the aliphatic structures in HA's and FA's consist of n-fatty acids esterified to phenolics. In addition we have smaller amounts of "loosely held" fatty acids and n-alkanes which appear to be physically adsorbed and are not likely to be structural humic components. As far as aliphatic dicarboxylic acids are concerned, which may range from oxalic to sebacic acids (Matsuda and Schnitzer, 1972), these may in part arise from the oxidative degradation of aromatic structures or may to some extent originate from aliphatic chains joining aromatic rings as Ogner (1973) has suggested. From these considerations it becomes apparent that phenolic and benzenecarboxylic structures are the major building blocks of all soil humic materials on the earth's surface. Whether this is also true for water humic substances remains to be investigated. While the methods described in this paper make it possible to uncover HA- and FA-building blocks, future work must be concerned with uncovering how these fit together and what type(s) of structural arrangement is produced.

It may be advantageous at this point to focus our attention on FA, a relatively low-molecular-weight humic material that we have investigated extensively by X-rays and electron microscopy (Kodama and Schnitzer, 1967; Schnitzer and Kodama, 1975). X-Ray analysis shows that the carbon skeleton of FA consists of a broken network of poorly condensed aromatic rings with appreciable numbers of disordered aliphatic chains around the edges of the aromatic rings; that is, FA has a loose, open structure. Similar information comes from electron microscopy which shows that FA has a sponge-like structure of variable thickness (100-300 Å), perforated by voids of widely differing sizes (200 to 1000 Å in diameter). Thus, both X-ray analysis and electron microscopy of FA point to relatively "open," flexible structure perforated by voids of varying dimensions that can trap or fix organic and inorganic compounds that fit into the voids, provided that the charges are complementary. Other considerations that enter into this discussion are long-term observations that indicate that humic substances are readily dispersed or aggregated by small changes in pH, salt concentration or valence of cations. If we want to develop a meaningful concept of the chemical structure of humic materials, we must take all available evidence into account. It becomes, therefore, more and more apparent that humic substances are not single molecules but rather associations of molecules of microbiological, polyphenolic, lignin and condensed lignin origins. These are the benzenecarboxylic and phenolic building blocks that we have isolated, and which appear to be held together by relatively weak linkages such as hydrogen-bonding, van der Waal's forces and possible π-bonding. This structural concept applies primarily to FA's. In higher-molecular-weight humic fractions, that is, HA's and humins, the building blocks, although made up of the same types of benzenecarboxylic and phenolic structures, appear to be more complex and more stable (Neyroud and Schnitzer, 1974). The increased stability of each building block in HA's and humins may arise from more energetic linkages of the types discussed above and/or from additional chemical bonding via C-O and C-C bonds. Thus, whereas the building blocks in the higher-molecular-weight humic materials may be more complex than those in lower-molecular-weight ones, the molecular forces holding the building blocks together are similar, consisting mainly of low-energy bonds such as hydrogen bonds and van der Waal's interactions.

A chemical structure that is in harmony with most of the requirements listed above is shown in Figure 7.7. This molecular arrangement may account for a significant portion of the FA structure. Each of the compounds that make up the structure was isolated from FA without chemical degradation (Ogner and Schnitzer, 1971). The FA was exhaustively methylated, which appeared to break the hydrogen bonds holding the major components together, and the phenolic and benzenecarboxylic acids were separated and identified by the methods listed in Figure 7.2. The chemical structure in Figure 7.7 is relatively "open," flexible, bonded exclusively by hydrogen bonds, and contains sufficient voids for trapping inorganic and organic molecules of appropriate molecular sizes.

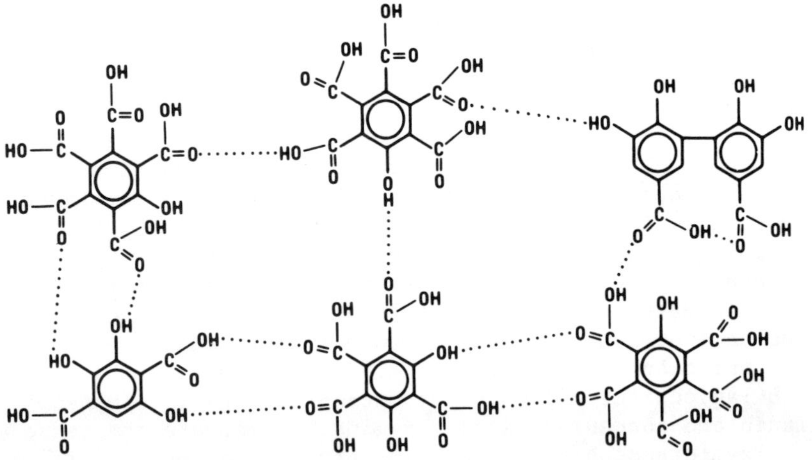

Figure 7.7 A partial chemical structure for FA.

SUMMARY

The chemistry of humic substances has been under continuous investigation for almost 200 years (Schnitzer and Khan, 1972), yet much remains to be learned about the synthesis, chemical structure and reactions of these materials which constitute the bulk of the organic matter in soils and sediments. There is increasing evidence that shows that water-soluble (FA's) and water-insoluble (HA's, humins) humic substances significantly affect the concentrations and reactions of inorganic as well as organic compounds in soil solutions and in surface waters. Furthermore, associations of FA's with metal ions, hydrous oxides and clay minerals in soils and waters may be active in mobilization, transportation and deposition phenomena in these systems. Mineral surfaces may catalyze structural changes in humic materials while, conversely, humic materials may induce crystallization of inorganic materials under conditions under which this normally would not be possible. Most of the runoff that enters streams and lakes has had contact with the soil and carries with it organic and inorganic soil materials. Among the latter, water-soluble FA's are most likely to be most conspicuous. Sedimentary rocks are another major source of dissolved materials in surface waters. Thus, soil chemists and organic geochemists (or biogeochemists) have many problems in common, but there is little contact between the two disciplines. A recent report by the National Academy of Sciences (1973) notes that organic geochemists are primarily concerned with four classes of compounds, that is, with hydrocarbons, fatty acids, porphyrins, and amino acids. The report gives the following reasons for these developments: These classes of compounds are readily extracted and isolated and sophisticated instruments are available for their identification. While these compounds can be readily analyzed, they constitute only small fractions of the total C in soils and sediments. The report goes on to state that the characterization of the more abundant, intractable, complex, organic polymers (apparently humic materials) has proceeded much more slowly but that the study of the chemical structure and constitution of these materials may be the most important and most difficult problem in organic geochemistry. While, admittedly, work on humic materials is often unrewarding and associated with many difficulties, this, however, is not sufficient reason to disregard the existence of these materials.

The purpose of this paper is to show that the recent availability of sophisticated and powerful analytical tools such as the gas chromatography-mass spectrometry-computer system has made possible significant advances in the structural chemistry of humic substances. Thus, important and exciting developments are now occurring in this field. It is hoped that organic geochemists will join soil scientists and others in taking advantage of these opportunities to develop a more accurate concept of the chemical structure of humic materials so that the carbon cycle in soils and waters can be more completely understood.

REFERENCES

Breger, I. A., 1966: Geochemistry of lipids. *J. Am. Oil Chemist Soc.* 43, 197-202.

Griffith, S. M. and M. Schnitzer, 1975: The oxidative degradation of humic and fulvic acids extracted from tropical volcanic soils. *Can. J. Soil Sci.* 55, 251-267.

Jones, J. G., 1969: Studies on lipids of soil microorganisms with particular reference to hydrocarbons. *J. Gen. Microbiol.* 59, 145-152.

Khan, S. U. and M. Schnitzer, 1971: Further investigations on the chemistry of fulvic acid, a soil humic fraction. *Can. J. Chem.* 49, 2302-2309.

Kodama, H. and M. Schnitzer, 1967: X-Ray studies of fulvic acid, a soil humic compound. *Fuel* 46, 87-94.

Kononova, M. M., 1966: *Soil Organic Matter.* (Oxford: Pergamon Press), 544 pp.

Matsuda, K. and M. Schnitzer, 1972: The permanganate oxidation of humic acids extracted from acid soils. *Soil Sci.* 114, 195-193.

National Academy of Sciences, 1973: *Orientations in Geochemistry.* (Washington, D.C.), 122 pp.

Neyroud, J. A. and M. Schnitzer, 1974a: The exhaustive alkaline cupric oxide oxidation of humic acid and fulvic acid. *Soil Sci. Soc. Am. Proc.* 38, 907-913.

Neyroud, J. A. and M. Schnitzer, 1974b: The chemistry of high molecular weight fulvic acid fractions. *Can. J. Chem.* 52, 4123-4132.

Ogner, G., 1973: Permanganate oxidation of methylated and unmethylated fulvic acid, humic acid and humin isolated from raw humus. *Acta Chem. Scand.* 27, 1601-1612.

Ogner, G. and M. Schnitzer, 1970: The occurrence of alkanes in fulvic acid, a soil humic fraction. *Geochim. Cosmochim. Acta* 34, 921-928.

Ogner, G. and M. Schnitzer, 1971: Chemistry of fulvic acid, a soil humic fraction, and its relation to lignin. *Can. J. Chem.* 49, 1053-1063.

Schnitzer, M. and S. U. Khan, 1972: *Humic Substances in the Environment* (New York: Marcel Dekker), 327 pp.

Schnitzer, M. and G. Ogner, 1970: The occurrence of fatty acids in fulvic acid, a soil humic fraction. *Israel J. Chem.* 8, 505-512.

Schnitzer, M. and H. Kodama, 1975: An electron microscopic examination of fulvic acid. *Geoderma* 13, 279-287.

Schnitzer, M. and J. A. Neyroud, 1975: Alkanes and fatty acids in humic substances. *Fuel* 54, 17-19.

Schnitzer, M. and S. I. M. Skinner, 1974: The peracetic acid oxidation of humic substances. *Soil Sci.* 118, 322-331.

Skinner, S. I. M. and M. Schnitzer, 1975: Rapid identification by gas chromatography-mass spectrometry-computer of organic compounds resulting from the degradation of humic substances. *Anal. Chim. Acta* 75, 207-211.

CHAPTER 8

INVESTIGATION OF EXTRACELLULAR
ELECTRON TRANSPORT BY HUMIC ACIDS

J. E. SCHINDLER
D. J. WILLIAMS
A. P. ZIMMERMAN

Department of Zoology, University of
Georgia, Athens, Georgia 30602

An interesting characteristic of humic materials is their ability to form stable free radicals. This characteristic and the behavior of humic materials in oxidation-reduction reactions has led to speculation on the possible dynamic electron donor-acceptor role of these compounds in nature (Steelink and Tollin, 1967; Ziechmann, 1972; Schindler and Alberts, 1974). If humic materials are capable of serving as electron donor-acceptors in anoxic sedimentary environments such as the profundal regions of lakes, they may be significant in directing the biogeochemical reactions of the system. This report deals with an investigation of the capacity and possible significance of electron transport by humic materials.
The humic acids used in this study were obtained from a small farm pond in Georgia. The techniques used in the extraction and purification have been reported elsewhere (Alberts *et al.*, 1974). Examination of the response of the humic acid to borohydride reduction and air oxidation indicates a characteristic, reversible bleaching in the 410 nm visible spectrum. ESR monitoring of oxidation-reduction reactions suggests structural alteration of the humic molecule. Lack of definitive knowledge of humic molecular structure makes these responses difficult to interpret directly, but their response is similar to that monitored for flavoproteins during electron transfers. Extracellular,

abiotic catalysts are not unknown. Fox (1969) has demonstrated several high-molecular-weight protenoids which mimic enzymes in nonspecific catalysis, some of which even show predictable Michaelis-Menton kinetic responses, heat denaturation, pH optima, inhibition and dependence on functional groups for activation.

Flavin electron transport involves the formation of semiquinone intermediates which may be stabilized through various resonance structures and electron sharing with metals; the quinone configurations which are characteristic of humic materials suggests that their redox reaction mechanisms could also involve semiquinone intermediates, perhaps also stabilized by metals. (Rather than a flavoproteins analogy, however, a better comparison is to the entire supramolecular respiratory assemblage.)

Actual transport by humic substances was demonstrated with experiments utilizing INT [2-p-Iodophenyl-3-p-nitrophenyl-5-phenyl-tetrazolium chloride]. The technique involves the reduction of tetrazolium to formazan by interaction with the electron transport assemblage (flavoproteins, quinones and cytochromes) and substitution of INT for the normal terminal electron acceptor. Since humic acids apparently display a reversible quinone-semiquinone configuration, tests were conducted to determine if INT was capable of serving as an artificial acceptor for humic molecules. Specifics of tetrazolium preparation and formazan extraction and determination have been reported previously (Zimmerman, 1974).

After determining that humic-tetrazolium electron transfers do occur, and that humics themselves do not contain a significant number of transferable electrons, various donors ($SnCl_2$, $NaBH_4$, ascorbic acid, C_2H_5OH) were chosen to test the transporting capacity of the humic material. Although all reducing agents eventually reduced INT in the presence of humic acid, ascorbic acid was chosen as the most satisfactory reducing agent. Figure 8.1 illustrates the typical responses of INT to reduction by ascorbic acid. The figure demonstrates the rate of the reaction with and without the humic material and also shows the effect of chemically reducing the humic material (as indicated by visible spectroscopy) before the experiments. The faster transfer rates after excess reducing agent was eliminated presumably reflect the effect of reducing various functional groups within the humic molecule.

Additional evidence of transporting capacity was obtained from experiments utilizing ^{203}Hg as a terminal electron acceptor. In this case, the mercury was reduced from the

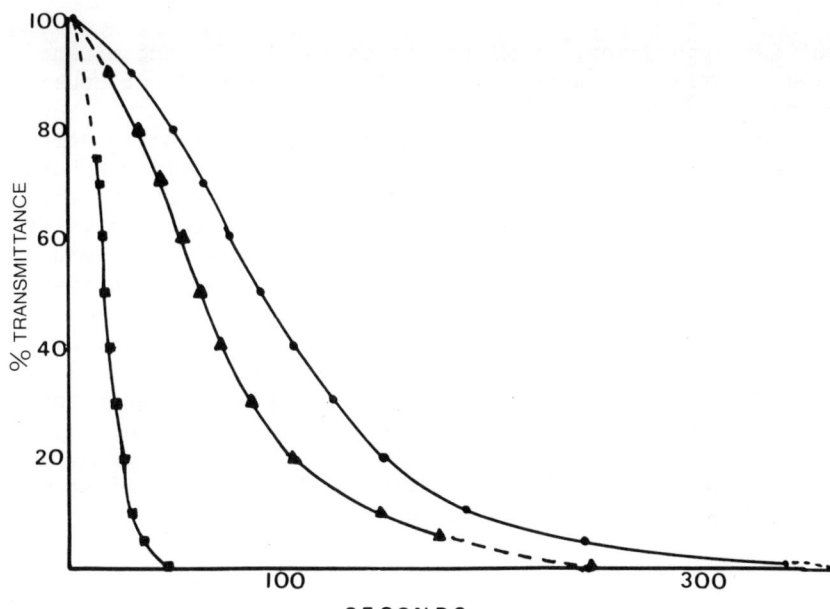

Figure 8.1 Rates of reduction of INT by: ●——● ascorbic acid (2mM); ▲——▲ ascorbic acid with oxidized humic acid; ■——■ ascorbic acid with reduced humic acid.

ionic form (Hg++) to the volatile, elemental form (Hg°). Any elemental mercury formed during the reaction was swept from the system with nitrogen gas, trapped in acid permanganate and monitored with an Ortec Gamma scintillation counter (Alberts et al., 1974). Initial results of these experiments with chemical reducing agents were identical to the INT tests, suggesting that humic materials facilitate Hg reduction by ascorbic acid. However, these tests with chemical reducing agents only demonstrate the capacity of the humic molecule to transport electrons and do not demonstrate any possible biological significance.

The existence of a humic acid electron pathway could allow anaerobic organisms to utilize a wider range of electron acceptors and continue metabolic oxidations even in the absence of their preferred (or reduction potential necessitated) terminal acceptor. Evidence exists for the possibliity that terminal electron acceptors for sulfate reducers in freshwater systems may be limiting (Ramm and Bella, 1974). Biologically, a humic acid pathway could allow continued or alternative activity of this particular group which may ultimately result in lessened heavy-metal sulfide deposition. Similar limitations with concomitant biogeochemical consequences could be postulated for other microbial groups.

Consequently, experiments were designed with biological systems to test the ability of organisms to utilize a humic transport system. Initially, whole sediment systems were set up with ^{203}Hg as the electron acceptor. Two levels of mercury were used with three levels of sucrose as a carbon source. Increased release of elemental mercury relative to controls suggests that the organisms were able to reduce mercury in some manner. Progressive carbon stimulation occurred only at the higher of the two mercury levels indicating that the systems were terminal electron acceptor limited rather than carbon limited.

Figure 8.2 demonstrates a reconstruction experiment using combinations of inocula from anoxic sediment systems, pure humic acids, sucrose and mercury.

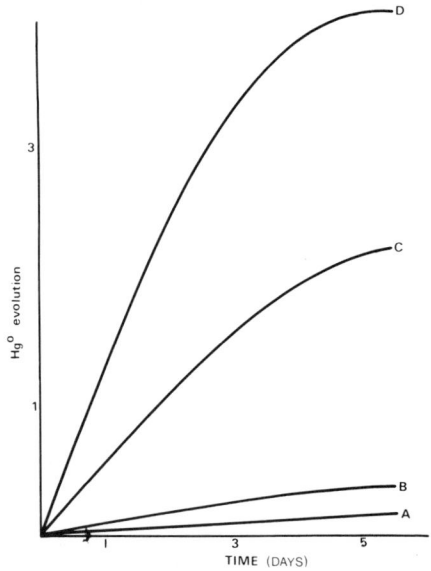

Figure 8.2 Average percent mercury evolution from four sets of aqueous systems containing anoxic sediment inocula:

    A - mercury
    B - mercury and carbon
    C - mercury and humic acid
    D - mercury, carbon and humic acid

Hg, 100 or 500 μg of Hg as $HgCl_2$ + 0.133 μg ^{203}Hg as ^{203}Hg $(NO_3)_2$ (no significant differences in release at different levels); carbon, 0.175 moles as $C_{12}H_{22}O_{12}$; and 1% humic acid.

The results indicate that the organisms are only capable
of reducing mercury in the presence of humic acid. Curve
C represents the control for abiotic mercury reduction
by humic acid (Miller et al., 1974), while the difference
between C and D represents additional mercury reduction by
organism activity. The experiments do not discriminate
between direct utilization of humic acid as an intermediate
electron acceptor and indirect transport of electrons from
microbially generated reduced substances. However, the
net environmental consequence would be the apparent utilization of cationic terminal electron acceptors and a
failure to develop pools of reduced by-products from microbial metabolism. The significance of such alternative
metabolic pathways obviously depends upon whether the
organisms utilize extracellular transport only in acceptor
limited situations or simultaneously with normal metabolism.

Although part of the mercury in both the biological and
chemical transport experiments was recovered as the elemental form, some remained in the reaction flask. Extraction
and gas chromatographic analysis of this organomercury
residual by the methods of Longbottom et al. (1972) lead
to the conclusion that an alkylmercury was formed during
the transport process. However, if the humics again mimic
the flavoproteins and disproportionate as metastable semi-
quinones, the reactions of mercury with the aromatic moieties
on the humic molecule would seem more likely to lead to
the formation of aryl derivatives. Therefore, we examined
the Longbottom moehod for its ability to separate alkyl-
and arylmercury compounds. We concluded that the method
was incapable of separating phenyl- and methylmercury compounds when the gas chromatograph is operated isothermally.
However, separations were achieved by operation of the gas
chromatograph with a temperature program.

Utilizing this method we concluded that the mercury
compound formed in the electron transport reaction of the
humic acid was an aryl derivative. Since arylmercury derivatives are apparently less soluble than methyl derivatives
in benzene, the Longbottom method could seriously under-
estimate the amount of organomercury compounds in
environmental samples. Furthermore, the dissociation of a
complex aromatic molecule like a humic acid would probably
yield a family of arylmercury derivatives, most of which
would not be extracted or chromatographed following the
current isothermal technique.

At this stage, we are not able to detail the magnitude
nor the direct environmental significance of the apparent
formation of these mercury compounds. Examination of fish

114 Environmental Biogeochemistry

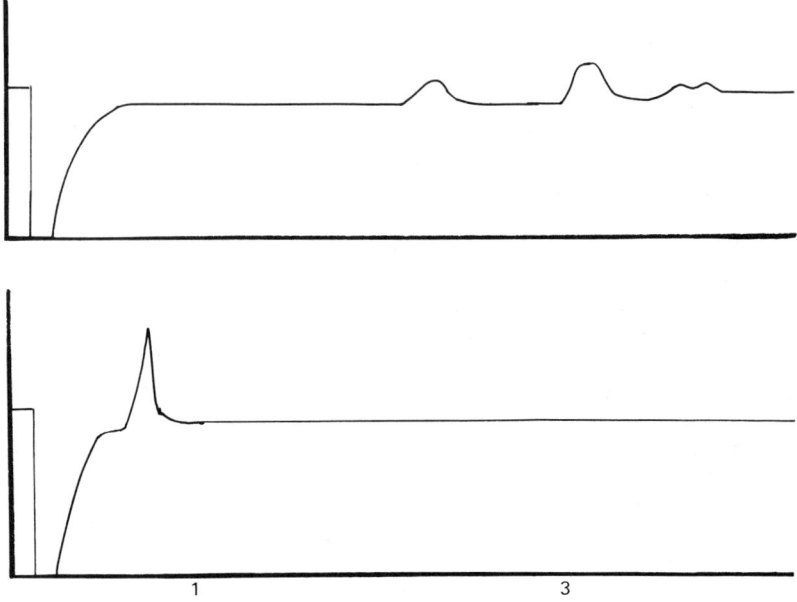

Figure 8.3 Gas chromatograph tracings of a 150° isothermal
operation yielding one peak with combined
CH_3HgI and C_6H_5HgI and a programmed operation
from 100-160° with an initial 2-min hold and
a 24°/min rise time yielding CH_3HgI at 2 min
17 sec and C_6H_5HgI at 3 min 10 sec.

tissues from a mercury-contaminated pond using programmed
separation reveals both arylmercury and alkylmercury compounds in the tissue. This suggests that the formation of
arylmercury compounds in the environment may be more significant than previously suggested in the literature.
   In conclusion, we have been able to demonstrate that
humic acids undergo reversible oxidation and reduction.
They are capable of catalyzing the transport of electrons
from reduced materials to electron acceptors that are
normally thermodynamically unavailable and are also capable
of enhancing the rate of already feasible reactions. In
experimental situations, anaerobic microbial populations
can utilize such a transport system. Consideration of
extracellular electron transport and abiotic arylation of
mercury by humic acids suggests the possibility of other
humic mediated reactions. Abiotic mimicry may help explain
anomalies in some biogeochemical cycles (Allison, 1955).

REFERENCES

Alberts, J. J., J. E. Schindler, R. W. Miller, and D. E. Nutter, Jr, 1974: Elemental mercury evolution mediated by humic acid. *Science* 184, 895-987.

Allison, F. E., 1955: The enigma of the soil nitrogen balance sheet. *Adv. Agron.* 7, 213-250.

Fox, S. W., 1969: Self-ordered polymers and propagative cell-like systems. *Naturwissenschaften* 56, 1-19.

Longbottom, J. E., R. C. Dressman, and J. J. Lichtenberg, 1972: USEPA Office of Research and Monitoring, National Environmental Research Center, Cincinnati, Ohio.

Miller, R. W., J. E. Schindler, and J. J. Alberts, 1974: Mobilization on mercury from freshwater sediments by humic acids. AEC Symposium on Mineral Cycling in Southeastern Ecosystems, Augusta, Georgia.

Ramm, A. E. and D. A. Bella, 1974: Sulfide production in anaerobic microcosms. *Limnol. Oceanogr.* 19, 110-118.

Schindler, J. E. and J. J. Alberts, 1974: An investigation of the role of organic materials in freshwater systems. *Verh. Internat. Verein Limnol.* 20.

Steelink, C. and G. Tollin, 1967: Free radicals in Soil. In: *Soil Biochemistry*. A. D. McLaren and G. H. Peterson, (eds.)(New York: Marcel Dekker Inc.).

Ziechmann, W., 1972: Uber die elecktronen-donator und-acceptor eigenshaften von huminstoffen. *Geoderma.* 8, 111-131.

Zimmerman, A. P., 1974: Electron transport analysis as an indicator of biological oxidations in freshwater sediment. *Verh. Internat. Verein Limnol.* 20.

CHAPTER 9

ISOLATION AND CHARACTERIZATION OF METAL-ORGANIC
COMPLEXES FROM TROPICAL VOLCANIC SOILS

S. M. GRIFFITH
M. SCHNITZER

Soil Research Institute, Agriculture
Canada, Ottawa, Ontario, K1A 0C6

INTRODUCTION

Most of the organic matter in soils and sediments is associated with inorganic constituents. While the chemistry of these inorganic-organic interaction products is not well understood, it is likely that these associations affect many reactions that occur in soils and waters. For example, reactive inorganic surfaces may catalyze the synthesis and degradation of organic substances adsorbed on them or, conversely, organic matter may favor the crystallization of inorganic compounds under unusually mild conditions. Inorganic-organic associations may also be active in transportation and adsorption phenomena. It has been suggested (National Academy of Sciences, 1973) that much of the organic matter in recent marine sediments has been transported to the site of deposition by clays.

Inorganic-organic associations are especially important in tropical soils of volcanic origin on a number of islands in the West Indies. These soils tend to accumulate large amounts of organic matter, which is extremely stable, but which appears to be associated with their infertility. After examining most of the available information on the subject, we came to the conclusion that the excessive stability or low turnover rate of the organic matter in these soils arose from it being strongly complexed by relatively large amounts of amorphous inorganics, and that the formation of metal-organic complexes of such a nature interfered with the biodynamics of these systems. The purpose of this paper is to describe the extraction and sequential purification of metal-organic complexes from these soils and to present data on a number of analytical characteristics of the complexes at different stages of purification as revealed by chemical, spectrophotometric and thermal methods. Our soil samples were collected on the Caribbean Island of Dominica.

MATERIALS AND METHODS

Soil Samples

A number of field and analytical characteristics of the soil samples that we used are listed in Table 9.1. The soils of Dominica have been described by Lang (1967). The Pont Cassé soil (samples 1 and 2) is referred to as rudimentary podzolic, occurring mainly on youthful volcanic piles, developed from andesitic agglomerate. Sample 1 was taken from a slightly higher elevation than sample 2. Samples 3 and 4 were collected from a profile belonging to the La Plaine series which is described as freely drained, mainly on young glacis, with a silica pan, developed from andesitic ash. Samples 5 and 6 were collected from a profile classified as Boetica and characterized as freely drained, on young glacis, without a silica pan, developed from andesitic ash. Sample 7 was collected from the Armadale profile, a poorly-drained Podzol in Prince Edward Island, Canada, and processed along with the soil samples from Dominica for comparative purposes. Since we have assembled a considerable amount of information on metal-organic reactions in the Canadian sample over many years, we used it as a check on the suitability of the methods that we developed during the course of this investigation.

Extraction of Metal-Organic Complexes

Three hundred grams of air-dry soil was weighed into a 4.5 liter polypropylene flask, 3000 ml of 0.5 N NaOH was added, the air in the flask and in the solution was displaced by N_2 and the system was shaken intermittently for 24 hr at room temperature. The dark-colored supernatant solution, after separation from the residual soil by centrifugation (850 x g for 30 min), was acidified to pH 2 with 6 N HCl and allowed to stand for 24 hr at room temperature to allow

Table 9.1
Origins and some Characteristics of Soil Samples

Sample No.	Horizon	Depth of Horizon (cm)	Soil Series	Order (Great Soil Group)	Geographical Origin	pH(H$_2$O)	% C[a]	% N[a]	C/N
1	Surface	0 – 15	Pont Cassé	Inceptisols (Hydrandepts)	Dominica	5.1	15.6	1.1	14.2
2	Surface	0 – 20	Pont Cassé	Inceptisols (Hydrandepts)	Dominica	5.4	25.2	1.6	15.8
3	Surface	0 – 15	La Plaine	Inceptisols (Dystrandepts)	Dominica	5.6	12.6	0.9	14.0
4	Sub-surface	15 – 25	La Plaine	Inceptisols (Dystrandepts)	Dominica	5.5	7.7	0.6	12.8
5	Surface	0 – 20	Boetica	Inceptisols (Dystrandepts)	Dominica	4.5	10.7	0.9	11.9
6	Sub-surface	20 – 40	Boetica	Inceptisols (Dystrandepts)	Dominica	4.2	7.3	0.7	10.4
7	Bh	15 – 20	Armadale	Spodosols	Prince Edward Island	4.1	1.9	0.07	27.1

[a]On oven-dry basis.

for the coagulation of the humic acid (HA) fraction. The
soluble material (mainly metal-organic complexes) was removed
from the HA by centrifugation, lyophilized, and dried in a
vacuum desiccator over P_2O_5 at room temperature. The dried
metal-organic complexes contained large amounts of whitish
crystalline materials which we identified as mainly NaCl,
and attempts were made to remove these materials.

Preparation of "Partly-Purified"
Metal-Organic Complexes

One hundred grams of lyophilized metal-organic complex
was shaken for 12 hr at room temperature in a 1.0 liter
ground glass stoppered flask with 750 ml of methanol. The
clear, dark-colored supernatant was separated from the gray-
ish, colloidal residue by filtration under suction. The
residue was washed repeatedly with small volumes of methanol.
The filtrate and washings were concentrated to a small volume on the rotary evaporator and allowed to stand overnight
to allow for the separation of insoluble materials. The
procedure outlined here was repeated until the methanol
solution remained clear after standing for 24 hr. Analysis
of methanol-insoluble materials showed that they contained
mainly NaCl, small amounts of organic matter, metals and
silica. The methanol-soluble material was then lyophilized.
Twenty-five grams of dry methanol-soluble material was
dissolved in 250 ml of distilled water, transferred to a
dialysis bag and dialyzed in a 2-liter graduated cylinder
against distilled water, changed at frequent intervals,
until the test for Cl^- in the reservoir became negative.
The total dialysis time was 44 hr. The dialysate (material
inside the bags) was first lyophilized and then dried in a
vacuum desiccator over P_2O_5. The ash content of the
"partly-purified" metal-organic complexes ranged from 14.5
to 24.8% and the loss-on-ignition (heating oven-dry material
at 750°C for 4 hr) from 75.2 to 85.5%.

Preparation of "Purified" Metal-Organic Complexes

To further purify the metal-organic complexes, each
"partly-purified" complex was dissolved in 200 ml of 0.1 N
NaOH solution, then shaken with 150 g of Rexyn 101 (H) for
5 hr, after which the resin was removed by filtration under
suction. The filtrate (pH 3.0) was transferred to an ultra-
centrifuge and centrifuged at 45,100 x g for 1 hr. The
clear supernatant was first lyophilized, then dried in a
vacuum desiccator over P_2O_5 at room temperature. The ash

contents of the complexes purified in this way ranged from 9.37 to 12.58% (see Table 9.4) for materials extracted from the tropical volcanic soils; the ash content of the "purified" metal-organic complex extracted from the Canadian Podzol Bh horizon, however, was only 1.13%.

The procedures employed for the extraction and purifications of the metal-organic complexes are summarized in Figure 9.1.

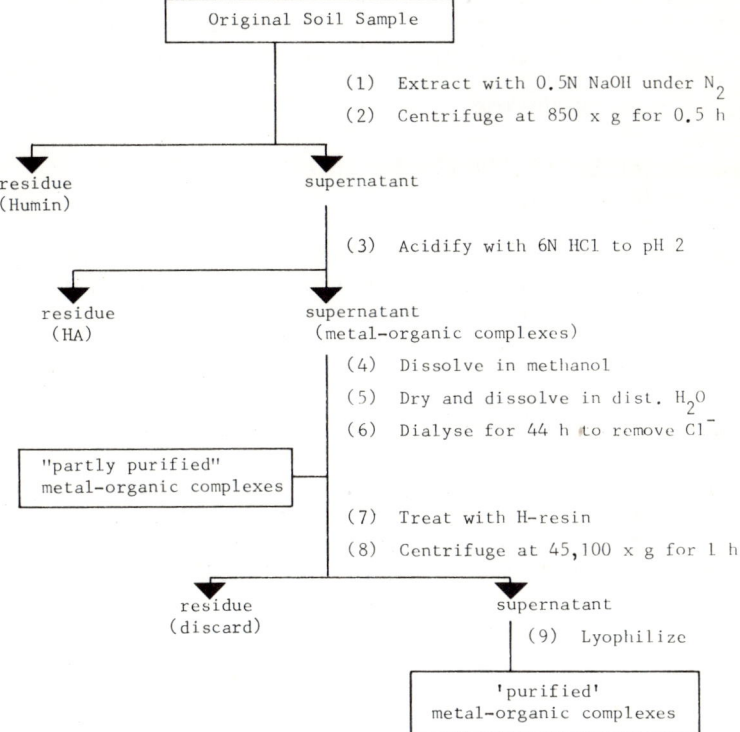

Figure 9.1 *Procedures used for the extraction and purification of metal-organic complexes.*

Analytical Methods

Moisture was determined by oven-drying air-dry material at 105°C for 24 hr, the ash content by heating oven-dried material at 750°C for 4 hr.

Carbon and hydrogen were determined by dry combustion, N by the automated Dumas method, S by oxygen-flask combustion, and O was calculated by subtracting %C + %H + %N + %S from 100.

Metal ions were determined on the ash, which was fused with Na_2CO_3 in a Pt-crucible. The fusion mixture was dissolved in distilled water, and after acidification was analyzed for Al, Fe, Cu, Ca and Mg by atomic absorption spectrometry. Analysis for Si was done by the method of Govett (1961) after fusion with NaOH in a Ni-crucible.

Each metal-organic complex was characterized by differential thermal analysis (DTA) under the experimental conditions described by Kodama and Schnitzer (1972), by IR spectrophotometry (Riffaldi and Schnitzer, 1973) and by X-ray diffraction analysis (Kodama and Schnitzer, 1971).

RESULTS AND DISCUSSION

Characteristics of "Partly-Purified" Metal-Organic Complexes

As shown in Table 9.2, yields of "partly-purified" metal-organic complexes ranged from 8.1 to 34.6 g per 300 g of air-dry soil. Between 22.8 and 70.4% of the carbon in the initial soil samples was found to occur in the "partly purified" metal-organic complexes. It is noteworthy that the highest proportions of soil-C were found in the metal-complexes isolated from subsurface horizons (Nos. 4, 6 and 7). The ash contents of the metal-organic complexes extracted from the tropical volcanic soils represented between 3.1 and 5.7% of the ash in the original soils as compared to only 0.2% for the Canadian sample No. 7.

The C-content of the "partly-purified" complexes ranged from 32.12 to 40.98%, which corresponded to an organic matter content of between 68.22 to 81.96% (see Table 9.3).

Table 9.2
Yields and Ash Content of "Partly Purified" Metal-Organic Complexes and Proportions of Soil-C and Soil-Ash in the Purified Complexes.

Sample No.	Yield (g)	% of Soil-C	% Ash	% of Soil-Ash
1	26.2	37.5	17.2	3.8
2	16.1	22.8	17.1	5.7
3	29.8	29.7	20.8	3.1
4	33.4	51.8	24.8	3.9
5	30.7	36.0	23.3	3.5
6	34.6	70.4	21.8	4.2
7	8.1	59.1	14.5	0.2

The most prominent metal in all complexes was Al, which constituted between 3.45 and 7.17% of dry weights of the different complexes. The Fe- and C-content of complexes 1, 2 and 7 was relatively high. The Si-content of the tropical complexes was considerably higher than that of the complex extracted from the podzolic soil. Amounts of Mg, Ca, and Cu in the complexes were <0.1%. The total metal + Si content of the seven metal-organic complexes ranged from 7.9 to 13.5%. The data in Table 9.3 indicate that Al, and to a minor extent Fe and Si, are the predominant inorganics in the complexes.

Kononova and Bel'chikova (1970) used Fe/C and Al/C ratios of model Fe- and Al-fulvic acid complexes prepared by Schnitzer and Skinner (1964) to characterize the forms in which Fe and Al were complexed by humic substances in Podzolic, Brown Forest and Chernozem soils of the USSR. If the (Fe + Al)/C ratio was <0.3, the metals were considered to be complexed as $M(OH)^{2+}$. Applying this approach to our data, we found (Table 9.3) that (Fe + Al)/C ratios of the purified metal-organic complexes that we had isolated ranged from 0.14 to 0.24, so that Al and Fe appear to be complexed as $Al(OH)^{2+}$ and $Fe(OH)^{2+}$; that is, each metal ion is bonded to 2 active sites or functional groups on the organic ligands. The occurrence in the complexes of Al in mainly the $Al(OH)^{2+}$ form is also indicated when we compare the analytical characteristics of the "partly purified" complexes with those of the "low" Al-fulvic acid complex prepared in the laboratory by Schnitzer and Skinner (1964).

Infrared spectra of the "partly purified" complexes were very similar to those shown by Schnitzer (1969) for metal-FA complexes. Absorbance at 1725 cm^{-1} and near 1225 cm^{-1} was relatively weak but increased in intensity in the 1630 cm^{-1} to 1610 cm^{-1} region, indicating a conversion of COOH to COO^{-} groups. Stronger absorbance near 1225 cm^{-1} in the "purified" complexes (Figure 9.2, "purified" complex 3) also indicates the likely involvement of C-O stretch of O-H deformations of CO_2H and/or OH groups of the ligands. Metal ions and positively-charged hydroxylated Fe- and Al-compounds were most likely bonded to these groups.

Characteristics of "Purified"
Metal-Organic Complexes

Yields of "purified" complexes were relatively small, ranging from 1.1 to 2.3 g. Between 2.0 and 3.7% of the C

Table 9.3

Analytical Characteristics of "Partly Purified" Metal-Organic Complexes (% Air-Dry)

Constituent	Sample No.						
	1	2	3	4	5	6	7
C	36.15	39.87	34.11	32.13	33.37	33.77	40.98
C × F^a = O.M.	72.30	79.74	68.22	75.51	78.42	79.36	81.96
Al	5.20	3.45	6.27	7.51	7.17	6.30	5.17
Fe	1.17	2.22	0.31	0.06	0.19	0.08	0.91
Si	2.97	3.02	4.77	5.80	3.66	4.25	1.66
Mg	0.01	0.01	0.01	0.01	0.01	0.01	0.01
Ca	0.07	0.08	0.07	0.08	0.05	0.07	0.07
Cu	0.02	0.02	0.01	0.03	0.01	0.02	0.03
Total inorganics in complex	9.44	8.80	11.44	13.49	11.09	10.73	7.85
$\frac{Fe + Al}{C}$	0.18	0.14	0.19	0.24	0.22	0.19	0.15

a_F = 100 divided by % C in "purified" metal-organic complexes (Table 9.5).

in the tropical volcanic soils was found to occur in the "purified" complex extracted from the Canadian soil. The ash in the "purified" complexes constituted in all instances less than 0.25% of the soil-ash. A comparison of the data in Tables 9.2 and 9.4 shows that between 90 and 95% of the soil-C and between 95 and 98% of the soil-ash were lost during the purification procedures, with the losses being especially high in the case of subsurface samples 4 and 6.

Table 9.4
Yields and Ash Content of "Purified" Metal-Organic Complexes and of Soil-C and Soil-Ash in these Complexes

Sample No.	Yield (g)	% of Soil-C	% Ash	% of Soil-Ash
1	1.5	3.1	9.4	0.12
2	1.1	2.0	9.9	0.23
3	1.5	2.2	12.6	0.10
4	1.4	2.6	12.1	0.08
5	2.3	4.1	11.1	0.13
6	1.2	3.7	12.2	0.09
7	1.9	19.5	1.1	<0.01

Each "purified" metal-organic complex was ashed and the composition of the ash determined by chemical and X-ray fluorescence analyses. The major component of all ash samples was Si, followed by smaller amounts of S, Ca, Ti, and traces of As and Zr. These data show that practically all of the Al and Fe in the "partly-purified" complexes was removed by the additional purifications.

The ultimate analysis of the organic matter in the "purified" metal-organic complexes is shown in Table 9.5. Except for sample 5, elementary analyses (on an oven-dry, ash-free basis) are quite similar to each other and are well within the range of those reported for fulvic acids (FA) (Schnitzer and Khan, 1972). Especially noteworthy is the relatively high ash content of the materials extracted from the tropical volcanic soils (Table 9.4), which we were unable to lower even with the extensive purification procedures that we employed. Also of interest is the relatively high N-content of the complexes extracted from the tropical soils (Table 9.5).

Infrared spectra of "purified" complex 3, after treatments with alkaline bromine solution and H_2O_2 and after ashing, are shown in Figure 9.2. As can be judged from the IR spectra, most of the organic matter in the complexes

Table 9.5
Ultimate Analysis of Organic Matter in "Purified" Metal-Organic Complexes

Sample No.	% (Oven-Dry, Ash-Free Basis).				
	C	H	O	N	S
1	49.23	5.30	41.20	3.02	1.25
2	49.61	4.84	40.84	3.10	1.61
3	50.62	4.68	39.68	3.29	1.73
4	46.85	4.60	44.61	2.15	1.79
5	42.83	3.84	47.75	2.02	3.56
6	46.93	4.52	44.57	2.28	1.70
7	50.49	3.62	43.42	0.53	1.94

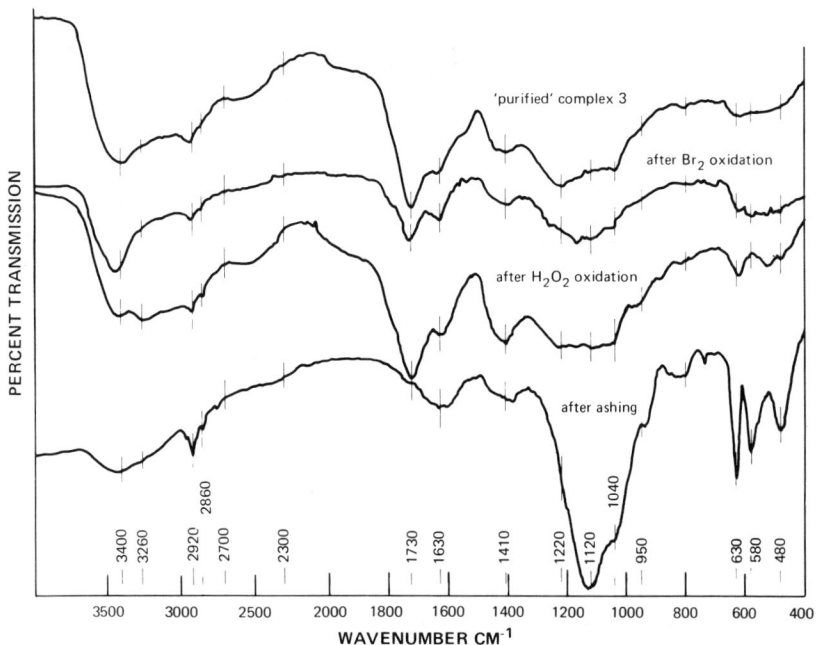

Figure 9.2 Infrared spectra of "purified" metal-organic complex 3, after bromine oxidation, H_2O_2 oxidation and ashing.

resisted oxidation by alkaline Br_2 solution and H_2O_2, although bands at 630, 525 and 480 cm^{-1} can be detected in the IR spectrum of the H_2O_2-treated material. Following ashing, very distinct, well-defined bands appear at 1120, 1040,

950, near 800, 735, 630, 580 and 480 cm^{-1}. All of these
IR bands are most likely due to Si-O valence and deforma-
tional vibrations (Stepanov, 1974). Figure 9.3 shows IR
spectra of "purified" complex 5 treated in the same manner
as complex 3. Again, most of the organic matter in the
complex was not oxidized by alkaline Br_2 solution nor by
H_2O_2. The IR spectrum of the ash shows well-defined bands,
most of which are at the same wave-numbers as those for
the ash from complex 3. There are a number of additional
bands at 910, 735 and 525 cm^{-1}. Figure 9.4 exhibits IR curves
for FA extracted from a Canadian Podzol Bh horizon and for an
interaction product of FA and silicic acid, which was prepared by shaking FA in distilled water with silicic acid for
one week at room temperature, removing excess reactants by dialysis and lyophilizing the product. The IR spectrum of the
mixture shows bands characteristic of both FA and silic acid.
Many of these bands appear also in Figures 9.2 and 9.3.

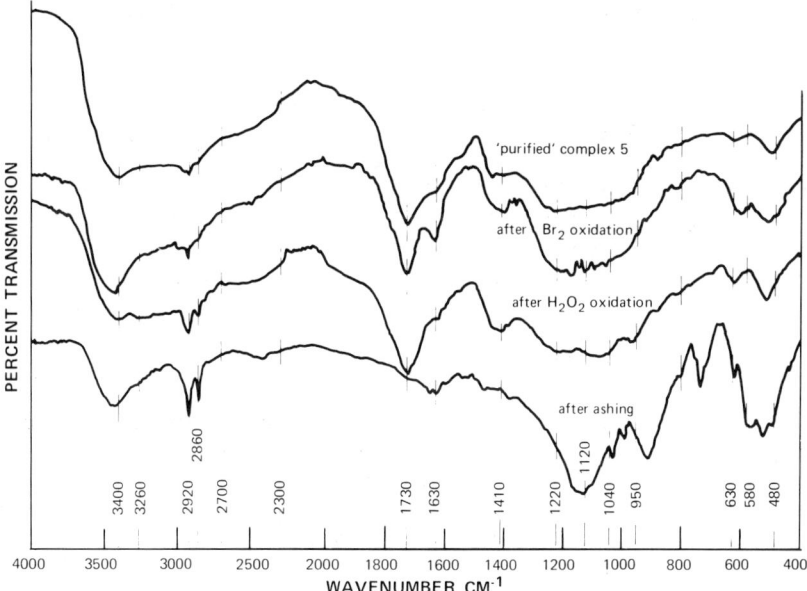

Figure 9.3 Infrared spectrum of "purified" metal-organic
complex 5, after bromine oxidation, H_2O_2
oxidation and ashing.

It may be relevant to mention at this point that the
"purified" complexes are completely soluble in water, acid,
base and methanol, and that their X-ray spectra show no
crystallinity. From the data presented here it appears
that the SiO_2 in the "purified" complexes is amorphous and

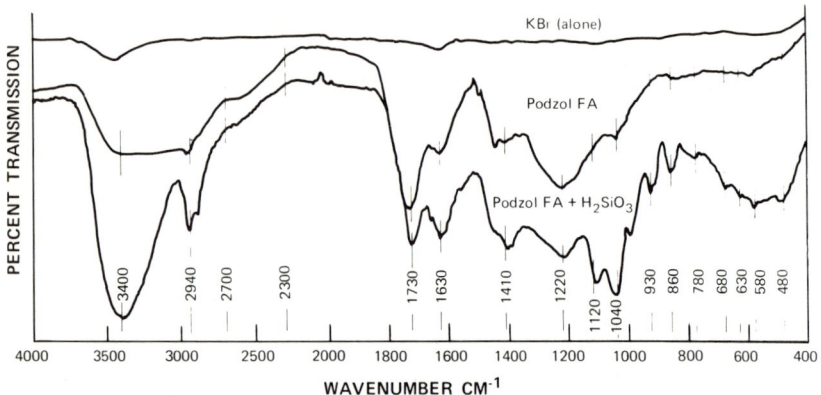

Figure 9.4 Infrared spectrum of KBr, Podzol Bh FA and a mixture of FA + silicic acid.

that it is completely enveloped by the organic matter so that the solubilities and spectroscopic characteristics are dominated by those of the organic matter, which is essentially FA. Our results can be explained on the basis of recent electron microscopic investigations (Schnitzer and Kodama, 1975) which show that FA has a polymeric sponge-like structure of variable thickness (100-300Å), punctured by voids, 200 to 1000 Å in diameter. If amorphous SiO_2 were firmly retained in the voids of FA, it would be surrounded and protected by the FA and its characteristics would be masked by those of FA as we have observed.

Characterization of Metal-Organic Complex 3 by DTA

The DTA curve of the "partly purified" complex (Figure 9.5) shows strong exotherms of approximately equal intensities at 354 and 471°C. Additional purification of the complex lead to the formation of a well-defined strong exothermic peak at 541°C and to lowering of the low-temperature peaks to 329 and 411°C, respectively. The DTA curve of the "purified " complex treated with H_2O_2 shows that the materials responsible for the low-temperature peaks had been oxidized but that the 540°C exotherm remained unchanged. Since sequential purification of the metal-organic complex involved primarily the removal of Fe and Al and of organic matter (FA) associated with these metals, it is likely that the 540°C exotherm is due to FA associated with SiO_2. Thus, it is possible to prepare this type of metal-organic complex in relatively pure form by oxidizing

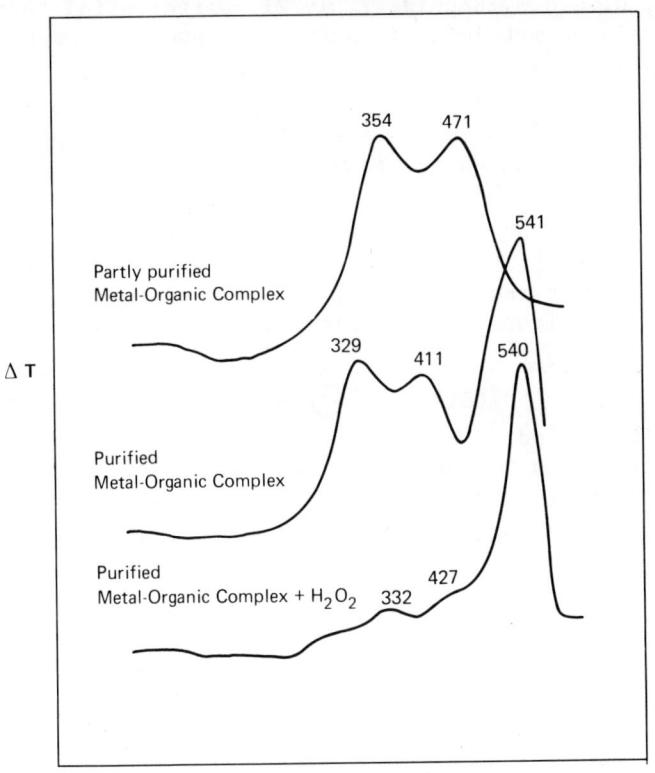

Figure 9.5 DTA analysis of "partly purified" metal-organic complex 3, after additional purification ("purified" complex) and after H_2O_2 oxidation.

the "purified" metal-organic complex with H_2O_2.

We are continuing our studies on these metal-organic complexes in order to obtain additional information on the nature, stability and behavior of $FA-SiO_2$ complexes.

REFERENCES

Govett, G. J. S., 1961: Critical factors in the colorimetric determination of silica. Anal. Chim. Acta 25, 69-80.

Kodama, H. and M. Schnitzer, 1971: Evidence for interlamellar adsorption of organic matter by clay in a podzol soil. Can. J. Soil Sci. 51, 509-512.

Kodama, H. and M. Schnitzer, 1972: Differential thermal analysis of metal-fulvic acid salts and complexes. *Geoderma* 7, 93-103.

Kononova, M. M. and N. P. Bel'chikova, 1970: Use of sodium pyrophosphate to separate and characterize organo-iron and organo-aluminum compounds in soil. *Soviet Soil Sci.* 351-365 (English translation).

Lang, D. M., 1967: Soil and Land Use survey. No. 21. Dominica. The Regional Research Centre of the British Caribbean at the University of the West Indies, Imperial College of Tropical Agriculture, Trinidad, W. I. pp. 26-27.

National Academy of Science, 1973: *Orientations in Geochemistry.* (Washington, D.C.), 122 pp.

Riffaldi, R. and M. Schnitzer, 1973: Effects of 6 N HCl hydrolysis on the analytical characteristics and chemical structure of humic acids. *Soil Sci.* 115, 349-356.

Schnitzer, M., 1969: Reactions between fulvic acid, a soil humic compound, and inorganic soil constituents. *Soil Sci. Soc. Am. Proc.* 33, 75-81.

Schnitzer, M. and S. U. Khan, 1972: *Humic substances in the environment.* (New York: Marcel Dekker), 327 pp.

Schnitzer. M. and H. Kodama, 1975: An electron microscopic examination of fulvic acid. *Geoderma* 13, 279-287.

Schnitzer, M. and S. I. M. Skinner, 1964: Organo-metallic interactions in soils: 3. Properties of iron- and aluminum-organic-matter complexes, prepared in the laboratory and extracted from a soil. *Soil Sci.* 98, 197-203.

Stepanov, I. S., 1974: Interpretation of infrared soil spectra. *Soviet Soil Sci.* 6, 76-88 (English translation).

CHAPTER 10

KEROGEN STRUCTURES IN RECENTLY-DEPOSITED ALGAL
MATS AT LAGUNA MORMONA, BAJA CALIFORNIA:
A MODEL SYSTEM FOR THE DETERMINATION OF
KEROGEN STRUCTURES IN ANCIENT SEDIMENTS

R. P. PHILP
M. CALVIN

Department of Chemistry, University of
California, Berkeley, California 94720

INTRODUCTION

Kerogen is a ubiquitous source of organic carbon on the earth's surface, containing over three orders of magnitude more organic carbon than all the other organic carbon reservoirs combined (Sackett et al., 1974). Kerogen has been defined in many different ways, but in this paper the term kerogen refers to organic material in sediments, shales and rocks which is insoluble in organic solvents. The deposition of plant and animal remains in marine and non-marine environments and the alteration of this debris during subsequent geologic periods by numerous reactions produces a wide variety of carbonaceous materials. Oil shales, kukersite, coorongite and torbanite represent different types of carbonaceous material containing various amounts of kerogenous organic material, from which high yields of oil can be produced at elevated temperatures.

The geochemical, environmental and economical significance of kerogen has led to many attempts to reveal its chemical and physical nature, its method of formation and its ancient precursors. These investigations have been hampered not only by the insolubility and macromolecular nature of the substance but also by the fact that kerogen is not a uniform molecule but most likely a conglomerate of various subunits of similar, but varying molecular

structures. Kerogen is both chemical and biologically very
inert and therefore can be used as a "chemical fossil" in
sediments. This term was used by Eglinton (1973) to describe
various classes of organic compounds which are relatively
stable under varying environmental conditions throughout
long periods of geological time. An example of kerogen's
potential as a chemical fossil was given by Sackett et al.
(1974) who analyzed amounts and isotopic composition of the
organic carbon, which is 95% kerogen, in sediments of
the Ross Sea, Antarctica. From this study it was shown
that much of the sedimentary organic carbon in the sediments
is derived from the rocks being eroded by glaciers on the
Antarctic continent.

It is now generally agreed that many oil shales were
formed by growth and deposition of algae, and that conditions
of sedimentation, compaction and lithification were
not too dissimilar form those existing at present. Thorne
and co-workers (1964) at the U. S. Bureau of Mines aptly
summarized the situation in the following way: "Oil shale
was formed by the deposition and lithification of finely
divided mineral matter and organic debris in the bottom of
shallow lakes and seas. The organic debris resulted from
the mechanical and chemical degradation of small aquatic
algal organisms." If indeed this is correct,then as Cane
(1969) has noted, with some surprise, it is unusual that
we fail to observe the initial stages of kerogen formation
in the contemporary environment, although, as he adds, this
may be due to the infrequency of the peculiar environmental
conditions for favorable vegetal growth. Bradley (1966,
1970) has suggested that the reason for failing to observe
these initial stages of kerogen formation is due to the
fact that the rate of deposition of kerogenous algal ooze
is less than 1 mm per century for the dry material. However,
under certain specialized conditions profuse algal growth
does occur and leads to the formation of Coorongite and
Balkaschite which represent an early stage in the formation
of algal kerogen. Cane (1969) has suggested therefore
that Coorongite can be regarded as the "peat" stage in
the "coalification"of algal shales and proposed that a
study of the nature of this intermediate should provide
important information on the nature of the more mature and
inert kerogen. Thus from an examination of living algae
and recent algal deposits it should be possible to get
additional information on the initial stages of kerogen formation.

Consequently a combined study of the soluble and insoluble organic matter present in recently-deposited algal
mats and oozes from Laguna Mormona, Baja California, has

commenced. The soluble lipid material in these mats and oozes will be discussed elsewhere in this volume by Eglinton and co-workers whereas we will concentrate on the insoluble or kerogenous fraction of the oozes.

Blue-green algae are especially interesting because of their ability to survive in environments which favor the preservation of organic matter such as hypersaline and reducing environments. Contemporary algae are also of interest since the search for the origins of unicellular life on earth has now reached the point where some of the microstructures refered to as exhibiting "alga-like" and "filamentous" morphologies occur in cherts from South Africa that are older than 3 eons (3 x 10^9 years, Schopf and Barghoorn, 1967). In order to get a clearer picture as to the origin and nature of these possible organisms, it is essential that the nature of modern prokaryotic algae is completely understood in the first instance.

The area at Laguna Mormona is characterized by extreme local variations in sedimentological, geochemical, and biological properties (Vonder Haar, 1973). Semi-arid climatic conditions and restricted water movement from the ocean have combined to produce an evaporite flat and hypersaline marsh environment. A barrier dune ridge separates the sea from a 100-m wide marsh which is characterized by algal mats. Mucilaginous algal laminae are disrupted by plant roots, early diagenetic growth of aragonite granules and also by desication cracks. Large quantities of organic matter are slowly accumulating because of the high salinity which inhibits organisms that normally consume and degrade the organic material. ("Algal mat," in general, refers to a cohesive fabric of filaments produced by a community of cyanophytes involving several species combined with different amounts of sediment (Logan et al., 1964). The main mat-producing genera are *Aphanocapsa, Aphanothece* and *Entophysalis* among the coccoid, and *Oscillatoria, Lyngbya, Microcoleus* and *Schizothrix* among the filamentous cyanophytes, although green and red algae, animals and abundant bacteria are often included in the community (Golubic, 1973).

The intertidal mats examined from Laguna Mormona are characterized by *Microcoleus chthonoplastes* and *Lyngbya aestuarii*. Sediment binding by *M. Chthonoplastes* is common to a wide variety of environments and no particular mat type can be consistently attributed to it. *Lyngbya aestuarii* has also been observed in arid subtropic bays such as Shark Bay and the Persian Gulf. In those environments the flat mats covering the floors of tidal pools and tidal channels are laminated structures with different proportions of trapped sediment. The mats often crack during draining, and shrinking of the algal fabric creates polygons of different sizes (Golubic, 1973).

Within the intertidal areas at Laguna Mormona several factors appear to influence the distribution and growth of the mats. They are characterized by the alteration of the water level and exposure to air. A chemical gradient between the limit of the tidal flooding and the evaporite sequence and freshwater influx leads to changes in mat distributions. Differences in mat morphology can be partially accounted for by the amount of sediment deposited over the mats by water movement.

The diverse nature of the organosedimentary structures within the mats can readily be appreciated by summarizing the three general areas of importance that should be considered when examining these structures: 1) the nature and species composition of the algal community; 2) the interaction of the community with major environmental variables, such as location in respect to tides, currents, sedimentation rates, light drainage and oxygen supply; and 3) the biological dynamics within the community balances between the rates of primary production and bacterial decomposition (Golubic, 1973).

As mentioned above there have been several studies directed at determining the structure of kerogen in ancient shales and sediments, in particular the Green River oil shale (Robinson et al., 1953, 1956, 1961, 1963; Burlingame et al., 1968a, 1969b; Djuricic et al., 1971). Djuricic and co-workers (1971) have also examined a collection of kerogens from ancient sediments of different geographic origin in an attempt to correlate variations in kerogen structures with the history and environment of the shale and thus give valuable information concerning the chemical formation of kerogen. All of these studies have involved degradation of the kerogen, either by oxidation, reduction, hydrolysis or pyrolysis techniques, followed by structural identification of the individual components of the degradation mixtures. From the data it is then possible to attempt the reconstruction of a structure for the original kerogen. The various reagents that have been used in the oxidation studies include chromic acid (Burlingame et al., 1969a), hydrogen peroxide (Downs and Himus, 1941), ozone (Jones, 1922) and potassium permanganate (Robinson et al., 1953). In the majority of these studies the mixtures of the degradation products obtained have been complex. Although it has normally been possible to analyze them by techniques such as computerized-gas chromatography-mass spectrometry after initial fractionation and derivatization, it is still difficult to construct anything but a preliminary model for the kerogen.

Thus it was anticipated that similar degradation studies of the less mature kerogenous material in these recently deposited algal mats would provide answers to the following questions:
(a) Is there any kerogen-like material in these recently deposited algal mats?
(b) Will degradation studies of young kerogens give information as to the structure and nature of ancient kerogens?
(c) What are the initial steps in the formation of kerogen?
(d) Are changes in complexity of the kerogen observed with changes in depth within these algal mats?
(e) Can kerogen-like material in recently deposited algal mats be used as a chemotaxonomic indicator?

EXPERIMENTAL

Sample Collection

A box coring device was used to collect a sample of the algal mats and ooze to a depth of approximately eight inches. This was divided into three sections which were transferred to clean glass bottles, capped and subsequently stored at 0°C until required for examination and extraction. No isopropanol was added to the samples immediately after collection as it was felt that changes due to microbial alteration would be minimal if the samples were stored at 0°C. In this study only the deepest and oldest sample of algal ooze was examined.

Sample Preparation

To determine whether or not there was any kerogen-like material produced by degradation and diagenesis of the algal ooze and also to get an idea of its structure, it was decided in this preliminary study to use an oxidative degradation technique similar to that used by Burlingame and Simoneit (1968b) in their oxidation studies of Green River oil shale kerogens. Almost every published work on the study of kerogen uses a different method for the preparation of the samples making it extremely difficult to compare results of various workers. Forsman and Hunt (1958) proposed an acid digestion of shales with consecutive treatments of

HCl, HNO3 and HF to remove carbonates, pyrites and silicates, respectively. Burlingame and Simoneit (1968a), however, used zinc dust and 6 M HCl to remove sulfides and free sulfur. Lawlor and co-workers (1963) advocated the use of lithium aluminum hydride for quantitative pyrite removal although this results in specific alteration of kerogen functional groups. One important factor to be taken into consideration with all these methods is the possibility of alteration of the organic kerogen by the action of strong mineral acids.

The mineral content of the algal ooze was found to be predominantly carbonate, and as a consequence of this was treated only with 6 M HCl. The method of isolation of the kerogen-like material is shown in Figure 10.1. After removal of supernatant water the sample was refluxed for 6 hr with 6 M HCl. It was centrifuged and the residue exhaustively extracted for a total period of two weeks with the following solvents: toluene/methanol (1:1), methanol and finally toluene.

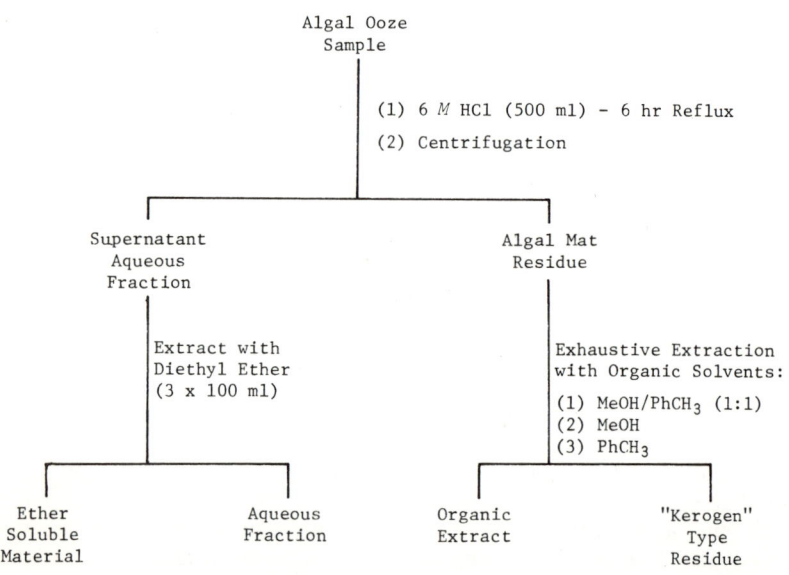

Figure 10.1 *Initial extract and fractionation scheme used to prepare kerogen residues from the recently-deposited algal ooze sample.*

The supernatant aqueous fraction was set aside for examination of its amino acid, lactone and more polar dicar-

boxylic acid content at a later date. A more detailed
study on the lipids in the total organic extract of the algal
mat and ooze has been carried out by the Organic Geochemistry Unit at Bristol; these results will be reviewed by
Professor Eglinton elsewhere in this volume.
 The resulting residue, hereafter referred to as kerogen,
was washed with triple distilled water and allowed to dry
under vacuo at room temperature. It should be mentioned at
this juncture that this residue was not subjected to any
further treatment, such as HF treatment or saponification.
However, subsequent work with this algal ooze sample has
shown that HF treatment does not have any noticeable effect
on the qualitative or quantitative nature of the results
described herein (R. P. Philp, unpublished results).

Oxidative Reactions

 The kerogen residue was subjected to successive chromic
acid oxidations (successive reaction times used: 3 and 6
hr). 4.3 g of kerogen residue (after hydrolysis and extraction) was refluxed 3 hr with 3 M chromic acid in 3 M
sulfuric acid. The residue was then filtered off, washed
with water and extracted three times each, first with heptane
and then diethyl ether, using ultrasonication to insure
thorough extraction (Figure 10.2). The spent chromic
acid solution was also extracted with heptane and then
ether. The respective extracts were combined and the acids
extracted from them with 1M KOH solution. The acids were
recovered and esterified with BF_3/MeOH and the esters of
normal acids were separated from those of branched-cyclic
acids by clathration with urea. The total acid extracts
from the 6-hr oxidation were not subjected to urea clathration.
 All analytical gas-liquid chromatography (GLC) was
carried out on a Perkin Elmer Model 900 gas chromatograph.
The total, normal and branched/cyclic fractions were gas
chromatographed on a 10 ft x 1/16 in. i. d. stainless
steel column, packed with 3% Dexsil 300 on Gaschrom Q and
programmed from 70 to 280°C at 4°/min with a helium flow
rate of 15 ml/min. (These same GLC conditions were used
throughout this study.) The various components of the
mixtures analyzed by GLC were identified by their retention
times, coinjection of standards, and low resolution mass
spectra obtained by combined gas chromatography-mass spectrometry (GC-MS). Combined GC-MS analyses were carried
out on a DuPont 492-1 instrument interfaced with a Varian
Aerograph Model No. 204 equipped with linear temperature
programmer. The column used for the GC-MS analyses was a

138 Environmental Biogeochemistry

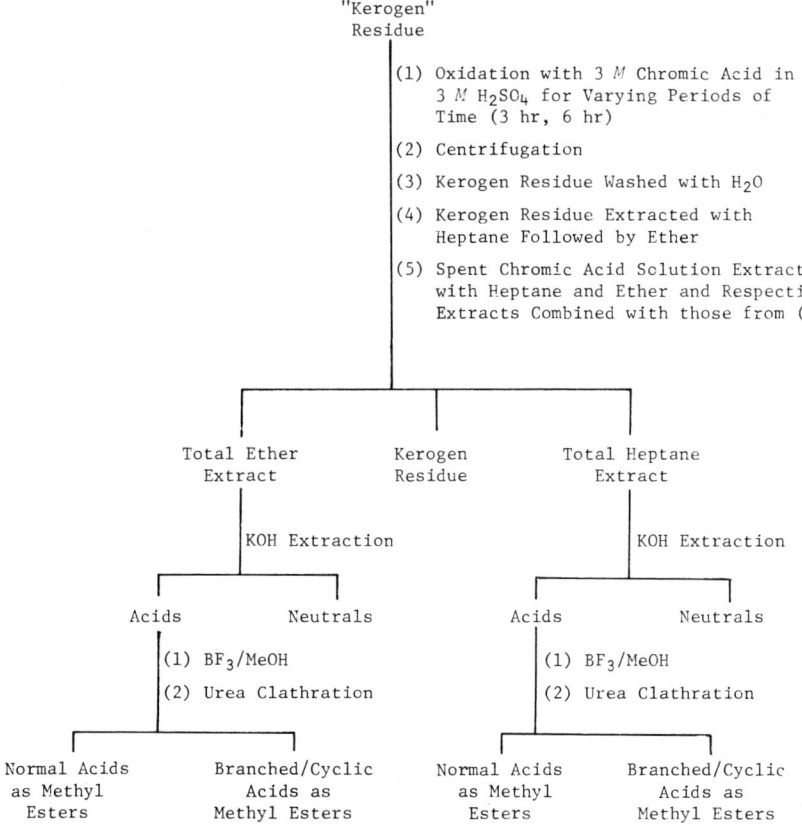

Figure 10.2 Oxidation scheme used on insoluble kerogen fraction from the algal ooze.

30 ft x 0.03· in. i. d. glass capillary column packed with 1% OV-1 coated on 80-100 mesh Gas Chrom Q. The mass spectral data were acquired and processed using a DuPont 21-094 data system.

RESULTS

The elemental analyses of the algal ooze, the kerogen residue and the residues after the various oxidations of the kerogen residue are given in Table 10.1. The yields of total and acid extracts for all extraction steps are listed in Table 10.2.

Table 10.1
Elemental Analyses of Algal Ooze and
Kerogen Residues at the Various Degradation Steps

	Dried Algal Ooze %	Kerogen Residue %	After 3-hr Oxidation %	After 6-hr Oxidation %
C	9.3[a]	10.31[b]	0.86[b]	0.54[b]
H	1.8	1.59	1.11	0.82
N	0.38	0.74	0.10	0.03
S	0.12	0.52	0.45	0.08
Residue	67.9	79.8	88.3	93.5
O[c]	20.5	7.04	9.18	5.03

[a] Total carbon content.
[b] Total organic carbon content.
[c] By difference.

Table 10.2
Extract Yields from Extraction and Oxidations
of the Algal Ooze Sample

Sample	Totals (Extracts or Residues)	Acid Fractions (mg) Heptane Soluble	Ether Soluble
1. Dried algal ooze	63 g	--	--
2. Hydrolyzed and exhaustively extracted ooze (i.e., kerogen residue)	4.3 g	--	--
3. Soluble lipid extract from hydrolyzed (1)	54 mg	39[a]	--
4. 3-hr Oxidation of (2)	149 mg	18	95
5. 6-hr Oxidation of residue from (4)	24 mg	2	9
6. Residue after 6-hr oxidation	1.9 g	--	--
7. Total solvent solubles and residues recovered	1.127 g	20	104

[a] Total solvent soluble acids

Exhaustively Extracted Fatty Acid Fractions

As mentioned, a detailed description of the soluble fatty acid extract can be found elsewhere in this volume (Cardosa et al., 1975). However, a preliminary examination of this fraction by computerized-gas chromatography-mass spectrometry (C-GC-MS) revealed the normal fraction to predominantly consist of a homologous series of normal saturated fatty acids in the range C_{14}-C_{30} with maxima at C_{16} and C_{26}. The presence of mono-unsaturated C_{16} and C_{18} fatty acids was also detected. Minor quantities of iso- and anteiso-C_{15} acids were observed in the branched/cyclic fractions along with relatively large quantities of a higher molecular weight acid tentatively identified as a C_{32}-triterpanoic acid. In view of the discussion below it is important to note that no dicarboxylic acids could be detected in this fraction by mass fragmentography (Hites and Biemann, 1970) or interpretation of individual spectra.

Three-hour Oxidation

Branched-Cyclic Fractions

The heptane fraction consisted mainly of a series of isoprenoid monocarboxylic acids dominated by C_{14}, C_{15}, C_{16}, C_{17}, C_{19}, C_{20} acids (peaks 1, 2, 3, 5, 7, 8 respectively in Figure 10.3) and their relative distribution is clearly illustrated in the lower gas chromatogram shown in Figure 10.3. Two other branched aliphatic carboxylic acids (C_{17} and C_{18}, peaks 4 and 6 respectively in Figure 10.3) were also present. The diethyl ether extract was dominated by a series of monomethyl branched dicarboxylic acids in the range C_4-C_8. No aromatic, cyclic or keto acids could be detected in either the heptane or the ether fraction.

Normal Fractions

The normal heptane extract was dominated by a series of normal monocarboxylic acids in the range C_{12}-C_{27} with a maximum at C_{16}. A series of saturated unbranched dicarboxylic acids were present in minor quantities in the range C_{13}-C_{27} with a maximum at C_{21}. A second series of saturated branched dicarboxylic acids (range C_{11}-C_{15}, max C_{13}) was also present in minor quantities. The other homologous series identified in this fraction was a series of α-methyl-branched monocarboxylic acids in the range C_{13}-C_{20} with a maximum at C_{17} (Figure 10.3).

The ether extract was dominated by the more polar unbranched dicarboxylic acids in the range C_8 to C_{27} with a maximum at C_9. The acids in the range C_{13}-C_{27} were present only in minor quantities. There was also another series of saturated branched monocarboxylic acids in this fraction in the range C_8-C_{13} with a maximum at C_{10} (Figure 10.4).

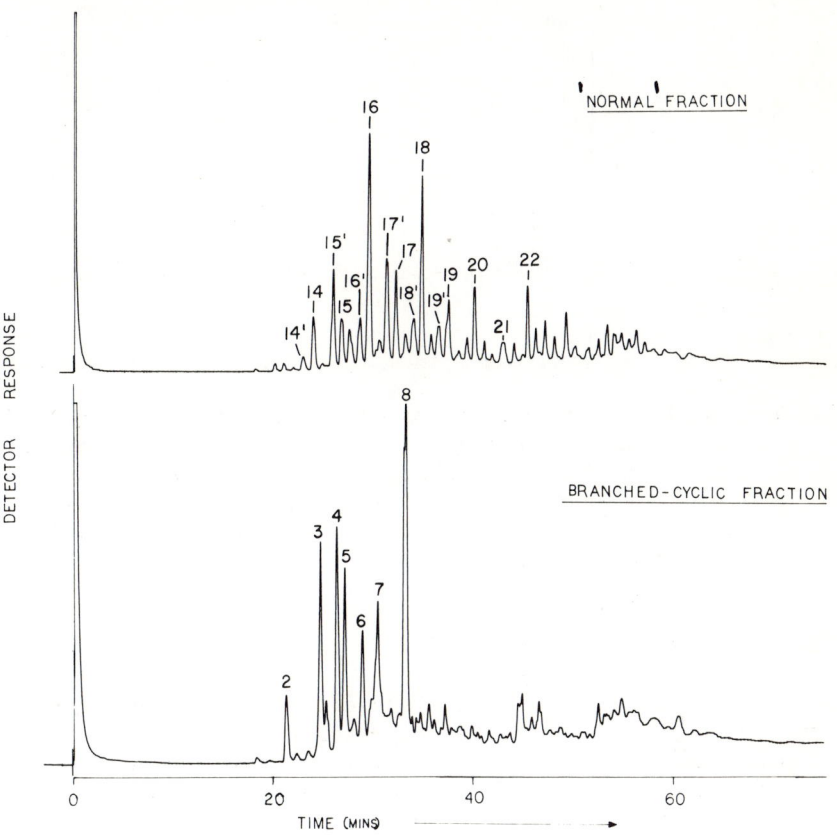

Figure 10.3   Gas chromatograms of the heptane-soluble normal and branched/cyclic acid esters isolated from the 3-hr oxidation of the algal kerogen (GC conditions are given in the text). Only the major components in each fraction are labeled in this figure. In the upper trace the carbon numbers of the normal fatty acids are indicated by the arabic numerals and the monomethyl branched acids indicated by the primed arabic numerals. In the lower trace peak numbers 1, 2, 3, 5, 7, 8 correspond to isoprenoid acids and numbers 4 and 6 to branched acids whose structures have not been assigned as yet.

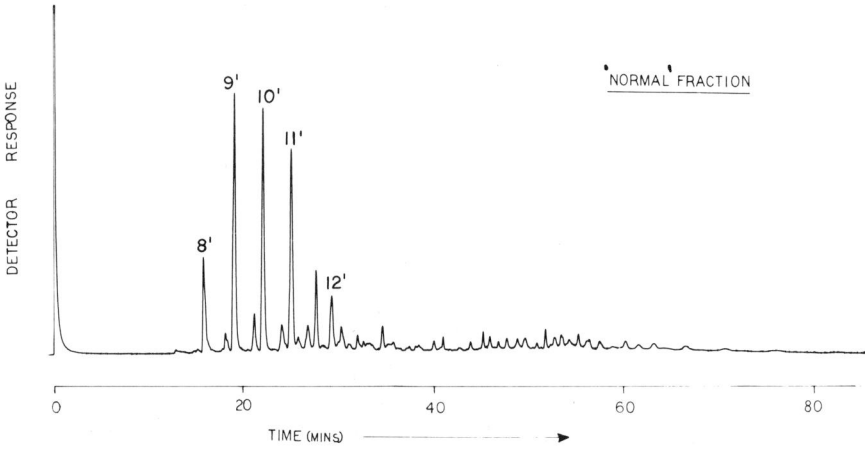

Figure 10.4 Gas chromatograms of the ether soluble normal acid esters isolated from the 3-hr oxidation of the algal kerogen (GC conditions are given in the text). The primed arabic numerals refer to the carbon number of the normal dicarboxylic acid esters.

Six-hour Oxidation

These fractions were not subjected to urea adduction due to the small quantities of material available. The heptane extract did not give any resolvable peaks by GC and was therefore not further analyzed by GC-MS. The ether extract basically consisted of two homologous series. The first was a series of normal acids in the range $C_{17}-C_{29}$ with a maximum at C_{24}, and the other a series of α,ω-dicarboxylic acids in the range $C_{12}-C_{24}$ with a maximum at C_{20}.

DISCUSSION

Before commencing the discussion on the qualitative aspects of these results it is valuable to examine the quantitative results shown in Tables 10.1 and 10.2. First, from Table 10.1 it can be seen that after only the 3-hr oxidation the organic carbon content of the kerogen residue has been reduced to 0.86% and is further reduced to 0.54% after 6 hr. First, this suggests that there is kerogenous material in this very young algal debris which can be degraded by oxidation. Second, this kerogen appears less resistant to oxidative degradation than the kerogen from Tasmanian tasmanite. In this case 90 hr were required for the organic carbon content to be reduced to the value obtained for the algal kerogen after a 3-hr oxidation period (Simoneit and Burlingame, 1973).

From Table 10.2 several important factors emerge, primarily the fact that 58.7 g of material has been removed in the initial hydrolysis experiment. This suggests the presence of large amounts of carbonate in the algal debris. A certain amount of lipid material will also be water-soluble and hence not accountable for in this particular work-up scheme. The majority of degradation products are produced after the 3-hr oxidation, once again suggesting that the kerogen nucleus is less resistant to oxidation than the kerogens from older shales. The residue remaining after the 6-hr oxidation is predominantly inorganic silicates as shown by X-ray fluorescence measurements. Again the discrepancy in the materials balance is due to the production of carbon dioxide and water-soluble products during the oxidation reactions. It is intended that future experiments will be directed towards permanganate oxidation of the algal kerogens using a method similar to that described by Djuricic and co-workers (1971) which includes removal of the products from the oxidative environment as soon as they are formed.

From these results it appears that the nucleus of this kerogenous material is predominantly a system of cross-linked polymethylene chains of varying chain lengths and various degrees of branching. The kerogen appears to be highly aliphatic, only trace amounts of aromatic acids were detected. It would also appear that condensed to this system on the periphery of the nucleus are unbranched hydrocarbon chains and also isoprenoid chains which when oxidized give rise to the normal and isoprenoid acids, respectively. The fact that less time is required to completely oxidize this immature kerogen-like material would suggest that it is not a very highly condensed system as some of the more mature kerogens appear to be. It also suggests that the cross-linkages are more susceptible to oxidation than in the older kerogens. In the soluble fractions there are relatively large amounts of cyclic acids, such as the C_{32} triterpanoic acid mentioned above. However, only trace amounts of cyclic acids could be detected in the products from the kerogen oxidation. This suggests that the first stage in the formation of the kerogen is the condensation of the functionalized aliphatic lipids to give the cross-linked polymethylene nucleus. The branched dicarboxylic acids and isoprenoid acids could either be condensed on

the system at its periphery or alternatively at this very
early stage in the kerogen formation could be trapped
within the polymethylene matrix prior to condensation. The
most probable source of the isoprenoid acids is the phytol
side of chlorophyll. Simoneit and co-workers (personal
communication) have shown by stereochemical determinations
that this indeed is the case with the Green River Shale.
The monomethyl branched dicarboxylic acids from the 3-hr
oxidation could possibly arise from the mixture of 7- and
8-methyl-heptadecanes which are known to be the major hydrocarbon constituents of blue-green algae (Han et al., 1968).
Again with this young immature kerogen these constituents
could be entrapped within the polymethylene matrix and on
oxidation give rise to the branched dicarboxylic acids
detected in certain fractions.

The results of this preliminary study have indicated
that there is kerogenous-like material present in these
recently deposited algal mats whose structure bears certain
similarities to that of more mature and inert kerogens.
However, the problem still arises as to the method of
formation of the kerogen nucleus in these algal mats.
It also remains to be shown whether or not this material
is formed by simple condensations of soluble unsaturated
lipids known to be present in the algal mats. Alternatively,
as suggested by Oehler and co-workers (1974), intracellular
chlorophyll molecules might become grafted onto cellular
macromolecules through ester and related linkages and these
grafted pigment complexes may become incorporated into the
insoluble kerogen fraction.

CONCLUSIONS

1. On the basis of the degradation products obtained
it can be said that a kerogen-like material is formed early
on in the deposition of the algal mats.
2. The main kerogen nucleus seems to consist of polymethylene chains with some cross-linking, a small amount
of aromatic material, plus unbranched material on the periphery.
3. The kerogen nucleus is probably derived either
from condensation of unsaturated oxygenated compounds or
even the chlorophyll moeity.
4. The presence of relatively abundant quantities of
phytanic acid suggests that this originates from the chlorophyll and is ester-linked to the kerogen nucleus.
5. To expand the idea that kerogen is being formed
rapidly, a depth study of kerogens from these algal mats

will be undertaken to see if the degradation products show increasing complexity with depth.

This initial study has provided some answers to the questions asked above but until further samples are examined it will not be possible to say whether the kerogen can be used as a chemotaxonomic indicator or whether there are changes in its complexity with changes in depth.

Two very important issues which still remain unanswered relate to 1) the method of kerogen formation, and 2) the types of linkage that hold the kerogen matrix together. To provide an answer to the first issue is difficult. One way to study the short-term fate of biolipids is to use ^{14}C-labeled lipids and incubate them in the sediment for a period of several months or more. However, in order to observe the incorporation of these biolipids into the kerogen matrix would take an impractical period of time. On the second point one can speculate and say that there are ether linkages or other types of heteroatom linkages in the kerogen which are attacked on oxidation with chromic acid. One important point to raise here is what are the differences between the linkages in this algal kerogen which appear to be readily attacked by chromic acid and those in ancient kerogens which are fairly resistant to oxidative degradation? Is it merely the type of cross-linkage, or is it due to that fact that the older kerogens are more highly condensed and thus potential sites of oxidation more sterically hindered?

It is anticipated that by looking at other algal kerogens and also other slightly older kerogens the answers to these questions and others may become apparent.

ACKNOWLEDGMENTS

We thank Mr. Stephen Brown for assistance in obtaining the GC-MS data and Ms. Adrienne Ross for typing the manuscript. We also wish to thank Dr. S. W. Awramik and Mr. B. Simoneit for providing critical comments on the manuscript This work was funded by NASA grant number NGL 05-003-003 and the DuPont data system was purchased on National Cancer Institute contract #NCI-FS-(71)-58.

REFERENCES

Bradley, W. H., 1966: Tropical lakes, copropel and oil shale *Bull. Geol. Soc. Amer.* 77, 1333-1337.

Bradley, W. H., 1970: Green River oil shale - concept of origin extended. *Bull. Geol. Soc. Amer.* 81, 985-1000.

Burlingame, A. L., P. A. Haug, H, K. Schnoes, and B. R. Simoneit, 1969a: Fatty acids derived from the Green River formation oil shale by extractions and oxidations- a review. In: Schenck, P. A. and I. Havenaar (eds.) *Advances in Organic Geochemistry 1968*, (Oxford: Pergamon Press), pp. 85-129.

Burlingame, A. L. and B. R. Simoneit, 1968a: Isoprenoid fatty acids isolated from the kerogen matrix of the Green River formation (Eocene). *Science* 160, 531-533.

Burlingame, A. L. and B. R. Simoneit, 1968b: Analysis of the mineral entrapped fatty acids isolated from the Green River formation. *Nature.* 218, 252-256.

Burlingame, A. L. and B. R. Simoneit, 1969b: High resolution mass spectrometry of Green River formation kerogen oxidations. *Nature* 222, 741-747.

Cane, R. F., 1969: Coorongite and the genesis of oil shale. *Geochim. Cosmochim. Acta* 33, 257-265.

Cardosa, J., P. Brooks, G. Eglinton, R. Goodfellow, J. R. Maxwell and R. P. Philp, 1975: Lipids in recently deposited algal mats at Mormona, Baja California. This volume.

Down, A. L. and G. W. Himus, 1941: A preliminary study of the constitution of kerogen. *J. Inst. Pet.* 27, 426-444.

Djuricic, M. V., D. Vitorovic, B. D. Andresen, H. S. Hertz, R. C. Murphy, G. Preti and K. Biemann, 1971: Acids obtained by oxidation of kerogens of ancient sediments of different geographic origin. In: von Gaertner, H. R. and H. Wehner (eds.) *Advances in Organic Geochemistry 1971* (Oxford: Pergamon Press), p. 305-321.

Eglinton, G., 1973: Chemical Fossils: A combined organic geochemical and environmental approach. *Pure and App. Chem.* 34, 611-632.

Forsman, J. P. and J. M. Hunt, 1958: Insoluble organic matter (kerogen) in sedimentary rocks. *Geochim. Cosmochim. Acta* 15, 170-182.

Golubic, S., 1973: The relationship between blue-green algae and carbonate deposits. In: Carr, N. G. and B. A. Whitton (eds.), *The Biology of Blue-Green Algae* (Berkley: University of California Press), 434-472.

Han, J., E. D. McCarthy, W. Van Hoeven, M. Calvin and W. H. Bradley, 1968: Organic geochemical studies II. A preliminary report on the distribution of aliphatic hydrocarbons in algae, in bacteria and in a recent lake sediment. *Proc. Nat. Acad. Sci. USA* 59, 29-33.

Hites, R. A. and K. Biemann, 1970: Computer evaluation of continuously scanned mass spectra of gas chromatographic effluents. *Anal. Chem.* 42, 855-860.

Jones, C. L., 1922: The problems of the shale oil industry. *Chem. Met. Eng.* 26, 546-553.

Lawlor, D. L., J. I. Fester and W. E. Robinson, 1963: Pyrite removal from oil shale concentrates using lithium aluminum hydride. *Fuel* 42, 239-244.

Logan, R. W., R. Rezak and R. N. Ginsburg, 1964: Classification and environmental significance of algal stromatolites. *J. Geol.* 72, 68-83.

Oehler, J. H., Z. Aizenshtat, and J. W. Schopf, 1974: Thermal alteration of blue-green algae and blue-green algal chlorophyll. *Am. Assoc. Petroleum Geol. Bull.* 58, 124-132.

Robinson, W. E., J. J. Cummins and K. E. Stanfield, 1956: Constitution of organic acids prepared from Colorado oil shale based on n-butyl esters. *Ind. Eng. Chem.* 48, 1134-1138.

Robinson, W. E., H. H. Heady and A. B. Hubbard, 1953: Alkaline permanganate oxidation of oil-shale kerogen. *Ind. Eng. Chem.* 45, 788-791.

Robinson, W. E. and D. L. Lawlor, 1961: Constitution of hydrocarbon-like materials derived from kerogen oxidation products. *Fuel* 40, 375-388.

Robinson, W. E., D. L. Lawlor, J. J. Cummins and J. I. Fester, 1963: Oxidation of Colorado oil shale. Bureau of Mines, Rept. of Invest. 6166, 33 pp.

Sackett, W. M., C. W. Poag and B. J. Eadie, 1974: Kerogen recycling in the Ross Sea, Antarctica. *Science* 185, 1045-1047.

Schopf, J. W. and E. S. Barghoorn, 1967: Alga-like fossils from the early precambrian of South Africa. *Science* 156, 508-512.

Simoneit, B. R. and A. L. Burlingame, 1973: Carboxylic acids derived from Tasmanian tasmanite by extractions and kerogen oxidations. *Geochim. Cosmochim. Acta* 37, 595-610.

Thorne, H. M., K. E. Standfield, G. U. Dinneen and W. I.

Murphy, 1964: Oil Shale technology: a review. Information circular 8216 U. S. Bureau of Mines, Washington.

Vonder Haar, S. P., 1973: Evaporite environment at Laguna Mormona, Pacific coast of Baja California, Mexico. *Geol. Soc. Amer. Abs.* with Programs 5, 117-118.

CHAPTER 11

LIPIDS OF RECENTLY-DEPOSITED ALGAL MATS
AT LAGUNA MORMONA, BAJA CALIFORNIA

J. CARDOSO, P. W. BROOKS, G. EGLINTON
R. GOODFELLOW, J. R. MAXWELL

Organic Geochemistry Unit, The School of Chemistry,
The University, Bristol BS8 1TS, England

R. P. PHILP

The Chemistry Department, The University of
California, Berkeley, California 94720

INTRODUCTION

One key to understanding the geolipid content of ancient sediments lies in the study of the lipid composition of present-day environments and recent sediments. However, a number of parameters must be considered involving the deposition of aquatic sediments. A previous survey of several temperate lacustrine and subtropical lagoonal sediments (Eglinton *et al.*, 1974) revealed some differences in distribution for various lipid classes, presumably as the result of differing conditions of deposition, trophic status, extent of microbial transformation, etc. The situation is often complicated by the simultaneous contribution to the sediment from both autochthonous and allochthonous (terrestrial biota) inputs. Furthermore, the increasing widespread extent of both domestic and industrial pollution of aquatic environments adds to the difficulty of understanding the lipid distributions of contemporary sediments. Clearly, one of the major problems in the study of recent sediments lies in the choice of a model environment. Thus, an environment having clearly defined input sources, a well-studied history of deposition, and relative seclusion from any major pollution source is required. Furthermore, the paleoenvironmental significance of the sediment should be such as to render the results of its investigation valid for extrapolation to geological formations. The contribution of algal deposits to several ancient sediments is well-documented (*e.g.*, Traverse, 1955; Thorne *et al.*, 1964) and indicates that the study of recent algal formations should prove useful.

A preliminary survey of the lipid composition of sediments of a subtropical, hypersaline coastal pond at Laguna Mormona (Baja California, Mexico) is described herein. A comprehensive description of this algal mat formation is given elsewhere (Philp and Calvin, 1975; Horodyski and Van der Haar, 1975) and only a diagrammatic illustration of the stratified layers present in the core taken for study will be given here (Figure 11.1). Samples from the mat and the organic ooze were chosen for study. (The gelatinous, intermediate layer was not examined. In addition, other mat and ooze samples as far apart as 2 km have been briefly examined.) It was thought that a comparison of the lipid composition of the mat and ooze layers might provide information about the early-stage changes attending the formation of algal mat deposits.

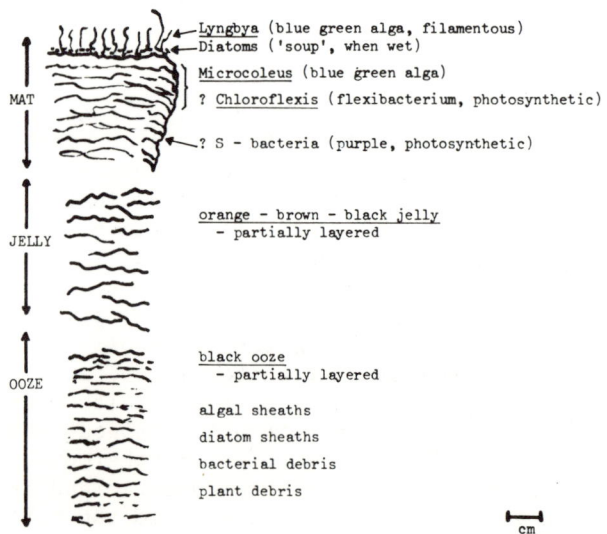

Figure 11.1 *Schematic representation of layering in the analyzed core. S. Awramik (U. C., Santa Barbara), Horodyski (U.C.L.A.), Vonder Haar (U.S.C.) personal communication.*

EXPERIMENTAL

The extraction and fractionation procedures were performed as described previously (Eglinton, 1973; Eglinton et al., 1974). All solvents used were redistilled. All glassware was pre-cleaned with hot chromic acid. Care was taken throughout to minimize contamination.

Thin Layer Chromatography (TLC)

Separation of saturated and unsaturated hydrocarbons and esters, respectively, was carried out on silica gel G impregnated with 10% silver nitrate. Developing solvents were hexane and hexane/diethyl ether (90:10), respectively. Alcohols were separated on silica gel G developing with chloroform into fractions containing the 4-desmethyl sterols (R_f = 0.2) and other alcohols (R_f = 0.3).

Gas Liquid Chromatography (GLC)

GLC analyses were carried out on a Varian 2700 instrument equipped with hydrogen flame ionization detector and linear temperature programmer. Open tubular coated capillary columns (glass 20 m x 0.25 mm) were used in most analyses, operated at 140-280° at 4°/min with helium carrier gas (ca. 2 ml/min) and sample split of ca. 20/1, unless otherwise stated. Column efficiencies were typically 35-45,000 theoretical plates measured with 5 α-cholestane at 270°C. For some of the sterol and alcohol analyses, packed columns (stainless steel 1.5 m x 0.3 cm) packed with 3% OV-17 on Gas Chrom Q were used and operated at 270° isothermal. Carrier gas flow rate (N_2) was 15-20 ml/min.

Computerized Gas Chromatography-Mass Spectrometry (C-GC-MS)

Low resolution spectra were recorded using a Varian Aerograph 1200-Varian MAT CH-7 gas chromatograph-mass spectrometer combination, with the total column effluent delivered to the ion source via an all-glass Watson Biemann separator. Spectra were recorded at a filament current of 300 µA and 70 eV ionization energy. The columns were either wall-coated (OV-101) capillary (glass 20 m x 0.5 mm), or packed stainless steel (3 m x 0.16cm) with 3% Dexsil on Gas Chrom Q. Operating conditions were as above. The spectrometer was linked to a 16K memory PDP 8e computer via a Carrick interface (Eglinton et al., 1974) and was operated in the continuous cyclic scanning mode.

Table 11.1
Hydrocarbons and Fatty Acid Distributions

Sample	Hydrocarbons (ppm)[a]				Fatty Acids (ppm)[a]			
	Alkanes		Monoenes		Saturated		Monounsaturated	
	N[b]	B/C[b]	N	B/C	N	B/C	N	B/C
Mat (104.3 g dry wt)	82.5	26.9	6.7	32.6	181.2	52.7	147.7	6.7
Ooze (75.2 g dry wt)	51.0	11.0	9.3	33.0	63.8	9.3	12.0	4.0

[a] All ppm values were calculated relative to the dry weight of sample after extraction.
[b] N = normal (straight chain). B/C = branched and cyclic chain.

Methylation and Acetylation

Conversion of free fatty acids to methyl esters was carried out using ethereal diazomethane (Schlenk and Gellerman, 1960). Alcohols (1-50 mg) were acetylated using acetic anhydride-pyridine (1:1) at room temperature overnight. After addition of water (20 ml), the acetates were extracted with hexane.

Urea Adduction

Separation of straight-chain and branched and cyclic lipids was carried out by urea clathration. A saturated solution of urea in methanol (1 ml) was added to a solution of the mixture in hexane (4 ml) and acetone (2 ml). After evaporation of the solvent, the non-adduct was extracted with hexane. The adduct fraction was recovered by dissolution of the urea crystals in water and extraction with hexane.

Sample Collection and Extraction

The sample of algal mat was collected using a box corer and each layer bottled separately. The samples were stored in sealed glass jars at 4°C until analysis. The two layers (Figure 11.1) were extracted by a modification (Rhead, 1971) of the method of Dole and Meinertz (1960), using heptane/isopropanol (1:4). The total extract was saponified, by heating under reflux (2 hr) with methanolic KOH (6 N) and the acids removed. The nonsaponifiable lipids were fractionated on an alumina column (15 cm x 2 cm), deactivated with 5% water. The total hydrocarbon fraction was eluted with hexane (250 ml) and divided into three classes by Ag^+ TLC according to the degree of unsaturation. The alcohol fraction was obtained by elution with either (250 ml). A more polar fraction was eluted with methanol (250 ml) but was not examined further.

RESULTS

A diagrammatic representation of the core is shown in Figure 11.1. Two layers were examined, *viz*. 1) the upper 2.0 cm-thick layer) comprising the fresh algal mat, predominantly the blue-green species *Microcoleus chthonoplastes* (Philp and Calvin, 1975), and 2) the black anaerobic algal ooze at the depth of *ca*. 10 cm.

Most of the interwoven filamentous material of the upper layer is made up of fibers of *M. chthonoplastes*. The presence of *Lyngbya aestuarii* and diatoms appears to be restricted to the mat surface, while there is evidence for the presence of purple photosynthetic bacteria in the lower mat layer.

The ooze shows little stratification, being made up essentially of partially transformed algal and plant debris.

n- Alkanes

The distributions are shown in Figure 11.2. About 75% of the n-alkanes in the mat are accounted for by n-C_{17}, with smaller amounts of higher homologues maximizing at n-C_{27}. The ooze is characterized by a bimodal distribution with maxima at n-C_{17} and n-C_{27}. A high CPI value was noted for both layers, although some loss of odd/even dominance (4.5 as compared with 5.9, Figure 11.2) is apparent in the ooze.

A similar situation was observed for other samples collected from different sites. The mats are characterized by a dominant n-C_{17} component with little contribution from homologues above n-C_{20}. On the other hand, a bimodal distribution, with the first maximum at n-C_{17} and the second one around n-C_{27} was apparent in all ooze samples.

n-Alkanoic Acids

The n-alkanoic acids (Figure 11.2) show a parallelism with the corresponding n-alkane distributions. Palmitic acid is thus the major constituent in the mat with smaller amounts of the >C_{20} components. A clear bimodal distribution is apparent in the ooze with n-$C_{24:0}$ being the most abundant. This change is accompanied by a decrease in CPI values from 9.8 to 5.7.

Similar distributions were also apparent in other samples analyzed. Thus a uniform but distinct distribution seems to characterize the fatty acid content of mat and ooze layers, respectively.

A marked decrease in the ratio of monounsaturated to saturated acids in the ooze relative to the mat was also observed (Table 11.2) as indicated by the ratio of n-$C_{18:1}$/n-$C_{18:0}$ decreasing from 2.25 to 0.69.

Branched and Cyclic Alkanes

This fraction from the mat is characterized by relatively few major components (Figure 11.3 and Table 11.3). Pristane (I), phytane (II), 7- and 8-methylheptadecanes (Han and Calvin 1969) and an unidentified $C_{20}H_{40}$ monocyclic hydrocarbon (Brooks, 1974) are present. This same range of compounds

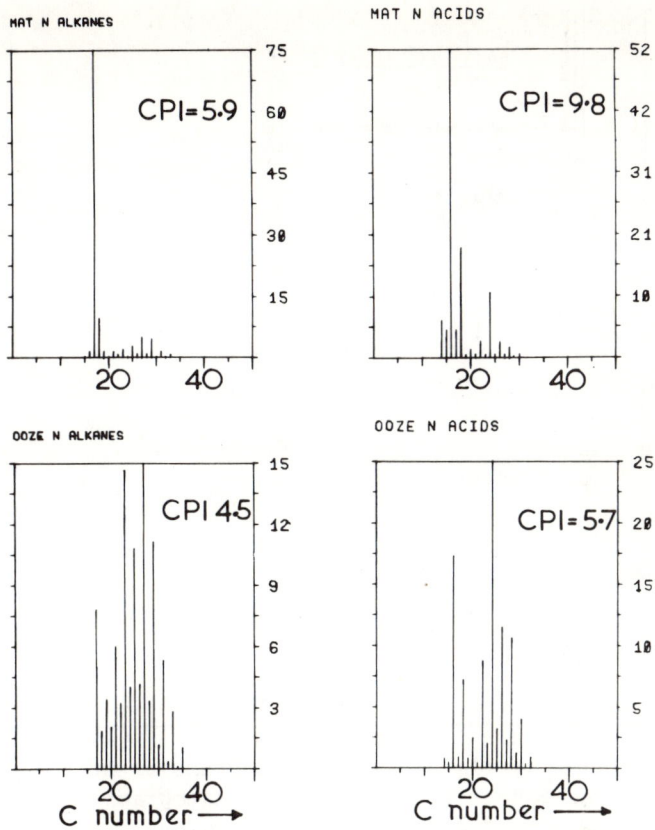

Figure 11.2 *n*-Alkane and *n*-alkanoic acid distributions. (Figures on the vertical axis show the relative percent abundances of each component.)

Table 11.2
Distribution of the C_{16} and C_{18}
Alkanoic and Monoenoic Acids

Sample	Concentration (ppm)				Ratios	
	$C_{16:0}$	$C_{16:1}$	$C_{18:0}$	$C_{18:1}$	$C_{16:1}/C_{16:0}$	$C_{18:1}/C_{18:0}$
Mat	60	50	40	90	0.83	2.35
Ooze	9	0.6	3.8	2.6	0.06	0.69

156 Environmental Biogeochemistry

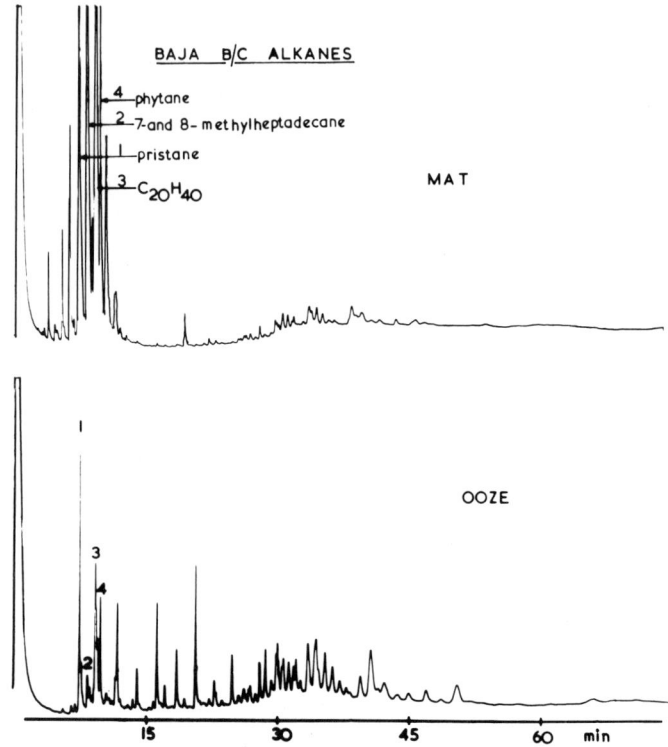

Figure 11.3 Capillary gas chromatograms of branched/cyclic alkane fractions. Chromatograms were recorded using a 20 m x 0.25 mm glass capillary, coated with OV-101 and temperature programmed from 140-280° at 4°C/min.

Table 11.3
Distribution of Branched/Cyclic Alkanes

Compound	Peak Number	B/C Alkanes	
		Mat (ppm)	Ooze (ppm)
Pristane	1	1.1	0.030
7- & 8-MeC$_{17}$	2	0.7	0.003
C$_{20}$H$_{40}$	3	0.4	0.014
Phytane	4	0.1	0.009
Others		<0.004	<0.003

is present in the ooze, although changes in relative concentration are apparent. A higher relative concentration of high-molecular-weight components is present in the ooze. This is accompanied by a smaller relative abundance of the 7- and 8-methylheptadecanes and the appearance of an unidentified series of equally spaced peaks, suggestive of an homologous distribution, with carbon number preference. A similar series of peaks was reported by Brooks (1974) for an algal mat and a marine sediment.

Branched and Cyclic Alkenes

Separation of the total B/C-hydrocarbons by Ag^+ TLC afforded a cut of R_f ca. 0-0.15 for both the mat and ooze. A major component of lower-molecular weight from both layers was identified as the C_{18} isoprenoid ketone, 6, 10, 14-trimethylpentadecan-2-one (III) (Table 11.4). This ketone was previously identified in other recent sediments (Ikan et al., 1973; Burlingame and Simoneit, 1974; Brooks, 1974) and has been shown to be a degradation product of phytol (IV) (Brooks, 1974).

Mono-and di-unsaturated hydrocarbons of carbon number C_{17}-C_{20} were detected in both layers but were not investigated further. Hydrocarbons of higher molecular weight were detected in the ooze but not in the mat. Preliminary mass spectral evidence suggests that they comprise a complex mixture of mono- and di-unsaturated sterenes (Table 11.4), ranging in carbon number from C_{27}-C_{29}. Furthermore, mass fragmentography suggests the presence of minor amounts of a possible C_{28} 4-methylsterene. A major component in this region of the chromatogram has also been identified by both co-chromatography and mass spectral comparison with an authentic standard as hop-22(29)-ene(diploptene, V). Trace amounts of this hydrocarbon were also tentatively identified in the mat by coinjection with authentic standard. Hop-22-(29)-ene has been reported as a major constituent of some species of blue-green algae, including *L. aestuarii* (Gelpi et al., 1970) and in a species of bacterium (Bird et al., 1971a,b).

Branched and Cyclic Alkanoic Acids

The lower-molecular-weight region of this fraction in the mat (Figure 11.4, Table 11.5) contains a number of components typically found in recently deposited sediments although relative abundances vary considerably (Eglinton, et al., 1974; Cranwell, 1973; Leo and Parker, 1966; Eglin-

158 Environmental Biogeochemistry

Table 11.4
Distribution of Branched/Cyclic Alkenes

Component	Concentration (ppm)	
	Mat	Ooze
6,10,14-trimethylpentadecan-2-one	2.43	0.20
C_{21} diterpene	–	0.11
Cholest-2-ene	–	0.38
C_{27} diunsat. sterene	–	tr*
C_{28} diunsat. sterene	–	0.06
C_{28} monounsat. sterene (4-Me?)	–	0.22
C_{28} monounsat. sterene	–	0.18
C_{29} diunsat. sterene	–	0.08
C_{29} monounsat. sterene	–	0.14
$C_{30}H_{50}$ triterpene	–	tr
Hop-22(29)-ene	tr	0.34

*tr = trace amounts

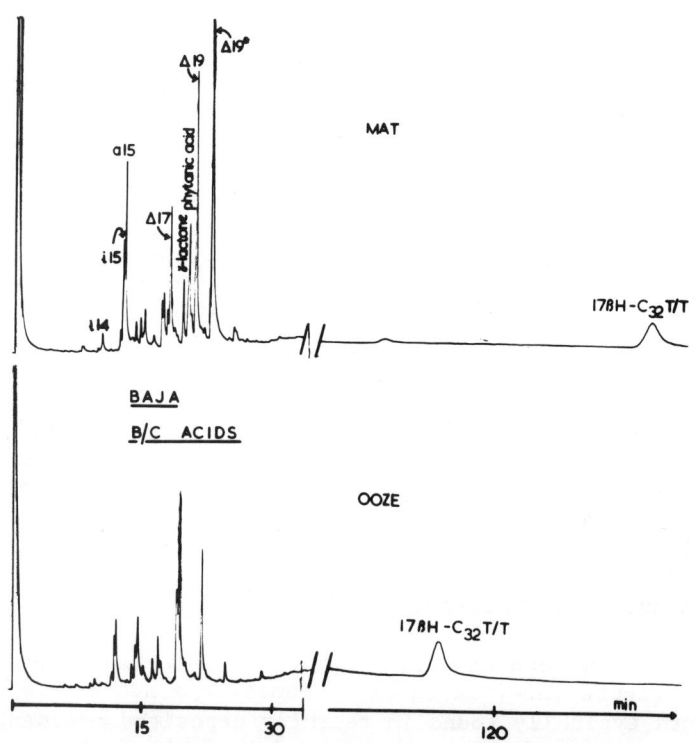

Figure 11.4 Capillary gas chromatograms of branched/cyclic alkanoic acids (as methyl esters). GLC conditions as for Figure 11.3.

ton, 1968). Thus *iso-* and *anteiso-*C_{15} together with cyclopropyl-C_{17} and C_{19} acids appear as dominant constituents. The corresponding fraction for the ooze could not be examined by GC-MS in view of the limited amount of sample available. A GLC trace for this fraction is shown in Figure 11.4 for comparison. The presence of a C_{32} triterpenoid acid (17βH-bishomohopanoic acid, VI) in the mat was confirmed by GC-MS and its presence tentatively inferred in the ooze layer by coinjection of the two fractions.

Sterols and Alcohols

The sterols of the mat and ooze were similar in composition, but the mat (25 ppm) contained a much higher concentration of sterols than the ooze (2.6 ppm). The sterols were subjected to GLC analysis as the acetates, which showed five major components corresponding in retention time to $C_{27}\Delta^5$, $C_{28}\Delta^{5:22}$, $C_{28}\Delta^5$, $C_{29}\Delta^{5:22}$ and $C_{29}\Delta^5$ steryl acetates. Greater resolution was obtained on capillary GLC (Figure 11.5) whereby each peak split into at least two components which corresponded, on the basis of coinjection data, to the Δ^5 steryl acetates and their ring-saturated analogues. This enabled the 5α-stanol:Δ^5 steryl ratio to be calculated by peak triangulation for each pair of compounds (Figure 11.6). To further strengthen the characterization of the sterol mixture, an aliquot of the steryl acetates from the algal mat was reduced to the free alcohols with lithium aluminum hydride and the sterols and stanols separated by Ag^+ TLC. GLC analysis of these fractions as their trimethylsilyl ethers confirmed the identifications based on the coinjection results.

The GLC traces of the alcohols (excluding 5α-sterols) as their acetates of the algal mat (122 ppm) and the ooze (62.5 ppm) were similar in composition, although minor differences in the proportions of some of the components were apparent. The major constituent in both the mat and the ooze was phytol with minor amounts of a series of *n*-alcohols (principally C_{22}, C_{24} and C_{26}). The remainder was a complex mixture of 5β-stanols and tetra- and pentacyclic (triterpenoid) alcohols (Table 11.6). 17βH- bishomohopanol(VII) along with a C_{30} pentacyclic triterpenoid alcohol [apparently 22- or 29- hydroxyhopane (VIII)] were identified by mass spectral comparison and, for the C_{32} compound, by coinjection with an authentic standard. The ratio of 5α:5β stanols was found to be approximately 4:1 in the mat.

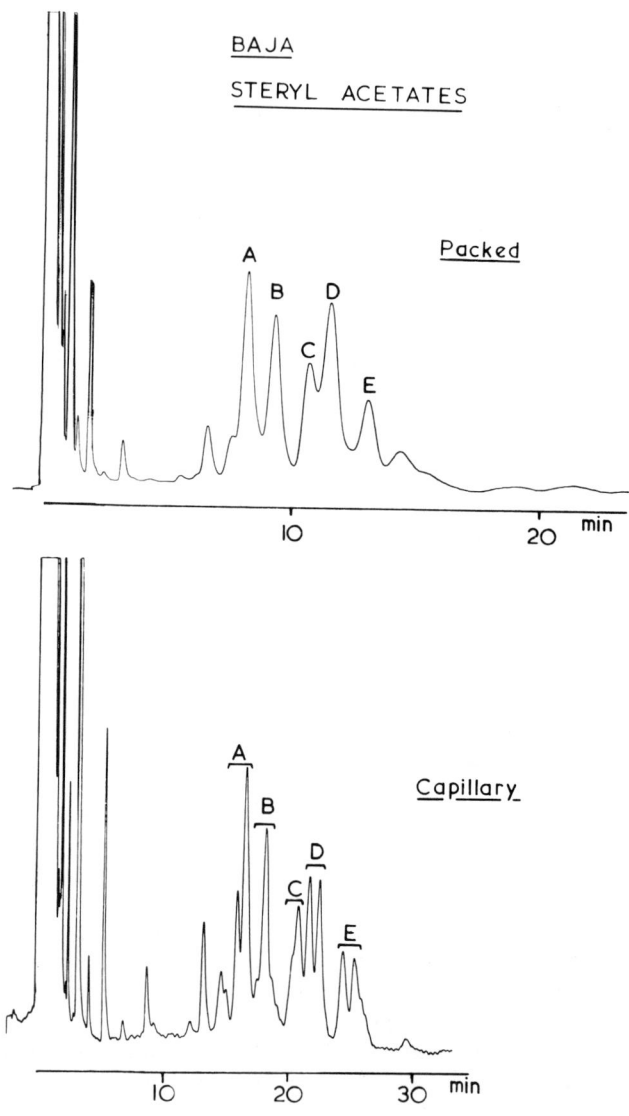

Figure 11.5 Gas chromatograms for the mat sterols (as acetates) on packed (1% OV-17 on Gas Chrom-Q, 80-120 mesh) and capillary columns (OV-101 glass column, 20 m x 0.25 mm) at 260°C.

Table 11.5
Distribution of Branched/Cyclic
Alkanoic Acids in the Mat Layer

Component	B/C Acids (ppm)
i-C_{14} [a]	0.38
i-C_{15} [a]	1.92
a-C_{15} [b]	2.87
Δ-C_{17} [c]	2.87
phytanic acid	1.53
Δ-C_{19} [d]	5.07
Δ-C_{19} [d]	8.42
17βH-bishomohopanoic acid	2.20

[a] i = iso-
[b] a = anteiso-
[c] Δ = cyclopropyl-
[d] = two isomers of the Δ-C_{19} acid were found but not fully characterized.

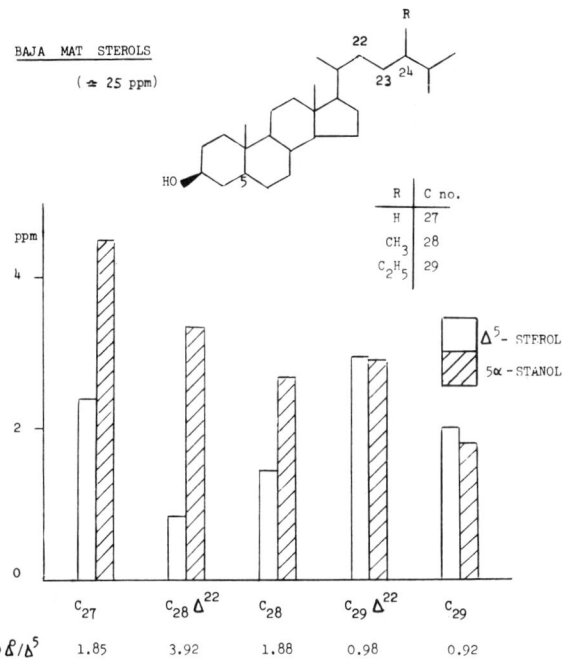

Figure 11.6 Δ^5-sterol and 5α-stanol distributions in the mat.

Table 11.6
Distribution of Alcohols
(excluding Δ^5 Sterols and 5α-stanols)

Compound	Structure No.	Conc. (ppm) Mat	Ooze
phytol	IV	75.80	31.27
n-C_{18} alcohol[a]		2.02	1.81
n-C_{20} alcohol[a]		0.71	1.15
n-C_{22} alcohol		7.12	3.56
n-C_{24} alcohol		2.97	3.56
n-C_{26} alcohol		1.90	2.30
5β-cholestan-3β-ol	IX		n.o.
24-methyl-5β-cholest-22-en-3β-ol	X		n.o.
24-methyl-5β-cholestan-3β-ol	XI		n.o.
24-ethyl-5β-cholest-22-en-3β-ol	XII	8.30	n.o.
24-ethyl-5β-cholestan-3β-ol	XIII		n.o.
cycloartanol	XIV		n.o.
cycloartenol	XV		n.o.
C_{30} pentacyclic triterpanol	VIII	0.24	1.45
17βH-bishomohopanol	VII	0.71	0.95
others		22.23	16.50
Total		122.00	62.50

[a] identified by GC retention time.
n.o. = not observed.

DISCUSSION

n-Alkanes and n-Fatty Acids

As expected, the mat shows (Figure 11.2) a high algal contribution typical of blue-green algae with n-C_{17} as the predominant alkane (Oro et al., 1967; Gelpi et al., 1970; Han et al., 1968; Blumer et al., 1971; Winters, et al., 1969). Lower concentrations of higher homologues are present, with a high odd/even predominance, maximizing at n-C_{27}. Although high-molecular-weight n-alkanes C_{27} to C_{30}, (Clark and Blumer, 1967) and n-alkenes, C_{23} to C_{33} (Gelpi et al., 1968) have been reported in several species of algae, only a slight odd/even predominance was noted for the alkanes. Our results show a strong odd/even dominance (CPI = 5.9) for the mat sample, which is more suggestive of the usual n-alkane distribution in higher plants (Eglinton et al., 1962). The presence of *Salicornia* and *Distichlis* close to the mats (Eglinton and Philp, 1974) tends to support this hypothesis. The main difference between the mat and ooze distributions appears to be the marked reduction in the n-C_{17} concentration in the latter. This could result from either a reduced algal contribution or from bacterial alteration. It has been shown that bacterial action on an alga and two species of higher plants resulted in marked changes in the relative distribution and carbon number preference of the original n-alkane distributions (Cranwell, 1975; Johnson and Calder, 1973). Similar changes are noticeable in a comparison of both layers in the present study. Apart from a decrease in CPI values from 5.9 to 4.5, there is a shift in the n-alkane pattern in the ooze (Figure 11.2) towards lower carbon numbers. These are, however, minor alterations which suggest that microbial degradation might not be the predominant effect in accounting for the observed changes.

A similar situation seems to occur for the n-alkanoic acids. Changes between the two layers (Figure 11.2) can hardly be rationalized as the result of bacterial activity. The typical bacterial fatty acids show a reasonably simple distribution in the range of C_{14} to C_{20} (Oro et al., 1967; Blumer et al., 1969). Also, the changes in fatty acid distribution following microbial decomposition of a species of alga (Cranwell, 1975) occasioned an increased abundance of the short-chain n-alkanoic acids. Higher-molecular-weight (>C_{20}) fatty acids seem, therefore, to be indicative of a terrestrial (i.e., higher plant) contribution to sediments (Cranwell, 1974). The high relative abundances of n-$C_{16:0}$ and n-$C_{18:0}$ in the mat are as expected on the basis of the reported fatty acid composition for L. aestuarii (Schneider et al., 1970) and M. chthonoplastes (Parker et al., 1967). Taken with the n-alkane distributions, the increased relative abundance of the >C_{20} components in the ooze and the high odd/even predominance indicate an increased higher plant contribution to the ooze.

As expected, there is a marked decrease in the ratio of monounsaturated to saturated acids in the mat and ooze, respectively. This is in keeping with the observation of Parker and Leo (1965) in a depth study of another blue-green algal mat community, and with the rapid conversion of radiolabeled oleic acid to saturated acids in an estuarine sediment (Rhead et al., 1971).

Branched/Cyclic Alkanes

Only a limited range of branched/cyclic alkanes has been reported in blue-green algae (Winters et al., 1969; Gelpi et al., 1970) and in bacteria (Tornabene et al., 1967;

Han and Calvin, 1969). As expected, the mixture of 7- and 8-methylheptadecanes thought to be a characteristic marker of blue-green algae (Han et al., 1968: Han and Calvin, 1969; Fehler and Light, 1970; Gelpi et al., 1970), is present. Pristane and phytane have not been reported as abundant constituents of blue-green algae (Han and Calvin, 1969; Oro et al., 1967) but have been reported as abundant constituents of several species of bacteria (Han et al., 1968).

The most marked change in the low-molecular-weight hydrocarbon distribution between the two layers is the pronounced decrease in the relative abundance of the 7- and 8-methylheptadecanes (Table 11.3) which is in keeping with our interpretation of a smaller algal input to the ooze.

Major differences are apparent, however, in the distribution of the higher-molecular-weight alkanes (Figure 11.3). The mat is characterized by extremely low amounts of high-molecular-weight compounds, whereas a relative increase of about 20 times is noticeable in the ooze. Structural characterization of these compounds in the ooze was not possible in view of the limited amount of sample available. Examination of other ooze samples from Laguna Mormona demonstrated however the presence of triterpenoid structures in the branched/cyclic hydrocarbon fraction. Mass spectral evidence suggests the presence of homohopane (XVI), a C_{31} triterpane, in one of these samples. This compound has been reported in a variety of recent and ancient sediments (Kimble et al., 1974; Ensminger et al., 1973; Eglinton et al., 1974; Van Dorsselaer et al., 1974). Mass fragmentography suggests the presence of other triterpane skeletons in the various ooze samples analyzed, which matches the situation for a number of other recent sediments where a range of such skeletons (C_{27}-C_{32}) has been observed. However, whether the higher abundance of these compounds in the ooze is the result of a short-term transformation of suitable functionalized precursors, possibly present in the mat, or finds its origin in a more significant higher plant contribution to the ooze layer, is not known at present.

It is noteworthy that there is an absence in the branched/cyclic alkane fractions of the complex, unresolved envelope of peaks found in several recent sediments (Eglinton et al., 1974). The presence of this "hump" in these sediments was proposed (Eglinton et al., 1974) to be suggestive of contamination from industrial and domestic wastes. If the "hump" does indeed reflect a pollution contribution, then the presence of a range of triterpanes in samples from Laguna Mormona probably indicates that they are indigenous and can be formed at an early stage.

Branched/Cyclic Alkanes

The presence of sterenes and triterpenes as major constituents of this fraction in the ooze, and their absence in the mat, is similar to the branched/cyclic alkane situation. The virtual absence of hop-22(29)-ene (V) in the mat is not surprising, although this triterpene has been reported as an abundant constituent of *L. aestuarii* (Gelpi et al., 1970) which is present as a minor contributing species to the sediment. The occurrence of the triterpene in bacteria has also been reported (Bird et al., 1971b) along with hop-17(21)-ene (XVII), the latter being found also in two Eocene sediments (Kimble et al, 1974; Arpino, 1973). The increased abundance of diploptene in the ooze where apparently there is a smaller algal input may suggest a bacterial derivation.

The occurrence of Δ^2 sterenes in recent sediments has not been previously reported,* although the presence of rearranged cholest-13(17)-ene homologues (XVIII) in a 180 million year old shale has recently been described (Rubinstein et al., 1975). The major component of the sterene series in the ooze has a mass spectrum which matched that of cholest-2-ene (XIX) (Rhead, 1971). Cholesta-3,5-diene (XX) has recently been described in the lipids of a crustacean (*Ligia oceanica*) (Hamilton et al., 1975). Rubinstein et al., (1975) have shown that cholest-2-ene is formed as the initial product of dehydration of 5α-cholestan-3β-ol (XXI) under acid conditions. Nonavailability of standards has precluded a full structural characterization of most of the sterenes present in the ooze but the mass spectra suggest Δ^2 sterenes to be the major constituents (Table 11.4).

A comparison of the sterene and stanol distributions (Tables 11.3 and 11.6) suggests a strong similarity in carbon number distribution. Gaskell and Eglinton (1975) have shown that stanols can be formed in a recent sediment through incubation experiments with radiolabeled cholesterol. The similarity between the stanol and sterene distributions in the ooze is therefore suggestive of the formation of the sterenes from the corresponding stanols, either by biochemical means or else through an acid or clay-catalyzed dehydration process.

---

*Mixtures of Δ^2 sterenes as reported herein were, however, observed in several recent sediments by M. Dastillung, I. Rubinstein and P. Albrecht (unpublished results).

Branched/Cyclic Alkanoic Acids

Unfortunately, no structural characterization of this fraction in the ooze was possible (Figure 11.4). The presence of *iso-* and *anteiso*-C_{15} acids along with major amounts of the C_{17} and C_{19}-cyclopropanoid acids seems to be indicative of a bacterial input to the mat (Table 11.5). The distribution of lower-molecular-weight components in this layer is similar to that reported by Cranwell (1974) for decayed algal material. The ratio of branched/cyclic to n-alkanoic acids is also higher in the mat than in the ooze sample (Table 11.1). This seems to correlate with the situation reported (Cranwell, 1974) for the surface sediment of productive lakes, in that a higher branched/cyclic acid concentration seems to be associated with sediments with a predominant autochthonous input. This is in keeping with the greater higher-plant type input to the ooze layer as suggested above.

The presence of the C_{32} triterpanoic acid (17βH-bishomohopanoic acid) (VI) in the mat and ooze relates to the reported occurrences of this and related triterpanoic acids in ancient and recent sediments (Van Dorsselaer *et al.*, 1974; Eglinton *et al.*, 1974). It has been suggested (P. Albrecht, personal communication) that the C_{32} acid derives from a C_{35} tetrahydroxy compound (XXII) which has been detected in a bacterium and a blue-green alga (Förster *et al.*, 1973; P. Albrecht, personal communication).

Sterols and Alcohols

Sterols are present in the majority of organisms apart from most bacterial and other lower forms of life and may serve as sensitive indicators of the depositional environment. Furthermore, their recent sedimentary transformation products could be important in the rationalization of the presence of certain steroidal compounds found in ancient oil shales (Steel and Henderson, 1972).

Although it was originally reported that the Cyanophyceae lacked sterols (Levin and Bloch, 1964), later reports have shown that certain members contain small amounts of cholesterol and phytosterol mixtures (de Souza and Nes, 1968; Reitz and Hamilton, 1968; Teshima and Kanazawa, 1972). The mat examined herein contains a complex mixture of sterols and stanols which may therefore derive from the major source organisms, the blue-green algae *M.chthonoplastes* and *L. aestuarii*. The alkane distribution in the ooze however indicates a possible additional contribution from higher plants and some of the ooze phytosterol may be derived from this source.

Several workers have found that the ratio of 5α-stanols to Δ^5 sterols increases with depth in cores from lake sediments (Henderson et al., 1971; Ogura and Hanya, 1973; Gaskell, 1974; Ogura, 1974), and this indicated that "rapid" sterol hydrogenation was occurring. Moreover, following injection of radiolabeled cholesterol into a lacustrine sediment, a small proportion of the recovered radioactivity was present in 5α-cholestanol (XXI) (Gaskell and Eglinton, 1975). The present data are in agreement with these observations. Thus, the presence of both Δ^5 sterols and 5α-stanols in the mat strongly suggests that hydrogenation of the Δ^5 bond is occurring. However the parallel Δ^5-sterol and 5α-stanol distributions relative to side-chain structure noted in Rostherne sediment (Gaskell, 1974) was not observed in this case. This may indicate that selective microbial reduction of sterols is taking place or that certain sterols are more accessible for reduction. Alternatively, the possibility cannot be excluded that the source organisms themselves contain a proportion of ring-saturated sterols. No significant increase in the 5α-stanol/sterol ratio was observed in the ooze. In addition the presence of the tetracyclic triterpenoid and phytosterol precursor, cycloartanol (XIV), in the algal mat is consistent with the sterol biosynthetic capacity of a photosynthetic organism.

Phytol is the major alcohol component of both mat and ooze. This high abundance is in keeping with its occurrence in nature as the side chain of chlorophyll and with its presence as a major component of several recently deposited sediments (Brooks and Maxwell, 1974; Brooks, 1974). 17βH-bishomohopanol (VII) was identified in both layers and major amounts of this compound together with the corresponding 17αH-isomer have been reported in other recent sediments (Brooks and Maxwell, 1975; Brooks, 1974). Together with 17βH-bishomohopanoic acid (VI), this C_{32} triterpenoid alcohol is believed to originate from a C_{35} tetrahydroxyhopane (XXII) precursor which has been detected in both a bacterium (Forster et al., 1973) and an alga (P. Albrecht, personal communication).

CONCLUSIONS

1. The n-alkane and n-alkanoic acid distributions in the mat and ooze layers in samples taken from the Laguna Mormona algal mats are explicable in terms of the expected contribution of blue-green algal components to the mat and the ooze and some higher plant contribution to the ooze.

2. The decrease in the ratio of monounsaturated to saturated acids in the mat and ooze, respectively, indicates a preferential removal of unsaturated components in agreement with the results obtained by other workers.

3. There is an absence in the branched/cyclic alkane fraction of the complex unresolved envelope of components present in a number of recent sediments, which may derive from a pollutant source.

4. Certain triterpenes of the hopane skeletal type are present in the mat and the ooze, thus extending the known occurrences of these apparently ubiquitous compounds in recent sedimentary situations.

5. The presence of stanols and sterenes in the ooze with similar carbon number distributions suggests that there may be a relationship between them.

6. The occurrence of both Δ^5 sterols and 5α stanols in both layers is evidence for the operation of a rapid reduction process, as shown by earlier investigation in other recent sediments.

I

II

III

IV

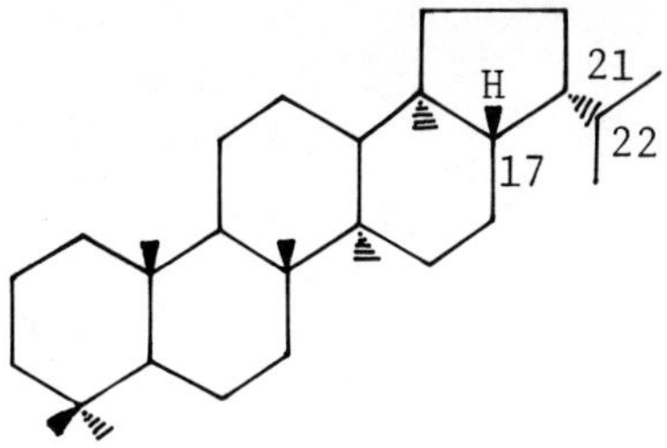

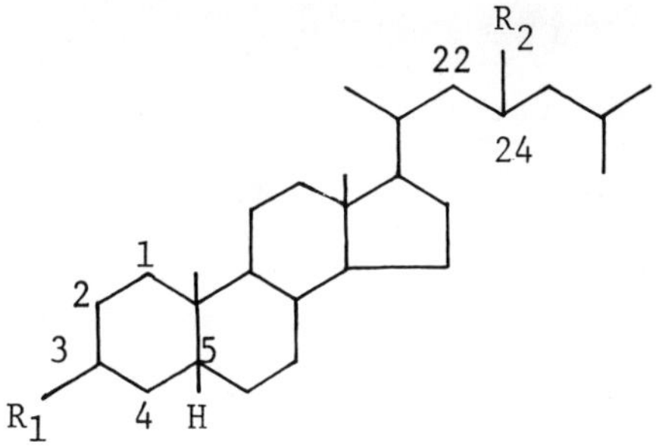

The Carbon Cycle 169

V	R = CH$_2$	IX	5β, R$_1$-OH, R$_2$-H
VI	R-(CH$_2$)$_2$COOH	X	5β, Δ^{22}, R$_1$-OH, R$_2$-CH$_3$
VII	R-(CH$_2$)$_3$OH	XI	5β, R$_1$-OH, R$_2$-CH$_3$
VIII	R-CH$_3$, 22-OH	XII	5β, Δ^{22}, R$_1$-OH, R$_2$-C$_2$H$_5$
or	R-CH$_2$OH, 22-H	XIII	5β, R$_1$-OH, R$_2$-C$_2$H$_5$
XVI	R-C$_2$H$_5$	XIX	Δ^2, R$_1$-H, R$_2$-H
XVII	Δ$^{17-21}$, R-CH$_3$	XX	Δ^3, Δ$^{5-6}$, R$_1$-H$_1$, R$_2$-H
XXII	R- (diol structure)	XXI	5α, R$_1$-OH, R$_2$-H

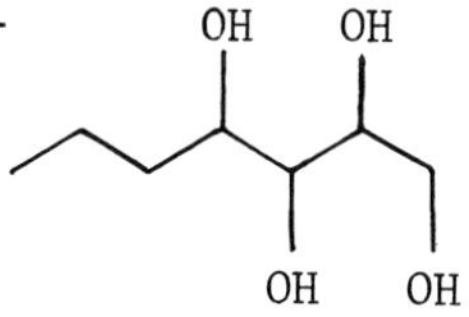

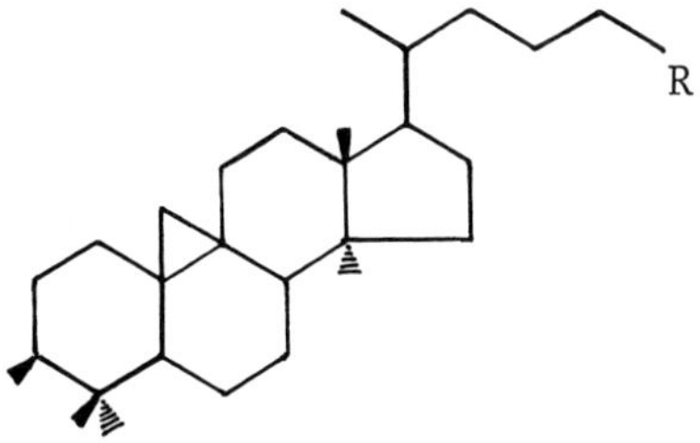

XIV R = CH(CH$_3$)$_2$

XV   R = C(CH$_3$)$_2$

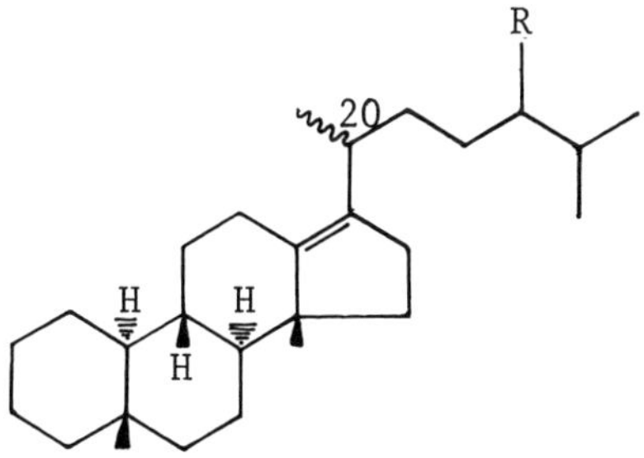

XVIII R=H, CH$_3$, C$_2$H$_5$, C$_3$H$_7$

ACKNOWLEDGMENTS

We thank the Nuffield Foundation, the Natural Environment Research Council (Grant Number GR3/2420), the Petroleum Research Fund and the National Aeronautics and Space Administration (Grant Number 05-003-003, through a subcontract from the University of California at Berkeley). PWB and JNC gratefully acknowledge the award of a NERC Fellowship and a British Council studentship, respectively.

REFERENCES

Arpino, P., 1973: PhD Thesis (Université de Strasbourg).

Bird, C. W., J. M. Lynch, S. T. Pirt, W. W. Reid, C. J. W. Brooks and B. S. Middleditch,1971a: Steroids and squalene in *Methylococcus capsulatus* grown on methane. *Nature*, 230, 473.

Bird, C. W., J. M. Lynch, S. T. Pirt and W. W. Reid, 1971b: The identification of hop-22(29)-ene in prokaryotic organisms. *Tet. Lett.*, 3189-3190.

Blumer, M., T. Chase and S. W. Watson, 1969: Fatty acids in the lipids of marine and terrestrial nitrifying bacteria. *J. Bacteriol.* 99, 366-370.

Blumer, M., R. R. L. Guillard and T. Chase, 1971: Hydrocarbons of marine phytoplankton. *Mar. Biol.* 8, 183-189.

Brooks, P. W., 1974: PhD. Thesis (University of Bristol)

Brooks, P. W. and J. R. Maxwell, 1974: Early stage fate of phytol in a recently-deposited lacustrine sediment. In: *Advances in Organic Geochemistry, 1973.* (eds) B. Tissot and F. Bienner, (Paris: Editions Technip), 977-991.

Brooks, P. W. and J. R. Maxwell, 1975: in preparation.

Clark, R. C. and M. Blumer, 1967: Distribution of n-paraffins in marine organisms and sediments. *Limnol. Oceanogr.* 12, 79-87.

Cranwell, P. A., 1973: Branched-chain and cyclopropanoid acids in a recent sediment. *Chem. Geol.* 11, 307-313.

Cranwell, P. A., 1974: Monocarboxylic acids in lake sediments; indicators derived from terrestrial and aquatic biota, of paleoenvironmental trophic levels. *Chem. Geol.* 14, 1-14.

Cranwell, P. A., 1975: Decomposition of aquatic biota and
sediment formation: organic compounds in detritus resulting
from microbial attack on the alga *Ceratium hirundinella*.
Chem. Geol. (submitted)

de Souza N. J. and W. R. Nes, 1968: Sterols: Isolation
from a blue-green alga. *Science*, 162, 363.

Dole, V. P. and H. Meinertz, 1960: Microdetermination of
long-chain fatty acids in plasma and tissues. *J. Biol.
Chem.* 235, 259-269.

Eglinton, G., R. J. Hamilton, and R. A. Raphael, 1962:
Hydrocarbon constituents of the wax coatings of plant
leaves: a taxonomic survey. *Nature* 193, 739-742.

Eglinton, G., 1968: Hydrocarbons and Fatty Acids in living
organisms and recent and ancient sediments. In: *Advances
in Organic Geochemistry, 1968.* (eds) P. A. Schenck and
I. Havenaar (Oxford: Pergamon Press),

Eglinton, G., 1973: Chemical Fossils: a combined organic
geochemical and environmental approach. *Pure and Appl.
Chem.* 34, 611-632.

Eglinton, G., J. R. Maxwell, and R. P. Philp, 1974: Organic
geochemistry of sediments from contemporary aquatic environments. In: *Advances in Organic Geochemistry, 1973.*
(eds) B. Tissot and F. Bienner (Paris: Editions Technip),
pp. 941-961.

Eglinton, G. and R. P. Philp, 1974: Report on field trip to
Laguna Mormona, Baja California. (University of Bristol).

Ensminger, A., A. Van Dorsselaer, C. Spyckerelle, P. Albrecht
and G. Ourisson, 1973: Pentacyclic triterpenes of the hopane
type as ubiquitous geochemical markers: origin and significance. In: *Advances in Organic Geochemistry, 1973.*
(Eds.) B. Tissot and F. Bienner (Paris: Editions Technip),
pp. 245-260.

Fehler, S. W. G. and R. J. Light, 1970: Biosynthesis of
hydrocarbons in *Anabaena variabilis*. Incorporation of
(methyl-^{14}C) and (methyl-^{2}H$_3$) methion into 7- and 8-
methylheptadecanes. *Biochemistry* 9, 418-422.

Förster, H. J., K. Biemann, W. G. Haigh, N. H. D. Tattire,
J. R. Cotrin, 1973: The structure of novel C$_{35}$ pentacyclic
terpenes from *Acetobacter xylinum*. *Biochem. J.* 135, 133-143.

Gaskell, S. J., 1974: PhD. Thesis (University of Bristol).

Gaskell, S. J. and G. Eglinton, 1975: Rapid hydrogenation of sterols in a contemporary lacustrine sediment. *Nature* **254**, 209-211.

Gelpi, E., J. Oro, H. I. Schnieder and E. O. Bennett, 1968: Olefins of high-molecular-weight in two microscopic algae. *Science* **161**, 700-701.

Gelpi, E. H. Schneider, J. Mann and J. Oro, 1970: Hydrocarbons of geochemical significance in microscopic algae. *Phytochemistry* **9**, 603-612.

Hamilton, R. J., M. Y. Raie, I. Weatherston, C. J. Brooks, and J. H. Borthwick, 1975: Crustacean surface waxes. Part I. The hydrocarbons from the surface of *Ligia oceanica*. *J. Chem. Soc. Perkin. I* 354-357.

Han, J. E. D. McCarthy, W. Van Hoeven, M. Calvin and W. H. Bradley, 1968: Organic geochemical studies.II-Preliminary report on the distribution of aliphatic hydrocarbons in algae, in bacteria and in a recent lake sediment. *Proc. Nat. Acad. Sci. USA* **59**, 29-33.

Han, J. and M. Calvin, 1969: Hydrocarbon distribution of algae and bacteria and microbiological activity in sediments. *Proc. Nat. Acad. Sci. USA.* **64**, 436-473.

Henderson, W., W. E. Reed, G. Steel, and M. Calvin, 1971: Isolation and identification of sterols from a Pleistocene sediment. *Nature* **231**, 308-310.

Horodyski, R. J. and S. P. Vonder Haar, 1975: Recent calcareous stromatolites from Laguna Mormona (Baja California), Mexico. *J. Sediment. Petrology.* (submitted).

Ikan, R., M. J. Baedecker and I. R. Kaplan, 1973: C_{18}-isoprenoid ketone in recent marine sediment. *Nature* **244**, 154-155.

Johnson, R. W. and J. A. Calder, 1973: Early diagenesis of fatty acids and hydrocarbons in a salt marsh environment. *Geochim. Cosmochim. Acta* **37**, 1953-1955.

Kimble, B. J., J. R. Maxwell, R. P. Philp, G. Eglinton, P. Albrecht, A. Ensminger, P. Arpino and G. Ourisson, 1974: Tri- and tetraterpenoid hydrocarbons in the Messel Oil shale. *Geochim. Cosmochim. Acta* **38**, 1165-1181.

Leo, R. F. and P. L. Parker, 1966: Branched chain fatty acids in sediments. *Science* **152**, 649-650.

Levin, E. Y. and K. Bloch, 1964: Absence of sterols in blue-green algae. *Nature* **202**, 90-91.

Ogura, K. and T. Hanya, 1973: The cholestanol-cholesterol ratio in a 200 meter core sample of Lake Biwa. *Proc. Jap. Acad.* 49, 201-204.

Ogura, K., 1974: Information of sterols in a core sample. Contribution on the Paleolimnology of Lake Biwa, 43, 194-201.

Oro, J., T. G. Tornabene, D. W. Nooner, and E. Gelpi, 1967: Aliphatic hydrocarbons and fatty acids of some marine and freshwater microorganisms. *J. Bacteriol.* 93, 1811-1818.

Parker, P. L. and R. F. Leo, 1965: Fatty acids in blue-green algal mat communities. *Science* 148, 373-374.

Parker, P. L., C. Van Baalen, and L. Maurer, 1967: Fatty acids in eleven species of blue-green algae: geochemical significance. *Science* 155, 707-708.

Philp, R. P. and M. Calvin, 1975: Kerogen structures in recently-deposited algal mats at Laguna Mormona, Baja California: a model system for the determination of kerogen structures in ancient sediments, this volume.

Reitz, R. C. and J. G. Hamilton, 1968: The isolation and identification of two sterols from two species of blue-green algae. *Comp. Biochem. Physiol.* 25, 401-416.

Rhead, M. M., 1971: PhD. Thesis (University of Bristol).

Rhead, M. M., G. Eglinton, and G. H. Draffan, 1971: Conversion of oleic acid to saturated fatty acids in Severn Estuary sediments. *Nature* 232, 327-330.

Rubinstein, I., O. Sieskind and P. Albrecht, 1975: Rearranged sterenes in a shale: occurrence and simulated formation (in preparation).

Schlenk, H. and J. L. Gellerman, 1960: Esterification of fatty acids with diazomethane on a small scale. *Anal. Chem.* 32, 1412-1414.

Schneider, H., E. Gelpi, E. O. Bennett and J. Oro, 1970: Fatty acids of geochemical significance in microscopic algae. *Phytochemistry* 9, 613-617.

Simoneit, B. R. and A. L. Burlingame, 1974: Initial reports on the deep sea drilling project, Vol XXI, Washington, U.S. Gov. Printing Office.

Steel, G. and W. Henderson, 1972: Isolation and characterization of stanols from Green River shale. *Nature* 238, 148-150.

Teshima, S. and A. Kanazawa, 1972: Occurrence of sterols in the blue-green alga, *Anabaena cylindrica*. *Bull. Jap. Soc. Sci. Fish.* **38**, 1197-1202.

Thorne, H. M., K. E. Standfield, G. U. Dinneen and W. I. Murphy, 1964: Oil shale technology: a review. Information Circular 8216, US Bureau of Mines, Washington.

Tornabene, T. G., E. Gelpi and J. Oro, 1967: Identification of fatty acids and aliphatic hydrocarbons in *Sarcina lutea* by gas chromatography and combined gas chromatography-mass spectrometry. *J. Bacteriol.* **94**, 333-343.

Traverse, A., 1955: Occurrence of the oil-forming alga *Botryococcus* in lignites and other tertiary sediments. *Micropaleontology* **1**, 343-348.

Van Dorsselaer, A., A. Ensminger, C. Spyckerelle, M. Dastillung, O. Sieskind, P. Arpino, P. Albrecht, G. Ourisson, P. W. Brooks, S. J. Gaskell, B. J. Kimble, R. P. Philp, J. R. Maxwell and G. Eglinton, 1974: Degraded and extended hopane derivatives (C_{27} to C_{35}) as ubiquitous geochemical markers. *Tet. Lett.* 1349-1352.

Winters, K., P. L. Parker and C. Van Baalen, 1969: Hydrocarbons of blue-green algae: geochemical significance. *Science* **163**, 467-468.

CHAPTER 12

NUCLEIC ACID BASE CONTENTS AS INDICATORS
OF BIOLOGICAL ACTIVITY IN SEDIMENTS

W. VAN DER VELDEN
ALAN W. SCHWARTZ

Department of Exobiology, University of
Nijmegen, Nijmegen, The Netherlands

INTRODUCTION

A knowledge of the distribution and diagenesis of the
nucleic acid bases in various environments is potentially
of great importance in organic geochemistry. Because of a
lack of data in this area, our laboratory is engaged in a
study of geologically recent sediments and soils with a view
toward understanding, eventually, the fate of individual
purines and pyrimidines. As part of this study we have
analyzed a number of core sections from the Great Lakes,
collected by the Canada Centre for Inland Waters (Burlington).
Previously, we have reported that cores from the central
and eastern basins of Lake Erie show sharply increased
purine and pyrimidine contents in the uppermost layers,
as well as a marked underabundance of uracil throughout the
cores (van der Velden and Schwartz, 1974). The data on total
purine and pyrimidine content were in excellent agreement
with other work on recent increases in organic C and total N
and P in Lake Erie sediments (Kemp *et al.*, 1972). An unresolved question was whether the observed increases represent
increased loading of organic material to the surface sediment, or merely a rapid degradative process within the
sediment. In order to attempt to clarify the processes
involved, we have analyzed additional core samples from
Lakes Erie, Ontario and Huron, as well as a number of sur-

face samples and a zooplankton sample from Lake Ontario.
This article constitutes a preliminary report on the purine
and pyrimidine contents of these samples. The sediments
studied have been the subject of previous analyses by Kemp
and co-workers (Kemp, 1971; Kemp and Mudrochova, 1972 and
1973). We have made use of data obtained by these workers
in order to relate the purine and pyrimidine contents to
total organic carbon and to calculate estimated dates of
deposition for given depths in the cores. (Kemp et al.,
1974). These estimates are based on average sedimentation
rates between the Castanea and Ambrosia pollen horizons
(Anderson, 1974). It should be noted that the existence
of these horizons within the cores indicates that the sediments
have not been stratigraphically disturbed since the time of
deposition.

MATERIALS AND METHODS

The locations of the sample areas are summarized in
Table 12.1. The zooplankton sample was collected at the
location of core E30 in the eastern basin of Lake Ontario.
The extraction procedure which was employed for isolation
of the purines and pyrimidines is shown schematically in
Figure 12.1.

Table 12.1
Sample Locations and Water Depths of Sediments

Sample Number	Sample Location	Station Location		Water Depth (m)
		Lat. N	Long. W	
SB1	Lake Huron - South Bay	45°37.5´	81°52.73´	56
G16	Lake Erie - Central Basin	42° 0.2´	81°36.2 ´	24
E30	Lake Ontario-Eastern Basin	43°30.5´	77°54.3 ´	225
KB1	Lake Ontario-Kingston Bay	44°04.71´	76°24.72´	26
CB3	Lake Ontario-Central Basin	43°33.0´	78°10.4 ´	186
WB4	Lake Ontario-Western Basin	43°24.10´	79°26.66´	101
855	Beaufort Sea	70°57.4´	135°03.4 ´	457
875	Beaufort Sea	70°31.8´	132°10.0 ´	35

Analysis and identification was performed by liquid
chromatography, utilizing anion and cation exclusion columns
operating at pH 10 and pH 4, respectively. A detailed
description of the procedures employed is reported elsewhere
(van der Velden et al., 1974; van der Velden and Schwartz, 1974).

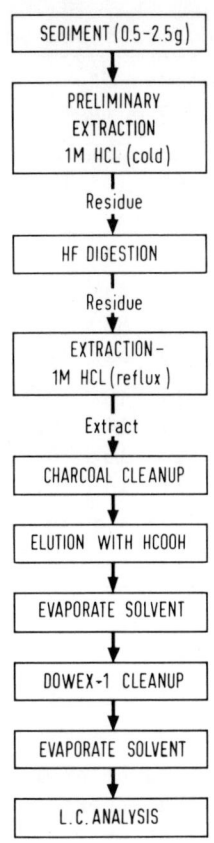

Figure 12.1 Procedure for the isolation of purines and pyrimidines from sediments and soils.

The recoveries of standard ^{14}C-labeled purines and pyrimidines added to sediment were: cytosine, 92%; uracil, 86%; thymine, 92%; adenine, 75%; guanine, 66%; and hypoxanthine, 53%. The reported values are uncorrected for losses.

RESULTS AND DISCUSSION

Analytical results for surface samples are summarized in Table 12.2. Also included in this table are analyses of four soils, derived from a number of vegetation types and two marine sediments from the Beaufort Sea (Peak *et.al.*, 1972). Evident are the low levels of uracil in all sediments and soils and the high hypoxanthine contents of Lake Ontario samples. Table 12.3 gives the mole % distribution of the five major bases in the same materials. The uracil content does not exceed 4 mole % in any sample. By comparison, the zooplankton sample contains 17 mole % uracil. The relative absence

Table 12.2
Purines and Pyrimidines in Surface Sediments and Soils

Location	Sample Depth (cm)	Concentrations of Individual Bases (ppm)[a]						Total
		C	U	T	A	G	HX	
Huron-SB1	0-1	38	2	24	45	38	10	157
Erie-G16	0-1	52	5	36	98	150	15	356
Ontario-E30	0-1	59	4	43	78	55	48	287
-KB1	0-5	31	3	20	47	63	32	196
-CB3	0-3	51	6	40	73	83	39	292
-WB4	0-3	32	3	25	43	51	24	178
Forest Soils[b] I	10-20	12	2	9	14	18	0.4	55
II	10-20	10	2	8	11	14	0.3	45
III	10-20	11	1	8	11	15	0.4	46
IV	10-20	11	2	8	11	14	0.3	46
Beaufort Sea 855	0-5	2.4	0.2	1.7	2.1	1.7	0.3	8.4
Beaufort Sea 875	0-5	2.1	0.3	1.7	2.2	2.5	0.3	9.1

[a] ppm = parts per million, based on dry weight of sediment, uncorrected for losses.
[b] vegetation types: I, Lamium; II, Pinus; III, Urtica; IV, Agrostis.

of uracil in the sediments and soils could be explained by a rapid turnover of RNA. It is known that there is generally very little RNA extractable from soils, while bases derived from DNA have been reported (Anderson, 1967). A similar, specific degradation of RNA may occur in aqueous environments and this is indeed suggested by the uniform, low uracil contents of both the soil and sediment samples.

The E30 core of Lake Ontario is particularly interesting, as it affords an opportunity of following the changes in purine and pyrimidine contents through a period of nearly 5000 years. (Complete data on individual purine and pyrimidine contents of this and other cores will be reported elsewhere.) Total purines and pyrimidines are plotted against estimated time of deposition in Figure 12.2. It is apparent that a rapid breakdown process, presumably microbial, extends quite deeply into the sediment. The curve only levels off at a depth of 20 to 30 cm and an age of 100-200 years. An interesting feature is the slow process, observable at ages greater than about 1000 years, which probably corresponds to chemical diagenesis of the purines and pyrimidines. In the deepest portion of the core, hypoxanthine, which as

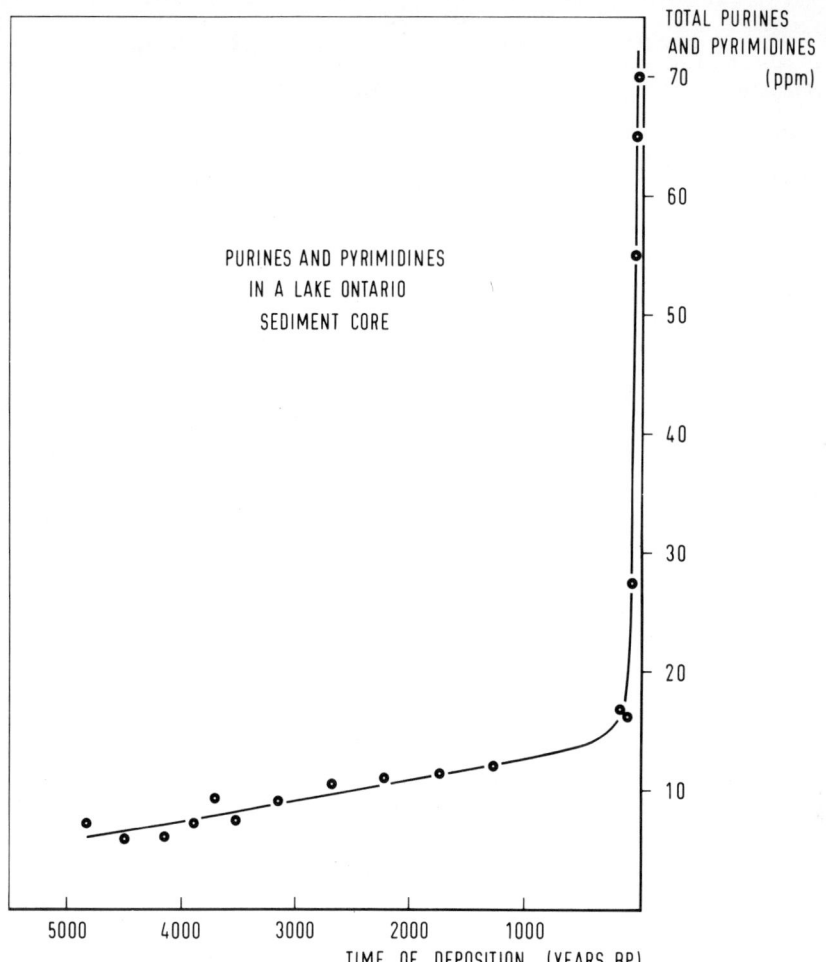

Figure 12.2 Total content of identified purines and pyrimidines in the Lake Ontario E30 core versus estimated time of deposition.

already noted is unusually high in Lake Ontario sediments, becomes dominant. Hypoxanthine is, of course, a hydrolytic product of adenine, which could contribute to the increase in its concentration relative to adenine (van der Velden et al., 1974).

Rapid microbial degradation of purines and pyrimidines appears to be a characteristic of relatively eutrophic locations in the Great Lakes. The most sensitive method of displaying this property is to plot the total content of purines and pyrimidines as a fraction of the total organic

Table 12.3
Mole % Distribution of the Five Major Bases

Sample Location	Mole % Distribution				
	C	U	T	A	G
Huron - SB1	30	2	17	29	22
Erie - G16	19	2	11	29	40
Ontario - E30	27	2	18	30	20
- KB1	23	2	13	28	34
- CB3	24	3	17	28	29
- WB4	25	2	17	27	29
Forest Soils I	25	4	18	24	29
II	26	4	19	23	27
III	28	3	18	23	28
IV	28	4	18	23	27
Beaufort Sea 855	34	3	21	24	18
Beaufort Sea 875	28	4	20	24	24
Zooplankton Sample	20	17	11	27	26

carbon against the estimated date of deposition of the core sections. This procedure has been applied to the data from Lakes Erie, Ontario and Huron and the results are presented in Figure 12.3. It can be seen that the nucleic acid base content decreases with time most rapidly in Lake Erie and least rapidly in Lake Huron. Similarly, the absolute value of the purine and pyrimidine content at the surface is highest in Lake Erie and lowest in Lake Huron. Both with regard to the rate of change of the base content with time and depth and the total content at the surface, Lake Ontario seems to be in an intermediate position. It is generally accepted that, while Lake Huron has remained oligotrophic, the central basin of Lake Erie has become eutrophic and that input of organic material to the sediment has increased dramatically in recent years (Beeton, 1969; Kemp, et al., 1973). Our data support this hypothesis, but also suggest that the increased microbial population in the sediment is adequate to recycle key metabolites such as the purines and pyrimidines. This is in accord with other observations that organic nitrogen compounds in lake sediments are degraded preferentially to the bulk of the organic matter (Kemp and Mudrochova, 1972). It is remarkable, however, that the increased rate of breakdown of the purines and pyrimidines

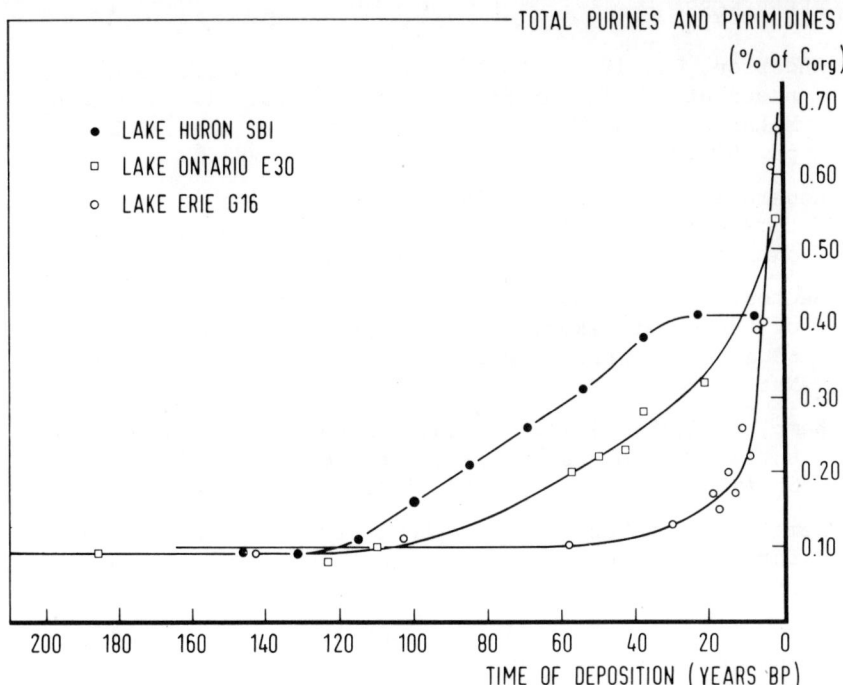

Figure 12.3 Purines and pyrimidines in cores of Lake Erie, Ontario and Huron. Total contents expressed as percentage of organic C versus estimated time of deposition.

appears to more than compensate for increased rates of organic sedimentation. Thus, all three curves ultimately reach a basal level of about 0.1%, but in an order which is quite unexpected, the area of highest organic sedimentation being reduced to 0.1% much more quickly than the area of lowest organic sedimentation. Additional studies are obviously needed to establish whether the pattern shown here is a general one.

ACKNOWLEDGMENTS

We thank A. L. W. Kemp for supplying the sediment cores and surface samples from the Great Lakes and G. W. Hodgson for the Beaufort Sea sediments.

REFERENCES

Anderson, G., 1967: Nucleic Acids, derivatives and organic phosphates. In: *Soil Biochemistry*, Vol. 1, eds. A. D. McLaren and G. H. Peterson. (New York: Marcel Dekker), pp. 67-90.

Anderson, T. W., 1974: The chestnut pollen decline as a time horizon in lake sediments in eastern North America. *Canadian J. Earth Sci.* 11, 678-685.

Beeton, A. M., 1969: Changes in the environment and biota of the Great Lakes. In: *Eutrophication: Causes, Consequences, Correctives*, ed., G. E. Likens (Washington: Nat. Acad. of Sci.), pp. 150-187.

Kemp, A. L. W., 1971: Organic carbon and nitrogen in the surface sediments of Lakes Ontario, Erie and Huron. *J. Sediment. Petrol.* 41, 537-548.

Kemp, A. L. W. and A. Mudrochova, 1972: Distribution and forms of nitrogen in a Lake Ontario sediment core. *Limnol. Oceanogr.* 17, 855-867.

Kemp, A. L. W., C. B. J. Gray and A. Mudrochova, 1972: Changes in C, N, P and S in the last 140 years in three cores from Lakes Ontario, Erie and Huron. In: *Nutrients in Natural Waters*, eds., H. E. Allen and J. R. Kramer (New York: Wiley), pp. 251-279.

Kemp, A. L. W. and A. Mudrochova, 1973: The distribution and nature of amino-acids and other nitrogen-containing compounds in Lake Ontario surface sediments. *Geochim. Cosmochim. Acta* 37, 2191-2206.

Kemp, A. L. W., T. W. Anderson, R. L. Thomas and A. Mudrochova, 1974: Sedimentation rates and recent sediment history of Lakes Ontario, Erie and Huron. *J. Sediment. Petrol.* 44, 207-218.

Peake, E., M. Strosher, B. L. Baker, R. Gossen, R. G. McCrossan, C. J. Yorath and G. W. Hodgson, 1973: The potential of Arctic sediments: Hydrocarbons and possible precursors in Beaufort Sea sediments. In: Proc. of the 24th Int. Geol. Congress, Montreal, August 1972, section 5, pp. 28-37.

van der Velden, W., G. J. F. Chittenden and A. W. Schwartz, 1974: Studies on the geochemistry of purines and pyrimidines. In: *Advances in Organic Geochemistry 1973*, eds, B. Tissot and F. Bienner (Paris: Technip), pp. 293-304.

van der Velden, W. and A. W. Schwartz, 1974: Purines and pyrimidines in sediments from Lake Erie. *Science* <u>185</u>, 691-693.

CHAPTER 13

CARBOHYDRATES IN LAKE ONTARIO SEDIMENTS

D. LIU

Environment Canada, Environmental Protection
Service, Canada Centre for Inland Waters,
Burlington, Ontario

INTRODUCTION

In the study of benthic ecology and productivity, some
comparative measurements of the nutritional value of sediment
are useful in the understanding of the effect of sediment
on the quality of the overlying water as well as the biological systems that predominate in the aquatic environment
(Ballinger et al., 1971; Buchanan et al., 1970). Organic
carbon, organic matter, carbonate carbon, total nitrogen and
chlorophylls have been used frequently in the characterization
of sediment (Hargrave, 1969; Kemp et al., 1968; Kemp, 1971 ;
Serruya, 1971; Vanderpost, 1972). However, such measurements
do not provide an accurate picture of the extent of pollution
of a sediment because of the inability of these analyses to
differentiate between the compounds which are readily biodegradable and those which are nonbiodegradable.

An intensive study of sediment from Lake Erie and Lake
Ontario during the period of 1971-1972 revealed that a good
relationship existed between carbohydrate content and organic
matter. To verify this relationship, sediment cores were
collected from the MOSES Station No. 14 (Lake Ontario)
on June 12, 1972 for the analysis of carbohydrate content.
The significance of the occurrence of carbohydrate in sediment will be discussed.

MATERIALS AND METHODS

Three sediment core samples were taken with a Benthos corer at MOSES Station No. 14 (Vanderpost, 1972), from Lake Ontario (43° 19' 59.637" N Lat., 79° 43' 38.622" W Long.) on June 12, 1972. The core samples were stored at 4°C and within 24 hours were subsectioned into 0-2, 6-9, 12-15, 15-18, 18-20, 20-25 and 40-45 cm fractions. Dry weights were obtained by drying the sediment samples at 105°C to a constant weight, and the carbohydrate contents in sediment were estimated by a modified phenol-sulfuric acid method (Liu et al., 1973).

RESULTS AND DISCUSSION

The most detailed study of carbohydrate content in sediment to date in this laboratory has been made on cores from MOSES Station No. 14 of Lake Ontario. Analyses were made at various depths of the core, and the reliability of the modified phenol-sulfuric acid method (Liu et al., 1973) for directly measuring carbohydrate content in sediment is clearly shown in Figure 13.1. From this graph and the water content (Table 13.1), it was possible to calculate the amount of carbohydrate (expressed as mg agar per gram of sediment dry weight) in the sediment core. The vertical distribution of carbohydrate content in the sediment cores of MOSES Station No. 14 is shown in Figure 13.2. For comparison, recent data (Vanderpost, 1972) have been incorporated into this figure. The most interesting finding in this study was that the distribution pattern of carbohydrate in the sediment column followed closely those of organic matter and organic carbon in the top 25 cm. However, below the 25-cm zone, the concentration of carbohydrate decreased rapidly indicating that some portions of carbohydrate had changed to humus material. Little, if any, relationship was observed between carbohydrate and inorganic carbonate carbon throughout the sediment column. An analysis of the ratio of carbohydrate to organic matter (Table 13.1) indicated that carbohydrate was one of the major constituents in the organic matter fraction of lake sediment.

Plant polysaccharides are known to be more resistant than amino acids and proteins towards biodegradation. Cheshire et al. (1973) demonstrated that plant polysaccharides could persist in aerated soil for 448 days. The anaerobic environment in bottom sediments would certainly facilitate the preservation of the carbohydrate in the sediment column. This is supported by the very recent work of Koyama et al.

The Carbon Cycle 187

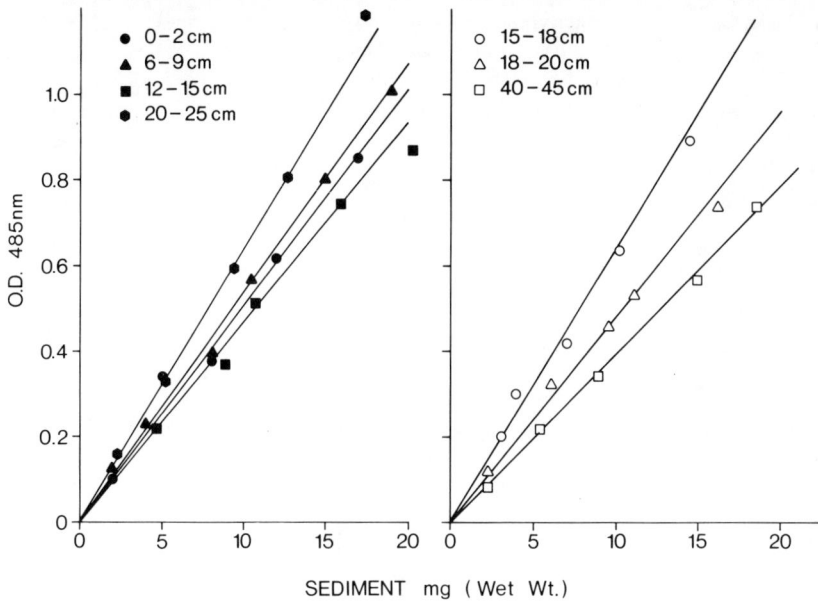

Figure 13.1 Proportionality between amounts of lake sediment and color development at 485 nm for carbohydrate by the modified phenol-sulfuric acid method.

Table 13.1
Some Physical and Chemical Characteristics of Sediment
at MOSES Station No. 14, Lake Ontario

Sediment Depth (cm)	Water Content (%)	Carbohydrate C Content (mg/g sediment dry wt)	Carbohydrate/ Organic matter Ratio
0-2	70.6	11.9	0.209
6-9	57.8	8.54	0.227
12-15	55.7	7.00	0.219
15-18	53.1	8.75	0.210
18-20	51.6	6.62	0.251
20-25	49.5	8.32	0.244
40-45	41.2	4.25	0.149

Sampling date: June 12, 1972.
Sampling site: 43 19' 59.637" N Lat., 79 43' 38.622"W Long.

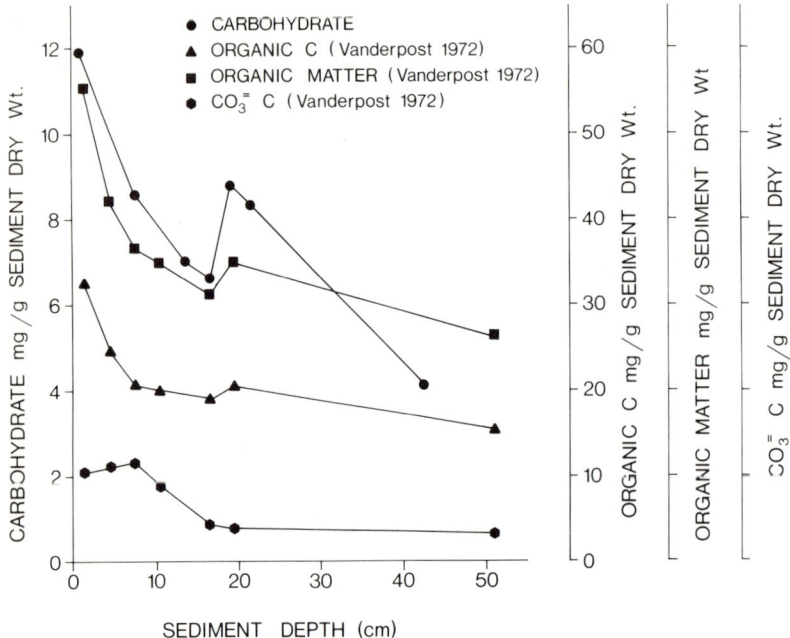

Figure 13.2 The vertical distribution of carbohydrate concentration in lake sediment from MOSES Station of Lake Ontario. Each point of the carbohydrate figure is an average of at least five replicates from the pooled sediment samples.

(1973), who studied the decomposition rate of various organic compounds in lake sediments. Their findings indicated that carbohydrate was far more resistant to degradation than pigments, fatty acids and proteins including amino acids. The carbohydrate components of lake sediment from Lakes Erie and Ontario have been found to exist as approximate equimolar mixtures of pentoses and hexoses indicating that the origin of the carbohydrate in the sediment was derived authochthonously from aquatic plant and algae (Liu et al., 1973). Recent studies on lake eutrophication have shown that the nutrient regeneration processes are controlled by the availability of organic nutrients in the sediment (Oppenheimer, 1960; Serruya, 1971; Ballinger et al., 1971). Thus, the analysis for organic nutrients such as carbohydrates in sediment could be a useful tool in the study of nutrient regeneration processes in lakes and rivers. This study shows that carbohydrate determination can be used to supplement the routine analyses for sediment characterization.

ACKNOWLEDGMENT

The author wishes to express his gratitude to Dr. N. W. Schmidtke for the review of this manuscript.

REFERENCES

Ballinger, D. G. and G. D. McKee, 1971: Chemical characterization of bottom sediment. *J. Water Poll. Control Fed.* 43, 216-227.

Buchanan, J. B. and M. R. Longbottom, 1970: The determination of organic matter in marine muds: the effect of the presence of coal and the routine determination of protein. *J. Exp. Mar. Biol. Ecol.* 5, 158-169.

Cheshire, M. V., C. M. Mundie and H. Shepherd, 1973: The origin of soil polysaccharides: transformation of sugars during the decompostion in soil of plant material labeled with ^{14}C. *Soil Sci.* 24, 54-68.

Hargrave, B. T., 1969: Epibenthic algal production and community respiration in the sediment of Marion Lake. *J. Fish Res. Board Can.* 26, 2003-2026.

Kemp, A. L. W. and C. F. M. Lewis, 1968: A preliminary investigation of chlorophyll degradation products in the sediments of Lakes Erie and Ontario. Proc. 11th Conf. Great Lakes Res., pp. 206-229.

Kemp, A. L. W., 1971: Organic carbon and nitrogen in the surface sediments of Lakes Ontario, Erie and Huron. *J. Sed. Petrology* 41, 537-548.

Koyama, T., N. Nikaido, T. Tomino and H. Hayakawa, 1973: Decomposition of organic matter in lake sediments. In: *Proceedings of Symposium on Hydrogeochemistry and Biogeochemistry*, Vol. II, E. Ingerson, ed., (Washington, D.C.: The Clarke Company), pp. 512-535.

Liu, D., P. T. S. Wong and B. J. Dutka, 1973: Determination of carbohydrate in lake sediment by a modified phenol-sulfuric acid method. *Wat. Res.* 7, 741-746.

Oppenheimer, C. H., 1960: Bacterial activity in sediments of shallow marine bays. *Geochim. Cosmochim. Acta* 19, 244-260.

Serruya, C., 1971: Lake Kinneret: the nutrient chemistry of the sediments. *Limnol. Oceanogr.* 16, 510-521.

Vanderpost, J. M., 1972: Bacterial and physical characteristics of Lake Ontario sediment during several months. Proc. 15th Conf. Great Lakes Res., pp. 198-213.

CHAPTER 14

SOIL POLYSACCHARIDES IN BURIED HUMIC HORIZONS
OF ASHITAKA LOAM FORMATION

R. HAMADA*, K. YOSHIZAKI,
K. SAKAGAMI AND T. KUROBE*

Tokyo University of Agriculture and Technology
Fuchu, Tokyo 183, Japan

INTRODUCTION

When the To-Mei Expressway (Tokyo-Nagoya Expressway) was under construction during the period of 1966-1967, a wide range of road cuts revealed exposed layers of volcanic ash falls at the foot of Ashitaka Volcano on the southeastern part of the piedmont region of Mt. Fuji. On that occasion, peculiar properties of these beds from the viewpoint of soil mechanics necessitated an integrated survey of the area. A voluntary group to undertake an integrated research on the so-called Ashitaka Loam was organized and named Ashitaka Loam Research Group (co-ordinator: Prof. Y. Kato of Shizuoka University, Shizuoka-shi, Shizuoka, Japan). The initial team consisted of 30 scientists from soil science, paleopedology, soil mechanics, clay science, geomorphology, stratigraphy, archeology and some other related fields.
Up to the present time, a series of papers has been published by the members of the group. The reports have dealt with stratigraphy, clay mineralogy, organic matter, amorphous mineral components, and ^{14}C age determination, in addition to some reports on the pedological features of the Loam.

---

*Members of the Ashitaka Loam Research Group (Co-ordinator: Prof. Y. Kato, Shizuoka University, Shzuoka, Japan)

191

The term "Loam" is a rather loose name for the Pleistocene tephra and volcanic ash fall and is the parent material of the soil in the area. The loam is low in solid phase (about 20-30%) and fine in particle size. Texturally, the loam varies from clay to silty clay.

The Ashitaka Loam Formation has been divided into three members: the Upper Loam which is 3 to 6 m thick, the Middle Loam which is 4 to 6 m thick, and the Lower Loam which ranges between 3 and 10 m in thickness. Above the Upper Loam, there is a Kuroboku soil (known as Ando soil). This soil is characterized by high organic matter or humus content of 10% or more. In some cases, the thickness of the A-horizon (or the uppermost part) of the humus-rich soil may be up to one meter. In the particular site of this study, there are alternate layers of buried humic horizons within the Kuroboku soil and in the upper part of the Upper Loam.

These alternate layers of humic horizons provide an excellent set-up to study the effects of time on the process of humification of the organic matter in the soil, and also the time-induced changes in the soil polysaccharides.

The Ashitaka Loam

As shown in Figure 14.1, Mt. Fuji is located almost at the center of the Japan archipelagoes, and Mt. Ashitaka is to the south of Mt. Fuji. The point of investigation was at the edge of the piedmont area of Mt. Ashitaka, shown in Figure 14.2 as Pt. X. The area is evergreen broad leaved forest region, has an average temperature of 15.9°C and mean annual precipitation of about 2000 mm. The climate, however, does not seem to significantly affect the soil genesis as is usually the case for zonal soils. The accumulation of humus is believed to be related to the presence of free aluminum in the soils derived from the volcanic ash.

Data have recently become available on the ages of the various horizons of the Kuroboku soil and the Upper Loam of the Ashitaka Loam Formation. Carbon-14 age-dating makes use of the charcoal fragments coexisting with the microliths and the organic matter in these horizons. Based on the archeological remains, horizons 2 and 4 are believed to be formed approximately 2000 and 4500 years before present (YRS. B.P. in the figures). According to Kato (1969), the horizon 5 corresponds to the so-called Fuji Black. From the ^{14}C method, Miyazaki (1968) determined the ages of the fulvic acid, humic acid and humin of this Fuji Black to be 5370 ± 200 B.P., 6540 ± 100 year B.P. and 7000 ± 100 year B.P. respectively. In considering the age of this horizon, we have tentatively adopted the value of 7000 years B.P.

The Carbon Cycle 193

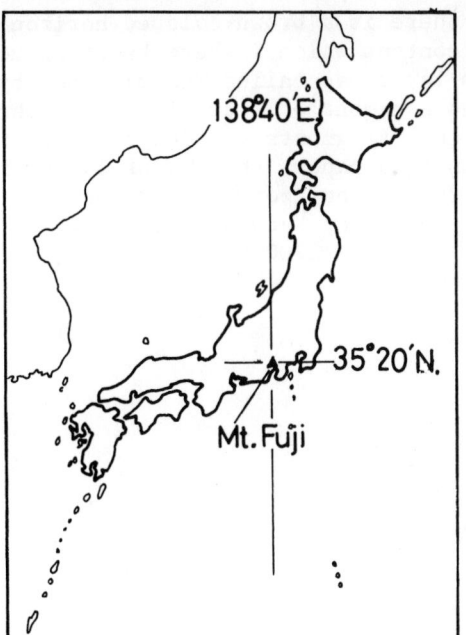

Figure 14.1 Map of Japan.

Figure 14.2 Area map and the sampling site.

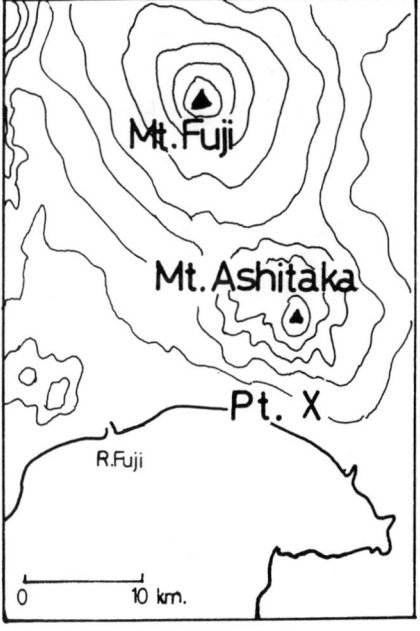

Below the Fuji Black, there is a brown-colored horizon with lower organic matter content which is here referred to as horizons 6 and 7. This layer was called Yasumiba bed by the archeologists who found a number of microliths from this bed and ^{14}C dating of charcoal fragments yielded the age of 14,000 ± 700 years B.P. In this paper, the age of 14,000 ± 700 years B.P. is adopted for the horizon 7. For the horizons 10, 17, and 18, ^{14}C data presented by Kato are adopted, namely 18,030 ± 450 years, 27,200 ± 2200 years and 28,100 ± 400 years B.P., respectively.

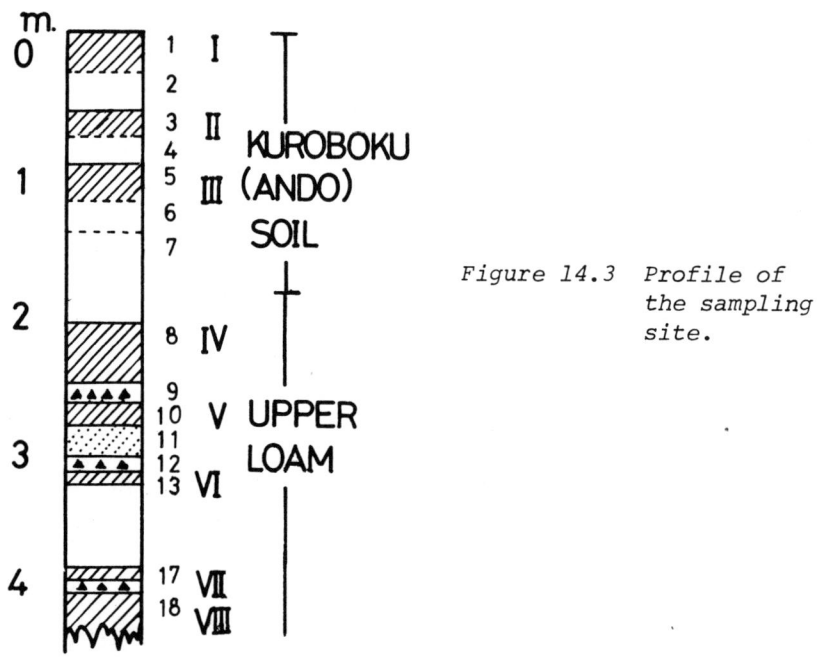

Figure 14.3 Profile of the sampling site.

The source of the volcanic ash was mostly from old Mt. Fuji and its surrounding volcanoes. Kato's study (1969), however, confirmed that some of the horizons came from a different source because their mineral compositions were different from those of old Mt. Fuji. Some interesting features of the clay mineralogy include the fact that the younger part of the volcanic ash (with the age of about a thousand years B.P.) contained small amounts of crystalline 14 Å clay minerals. This observation was unexpected since recent volcanic ashes are commonly believed to contain only amorphous clays or allophane and no crystalline clay minerals.

The Carbon Cycle 195

The degree of humification of organic matter was also investigated. From the humic acid concentrations, it would appear that the degree of humification in the Kuroboku soil increases from the top to the bottom except for horizon 3. Based on Kato's previously mentioned work, this horizon was not derived from the old Mt. Fuji. Down to the Upper Loam portion, the degree of humification was rather well advanced as was also mentioned above. However, in comparison to those of Kuroboku soil, the degree of humification was slightly lower in the Upper Loam. Organic matter content itself was higher in Kuroboku soil, ranging from 16.0 to 23.4% compared to the range of 3.6-6.4% in the Upper Loam. In age, Kuroboku soils are approximately 10,000 years or less, whereas the Upper Loam ranges from 10,000 to 30,000 years.

A STUDY OF THE POLYSACCHARIDE EXTRACTED FROM SOILS

Methods

The soil sample was air dried and passed through a 2-mm sieve. For extraction, a slurry consisting of 20 ml of 0.5 N sodium hydroxide solution and 2 g soil was placed in a centrifuge tube and left at room temperature for one hr, then centrifuged and the supernatant recovered. To this extract, sulfuric acid was added and the pH adjusted to 1.5. After bringing pH to 1.5, a precipitate or humic acid fraction was removed by centrifugation. Then to one volume of supernatant or the fulvic acid fraction, three volumes of 99.5% ethanol were added and the precipitate obtained considered to be the polysaccharide. This precipitate was separated through filter paper and dissolved again with 0.5 N sulfuric acid solution.

The determinations of sugars and uronic acids in the solution were carried out by the anthrone method (for sugars) and by the carbazole method (for uronic acids). The amounts of sugars and uronic acids were calculated and shown as mg glucose or uronic acid in 100 g soil on oven-dried basis. Total carbon determinations on the soil samples obtained from each horizon were carried out by the manometric method originated by Van Slyke and Folch.

Results

The concentrations of organic carbon in percent, glucose and uronic acid in mg/100 g soil are shown in Figure 14.4. Above horizon 6, the carbon contents range from 5% to 13.6% in horizon 1. Below horizon 7, the values remain constant

196 Environmental Biogeochemistry

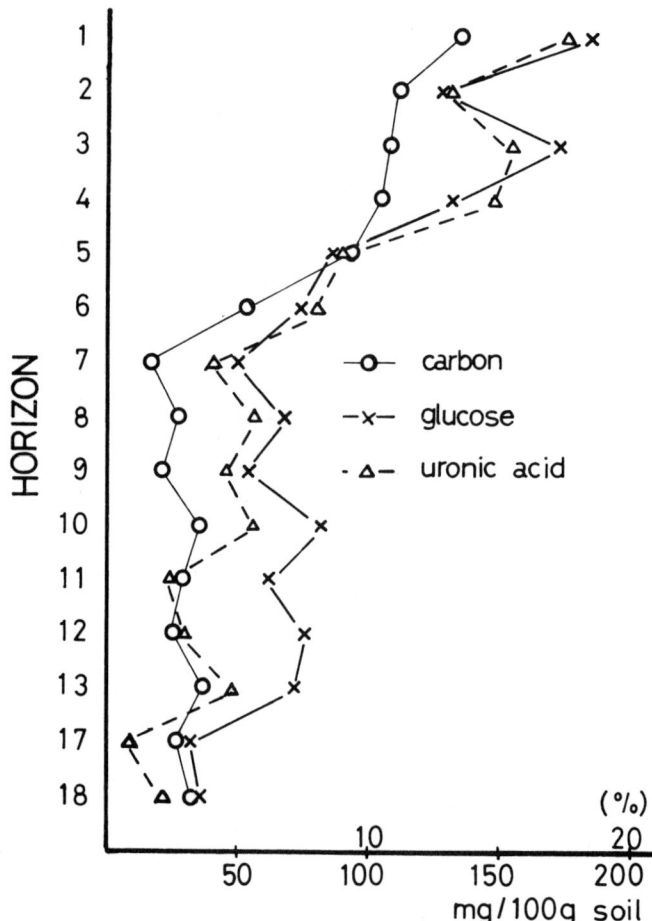

Figure 14.4 Amounts of carbon, glucose and uronic acid.

around 3%. Glucose and uronic acid contents decrease with the decrease in the value of the carbon content; however the tendency is somewhat different between the horizons above and below horizon 5. This difference is more clearly illustrated in Figures 14.5 and 14.6, where G/OM indicates glucose to organic matter ratio and U/OM indicates uronic acid to organic matter ratios, and the amount of organic matter was calculated by multiplying the carbon content by 1.72. The profiles for the two ratios are nearly identical. Above horizon 6, the ratios are about the same, but below horizon 7 the uronic acid to organic matter ratios become less than the glucose to organic matter ratios. In general, the G/OM and U/OM ratios are higher below horizon 7.

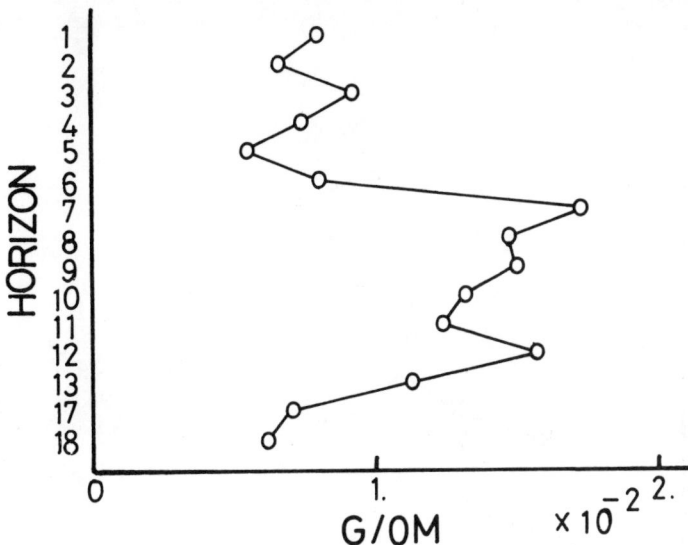

Figure 14.5 Glucose to organic matter ratios.

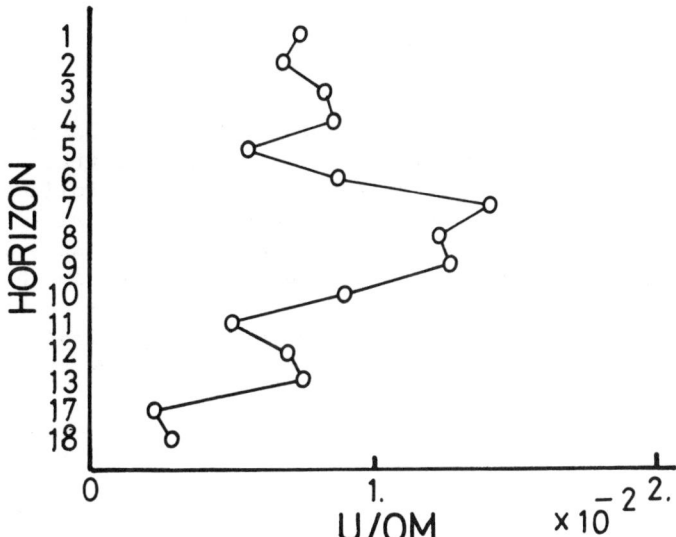

Figure 14.6 Uronic acid to organic matter ratios.

In Figure 14.7, uronic acid to glucose ratios (U/G) are shown. There is a rather monotonous decrease in the ratios along the profiles from the top to the bottom.

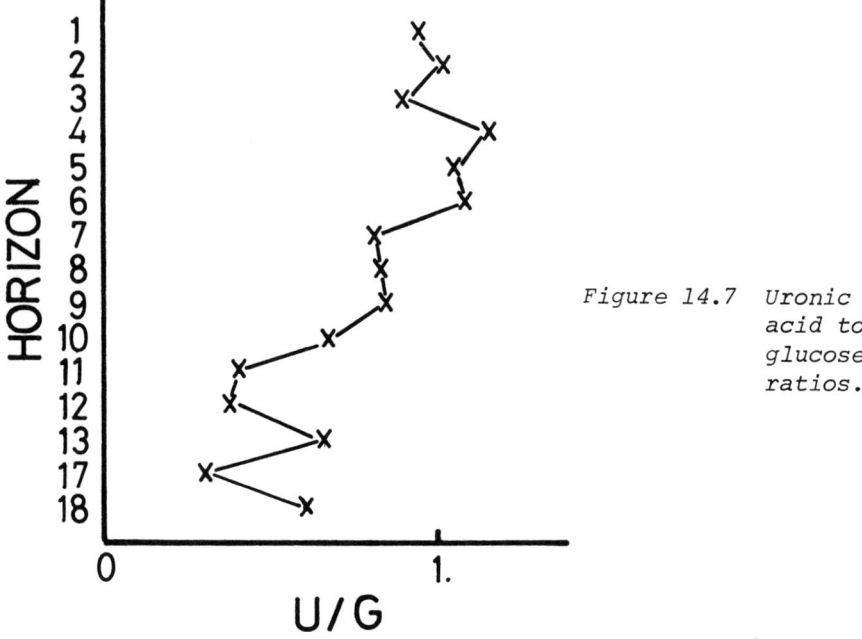

Figure 14.7 Uronic acid to glucose ratios.

Figures 14.8 and 14.9 show the relationship between age and polysaccharide concentrations. Ages are plotted on the logarithmic scale and indicated as YRS. B.P. Solid circles pertain to the samples whose ages have been determined in one way or the other. Circles with broken lines show ages approximated on the basis of the sample position in the soil sequence. Bars beneath the circles indicate that the samples are the humic horizons. Apparently there are differences in the grouping of the data above and below the 10,000 years B.P. time horizon.

Discussion

The soil and the buried humic horizons in Ashitaka Loam Formation which have been derived from volcanic ash falls belong to Kuroboku soil (or Ando soil). The high content of organic matter is quite common among these soils. This is commonly understood to imply luxuriant grass vegetation and unfavorable conditions for the degradation of organic matter or humus.

In considering the degradation of the (incorporated) plant materials, humus formation and gradual degradation

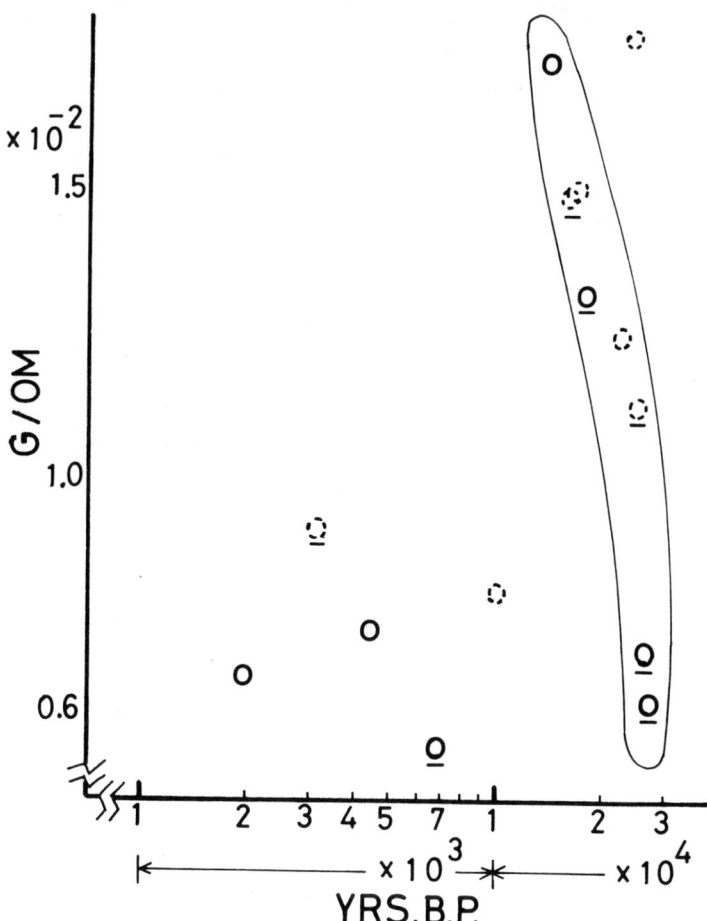

Figure 14.8 Relationship between age and glucose to organic matter ratios.

of humus, the profiles of the Ashitaka Loam Formation provide an excellent opportunity to study a naturally set up model. In particular, no experiments can match this set-up on the same time frame.

From Figure 14.1, as was mentioned before, one notices that carbon contents are low in the horizons of the Upper Loam. This would suggest that little, if any, incorporation of plant residues was made following burial of each horizon and that the degradation of humus proceeded for a long period of time (probably 10,000 years). Kato (1973) has also observed that in the recently-formed soil, the amount

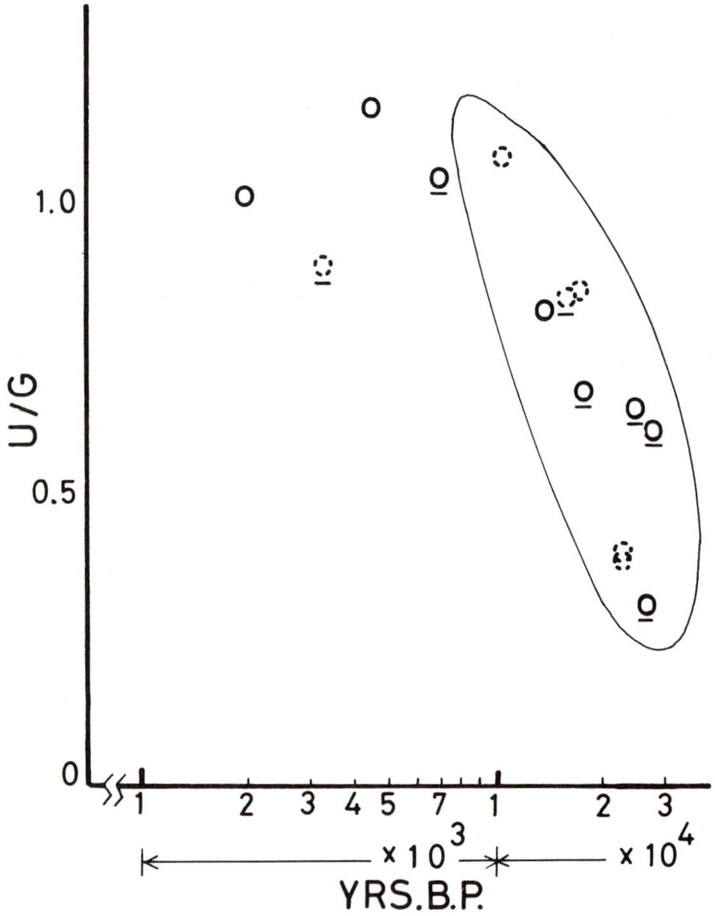

Figure 14.9 Relationship between age and uronic acid to glucose ratios.

of plant opal was similar to the amount of humus. This provides concrete evidence that the amount of humus present was dependent on the amount of plant residues incorporated originally into the horizon. Yamane (1973) also supports this observation through his study on Kawatabi grassland. As to the buried humic horizons in the Upper Loam, Kato (1973) noted that the ratio of plant opal to the organic matter was higher than the ratios found in recent soil. This may be considered as other evidence that there was abundant humus initially, by analogy with the characteristics of the surface horizons of recent soils. (It should be noted

that the term recent may tentatively be considered to be
in the order of a thousand years). Part of the reason for
that is the crossing of the line of carbon and glucose as
well as uronic acid in the vicinity of horizon 7. This may
also be interpreted to mean that polysaccharides are more
resistant during the final stages of the organic matter
degradation, particularly when no fresh plant residues
are incorporated into the soil. This is clear from Figures
14.5 and 14.6. In both cases glucose and uronic acid content
in the soil organic matter increase with the increase in
the duration of degradation, especially when the age exceeds
approximately 10,000 years. However, around the age of
27,000 to 28,000, both glucose and uronic acid to organic
matter ratios decrease substantially. There is a gradual
decrease in the uronic acid to glucose ratio (see Figure
14.7), but the rates of decrease are slightly different.
Above horizon 6 the rate is slow and could be called steady,
but below horizon 7 the rate of the decrease in the uronic
acid to glucose ratio becomes more pronounced. These variations apparently may have a paleoclimatic significance.

From his age-dating and field observations, Kato (1970)
attempted to correlate the horizon with the age of about
18,000 years B.P. with the upper black band (buried humic
horizon) and the horizon with the age of about 27,000 and
28,000 years B.P. with the lower black band of Tachikawa
Loam. Matsui (1968) was possibly the first to try to relate
the presence of buried humic horizons to the time of glaci-
ation and stated that the upper black band corresponded to
some time of Woodford stage and the lower black band corresponded to Farmdalian substage (28,000-22,000 years B.P.)
in the North America or the Paudorf-Archy interstadial
(33,000-25,000 years B.P.) in Europe.

At present, it is rather difficult to relate the nature
and the amount of the polysaccharides to the past climate.
However, the paleogeographic implications of the humic layers
are worth considering. One interpretation seems to be most
plausible, namely that the periods of 10,000 years B.P.
and 33,000 years B.P. represent abnormally cold conditions.
Thus the difference in the nature of polysaccharides before
and after the year around 10,000 years B.P. may possibly
be due to climate factors. Figures 14.8 and 14.9 show
the relationship between age in logarithmic scale and the
nature of polysaccharide. Again, the difference in the
nature of polysaccharides before and after 10,000 years
B.P. is observed.

REFERENCES

Kato, Y., and Ashitaka Loam Research Group, 1969: Pleistocene Tephras ("Loam") at the foot of Ashitaka volcano. *Quaternary Research* 8, 10-21.

Kato, Y., and Ashitaka Loam Research Group, 1970: ^{14}C age dating of the buried humic horizons of the Upper Loam in the Ashitaka Loam Formation. *Chikyu-Kagaku (Earth Science)* 24, 73-75.

Kato, Y., 1973: Amorphous components of buried humic soils among the upper member of Ashitaka "Loam" Formation, latest Pleistocene Tephras, Central Japan. *Quaternary Research* 12, 11-18.

Matsui, T., Naruse, H. and T. Kurobe, 1968: ^{14}C age dating of the buried humic horizons in the Tachikawa Loam. *Chikyu-Kagaku (Earth Science)* 22, 40-41.

Miyazaki, A., 1968: Problems on the ^{14}C age dating of the buried humic horizon. *Kodojo (Paleosol)* 4, 1-3.

Yamane, I., 1973: Soil and plants in Kawatabi grassland. *Pedologist* 17, 44-129.

CHAPTER 15

BIOGEOCHEMISTRY OF PCB AND DDT
IN THE NORTH ATLANTIC

GEORGE R. HARVEY
WILLIAM G. STEINHAUER

Woods Hole Oceanographic Institution
Woods Hole, Massachusetts 02543

INTRODUCTION

A study of the biogeochemistry of the industrial polychlorobiphenyls (PCBs) and the insecticidal DDT family (t-DDT = p,p' - DDT, p,p' - DDE and p, p' - DDD) in the ocean can serve as a useful way to probe into the movements of organic compounds in the marine environment. Neither PCB nor t-DDT is generated within the sea. They are totally manmade, thus there is no natural background to estimate at any particular site. Both classes of compounds are used almost exclusively on land but the manner in which they enter the open environment is different; p,p' - DDT is purposely sprayed into the atmosphere over agricultural lands and insect-infested areas. PCBs must leak into the environment or be accidentally introduced (Nisbet and Sarofim, 1972). Thus, one might expect t-DDT and PCBs to be distributed differently in the ocean even if they were equally stable in the environment.

We began studying the distribution and transfer of t-DDT and PCBs in the Atlantic during the winter of 1970-1971. At that time public and government interest in the distribution and effects of DDT and PCB in the marine environment were at a peak. Data was urgently needed on the environmental quality of the ocean so that policy decisions could be made. Unfortunately, there was no data to answer the questions in 1970, so the decisions had to be

made without it. Gloomy predictions were made by various
celebrities on the death of the seas brought about by petroleum, DDT and PCBs. Fortunately, some governments (especially the U.S. and Canada) committed themselves to collecting
environmental quality data on the oceans, and set out to
fund research in answer to some of the difficult questions.
Since 1970 marine research groups have provided answers to
some of the questions posed about the pollution of the sea.

Now in 1975 DDT has been banned in many countries and
PCB use has been curtailed in the northern hemisphere. The
shortage of food and energy is now the world's greatest
concern and not the quality of marine water. However, these
two concerns are very much related. As world energy needs
increase, more petroleum will be spilled into the sea from
tankers and offshore drilling rigs, and more nuclear fuel
wastes will find their way to the sea. DDT usage will be
reinstated, or a suitable cheap and *persistent* substitute
will have to be used for obvious economic reasons in developing countries. PCB, or a similarly heat-stable nonflammable
dielectric fluid, will continue to increase in production
because it is basic to the electric power industry. New
agricultural and industrial chemicals will be introduced
in megaton quantities around the world. In summary, one
could reasonably predict from our previous experiences that
from current demand for more food and more energy, the
increase in chemical pollution of our oceans will be an
unavoidable consequence unless rational alternatives, advice
and predictions based on research data are available so
nations can manage potentially harmful chemicals wisely.

RESULTS AND DISCUSSION

Organisms

Our observations of the PCB and t-DDT concentrations
of surface and midwater organisms have been described
elsewhere (Harvey *et al.*, 1974a). In that work we found no
evidence to support the popular food chain magnification
theory among the fish studied. The observed high PCB or
t-DDT concentrations found in some species, or the relatively
low concentrations found in others bore no relation to their
relative trophic level. The ratio of PCB/t-DDT was 30 or
greater in net plankton and decreased to about 3 in organisms
living in the 300-900 m range of the water column. The
DDE/DDT ratio increased with depth of habitat of the organisms
or with increasing trophic level. The t-DDT in the deeper
animals, or in the high predators, *e.g.*, sharks, had passed

through an unknown number of organisms previous to capture, each one being able to convert some of the p,p' - DDT to p,p' - DDE. Thus, the DDE/DDT ratio in the shark was >100.

More recently we have analyzed a collection of benthos captured on the Nares Abyssal Plain in 5800 m of water. Some of the results shown in Table 15.1 were quite surprising to us, as well as very revealing of the different biogeochemistries of PCBs and t-DDT.

Table 15.1
PCB and t-DDT in Deep Atlantic Benthos[a]

Specimen	DDT	DDE	PCB	PCB/t-DDT
	(in parts per billion wet weight)			
Rattail Stomach	b	0.6	0.0	
Rattail Fillet	b	21.4	0.5	0.02
Rattail Liver	b	381.0	340.0	0.9
Brotulid Stomach	0.5	4.4	11.5	2.3
Brotulid Fillet	1.7	2.2	36	9
Brotulid Liver	56	1800	1200	0.6
Holothurian #1 Stomach	b	0.2	0.2	1.0
Holothurian #1 Body	b	0.2	0.0	0.0
Holothurian #2 Stomach	1.1	0.2	0.6	0.46
Holothurian #2 Body	0.4	0.1	0.5	1.0
Core A-II-85-3-6 (0-2 cm)[a]	b	0.5	0.3	0.6

[a] Collected near 25°N, 62°W in 5500-5800 m water depth on Atlantis II cruise 85 (September, 1974).

[b] Saponified before analysis.

   1. t-DDT concentrations in the two benthic fish are not lower than those found in organisms in the upper regions of the water column.
   2. The PCB/t-DDT ratios are very much lower than those found near the surface.
   3. In most cases where it was possible to distinguish DDT was a significant fraction of the t-DDT; the PCB present was significantly altered in composition.
   4. All the organisms show very selective uptake and concentration of the two classes of compounds from the sediment and into liver and flesh.
   5. The holothurians do not seem to be significant in remobilizing sedimentary PCB or t-DDT.
   6. The data in Table 15.1 reconfirm our previous rejection of the concept of increasing concentration of PCB

and t-DDT up marine food chains (Harvey et al., 1974a). The detritiphoric holothurians depend entirely on the sediments for their food but there is no concentration factor. The fish, brotulid, had much higher body concentrations of these pollutants than the holothurians but more PCB than the rattail which was twice the size and had two large squid beaks (2 cm) and several squid eggs (tentative identification) in its stomach, obviously a very high predator in the food chain.

PCB and DDT in North Atlantic Water

During the course of this work we observed a dramatic decline of PCB concentrations in certain parts of the North Atlantic surface waters (Harvey et al., 1974b). The data supporting this observation are illustrated in Figures 15.1, 15.2 and 15.3, showing the prevailing concentrations in

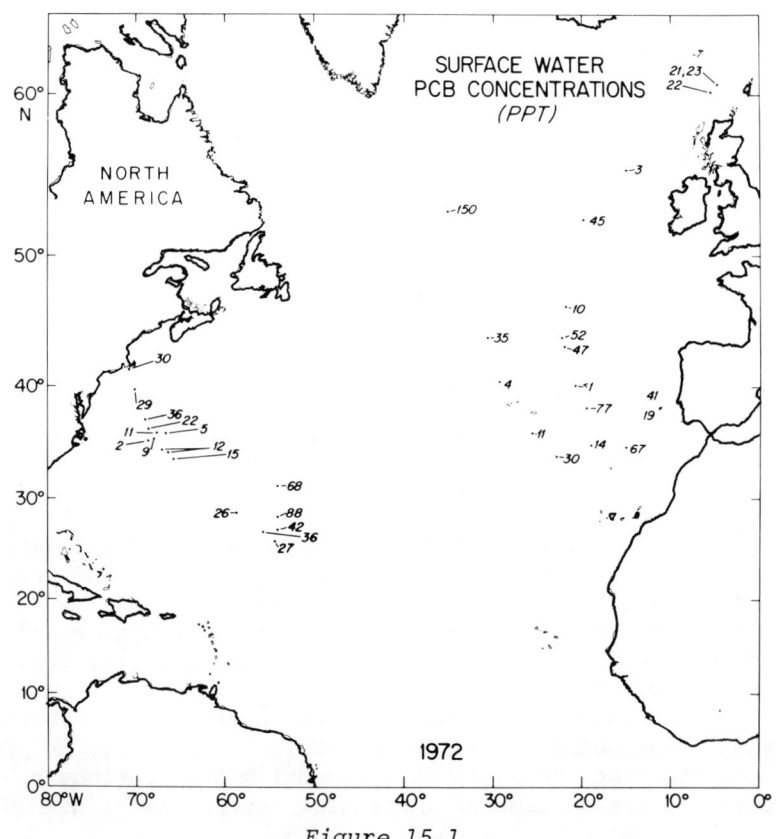

Figure 15.1

The Carbon Cycle 207

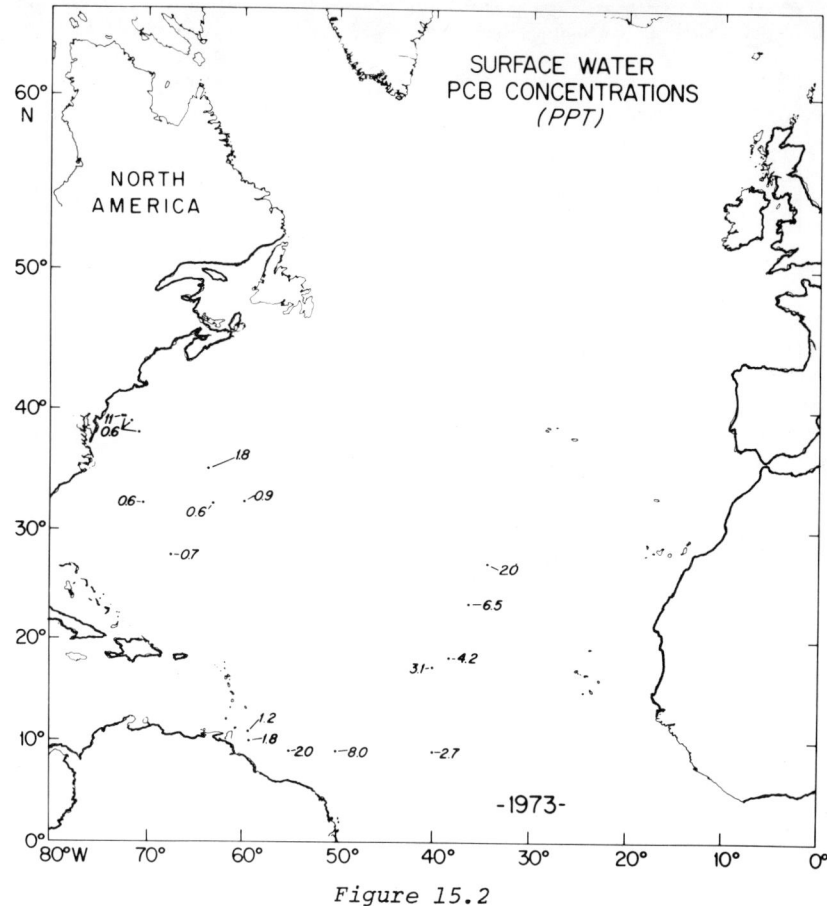

Figure 15.2

1972-1975. We believe that the observed decline of PCB in the surface waters was due to the cessation of certain dispersive industrial and commercial uses of PCB in the U.S. (1970-1971) and Europe (1972-1973).

Note, however, that we have not sampled north of 40°N since 1972 or south of the Sargasso Sea until 1973. Thus, only the Sargasso Sea has been periodically sampled during the period 1972-1975. However, since 1972 no PCB concentrations in the 10-100 ppt range have been found in the surface waters of any part of the North Atlantic. At no time during the sampling period was t-DDT detectable in the water column by our methods. We estimate the maximum concentration of DDT or DDE which could have been present at 0.5 ng/l.

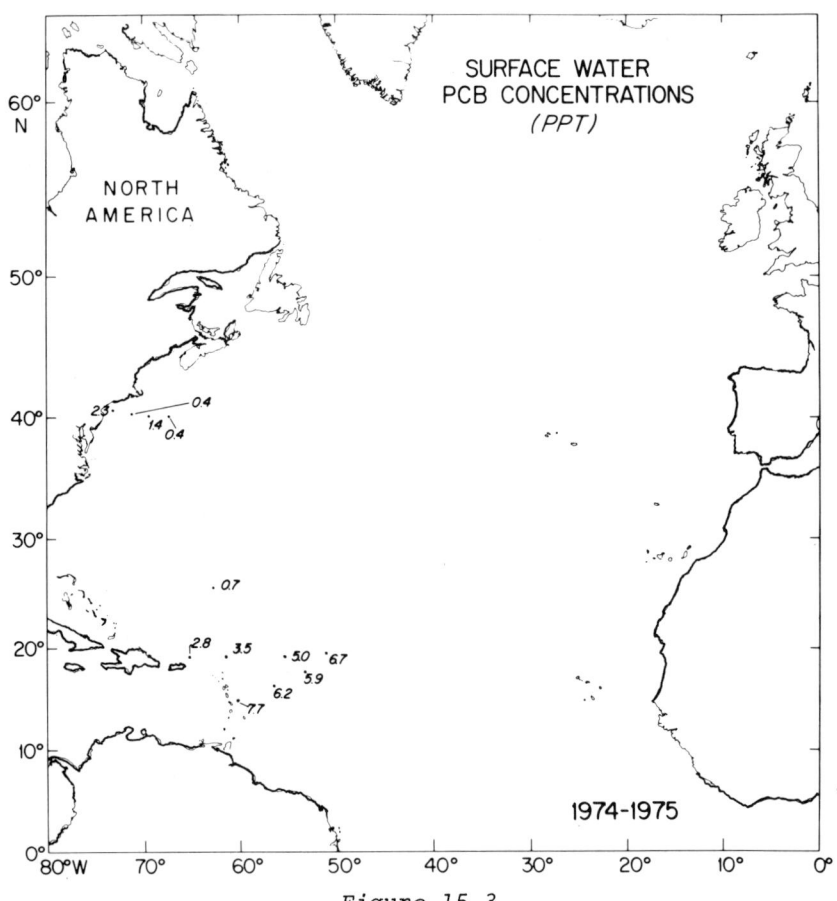

Figure 15.3

The PCB data require that, at least in the Sargasso, about 2×10^{-3} g/m^2 were lost from the upper 100 m of the water column in 1 year (1 m x 1 m x 100 m x 10^3 1/m^3 x 2×10^{-8} g/l). Five possible explanations for this loss have been seriously considered. They are:

1. Dilution of PCB by vertical and horizontal advection and diffusion.

Relevant studies on the depletion of dissolved radionuclides from the mixed layer after the 1963 test ban treaty show that advection and diffusion are too slow to have been responsible for the observed effect. For example, Sr-90 required more than five years to decline to one-half the surface concentrations of 1963 (Bowen, et al., 1969).

2. Association of PCB with descending particulate matter.

The generally known affinity of PCB for solid surfaces makes removal on falling particulates an attractive explanation for the removal of a portion of the 10^{-3} g/m^2 yr. In support of this explanation the plutonium nuclides, Pu-239-240, which are believed associated with particulates, decreased in surface waters considerably faster than Sr-90 (Noshkin and Bowen, 1973). Indeed, we have performed a series of equilibration experiments to determine the partitioning of PCB between seawater (20 liters) and suspended matter (10-20 mg dry wt.). The suspended matter used was from the top sections of near-shore and abyssal plain cores, and from deep sections of the latter cores. The PCB was introduced to the system in three ways: 1) as found naturally in the sediment, 2) as an added spike dissolved in the seawater and 3) as an added spike to a PCB free system (*i.e.*, extracted seawater and deep pelagic core material). The mixtures (in triplicate) were shaken periodically during five days before the subsequent analysis. Equilibrium was actually established after a few hours. The results are presented in Figure 15.4.

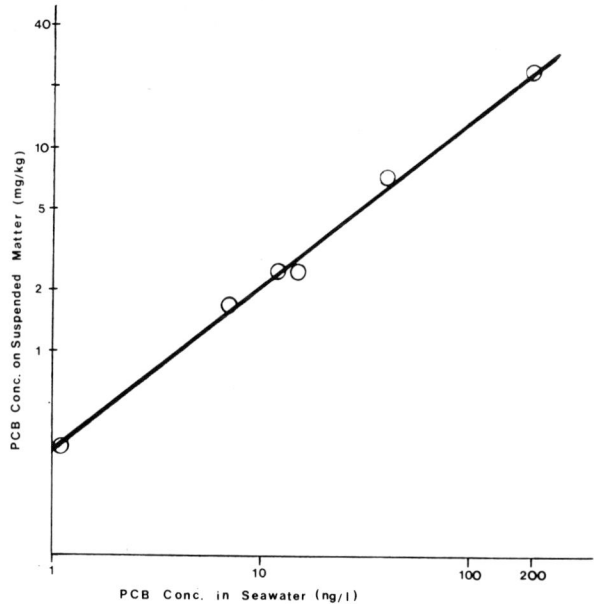

Figure 15.4 Distribution of PCB between seawater and particulates.

Considering the diverse sources of the particulate matter used in these experiments the correlation is remarkable. The concentration factor varies very little within the range of PCB concentrations used, $i.e.$, 3.4×10^5 to 1.1×10^5.

If we can assume from Figure 15.4 that in 1972 when the surface water contained 20-40 ng/l of PCB the concentration factor onto particulate matter was about 2×10^5, then the PCB concentration would have been about 2 ppm (2 mg/kg), a value we actually measured in phytoplankton in 1970-1972 (Harvey $et\ al.$, 1974a). Therefore, the 2×10^{-3} g/m^2 lost from the upper 100 m would have required about 500 g/m^2 of falling solid particles as adsorptive scavengers. In our present state of relative ignorance of particulate fluxes in the open ocean this number (500 g) appears impossible to achieve. The primary productivity of the open ocean is only 50 g C/m^2/yr (100 g organic matter) (Menzel, 1974), most of which (96%) is recycled within the mixed layer. Also, only about 3% of organic carbon which escapes the euphotic zone will be particulate, $i.e.$, 3 g/m^2. The sedimentation rate on the Sargasso floor is 13 g/m^2/yr (0.5 cm/ 10^3 yr $\times$ 10^4 cm^2 $\times$ 2.6 g/cm^3). Thus, even if all 13 g came directly from the surface of the sea, which is most unlikely, there are still not enough falling particles in 1 m^2 to scavenge 2 mg of PCB. From these data we must conclude that there exists at least 100 times less particulate matter available than would be necessary to invoke particulate scavenging as the major mechanism of the PCB flux from the euphotic Zone.

3. Evaporative codistillation of PCB from the surface waters of low and mid latitudes.

We do not know the major direction of flux of PCB through the surface microlayer. We do know that the concentrations above and below the surface microlayer are lower (Bidleman and Olney 1974). It seems reasonable to postulate that a large fraction of the PCB deposited on the sea surface by wet and dry fallout, and by solvation from the atmosphere, should be re-entrained with the air mass during periods of high evaporation. This condition is prevalent in the Sargasso Sea most of the year. The PCB could then be transported to areas of higher precipitation (land or sea), or be condensed in the colder high latitudes, $i.e.$, a giant distillation system. The PCB deposited at high latitudes could then be entrained by the cold, dense sinking water into the deep Atlantic and flow south with the North Atlantic deep water without passing through intermediate depths at the mid latitudes. Unfortunately, we have not had the opportunity to sample north of 40° since 1972 to test this

The Carbon Cycle 211

hypothesis. However, in support of the hypothesis, we have always seen lower PCB concentrations in the surface waters of the Sargasso Sea (compare Figures 15.1, 15.2 and 15.3).

In contrast, the southern North Atlantic lying in the influence of the Northeast Tradewinds reveals higher PCB concentrations in the water column than in the Sargasso Sea. In Table 15.2 we present two profiles determined in September of 1973 and two from February of 1975 (Atlantis II, Cruises 78 and 86). However, note that the PCB concentration maximum was in the surface water in September, but subsurface in February when the tradewinds are strongest. The rather gradual decrease in PCB concentration with depth may be explained by the high particle load of the Northeast Trades (Folger, 1970) carrying deposited PCB further into the water column in less time than is possible with simple wind mixing. Also, in this region of the Atlantic a permanent high salinity layer (> 37 ‰) exists between 80 and 130 m forming a barrier to normal diffusion.

Table 15.2
PCB Concentrations in Atlantic Water within Northeast Trades

Station	Date	Depth	PCB x 10^{-9} g/. (± 20%)
09°00N 40°00W	9/73	1	2.7
		80	1.7
		100	0.9
		110	1.8
		130	1.4
		200	2.0
		300	0.8
09°00N 49°57W	9/73	1	8.0
		80	1.8
		80	1.7
		100	1.1
		200	1.6
		250	1.2
19°00N 61°15W	2/75	1	3.5
		50	4.3
		120	3.9
		140	4.5
		180	3.8
		200	3.9
18°58N 55°03W	2/75	1	4.5
		50	8.4
		130	8.2
		140	8.7
		200	6.1
		850	4.2

4. Biogeochemical degradation of PCB in the mixed layer.

Very little can be said about the importance of biogeochemical degradation of PCB. We have noted before that DDT must be very susceptible to degradation since its concentration is so much lower than PCB in the open sea even though more has been made and all of it purposely injected into the open environment. Recently we discovered that ambient marine microorganisms were capable of converting a radiocarbon-labeled tetrachlorobiphenyl to water-soluble metabolites and CO_2. A carboxylic acid metabolite has been partially characterized, but its precise structure must be determined before we can ascertain the importance of this pathway for PCB removal.

5. A change of analytical methods resulted in lower measured PCB concentrations in 1973, or higher in 1972.

We have taken every precaution and followed up every possibility that may have led to different results in 1972, 1973 and 1974. The following comments are relevant to point 5 above.

--All analyses were completed at sea.

--Twenty and fifty-liter water samples were extracted in the same manner with Amberlite XAD-2 resin during the three years (see Methods).

--On each collecting cruise the XAD method (Harvey *et al.*, 1973) was checked at least once by solvent extracting by hand 20 liters out of a 60-liter water sampler. The remainder was extracted by XAD. The two results always compared within 20%.

--Surface water samples taken by bucket and water sampling bottles lowered to just below the surface have always compared well.

--Three Woods Hole Oceanographic research vessels were used for the water sampling. Oil, paint, grease, laboratory air and tap water always gave satisfactory blanks on the ships.

--The same personnel performed the analyses during the past three years.

--Three sets of reference and working analytical standards used in the work were used in several intercalibration experiments.

--Our results have always been within ±20% of other laboratories.

--The procedural blank was improved from 0.5 ng/l in 1972 to 0.3 ng/l in 1973, well below the magnitude of the changes observed.

PCB and DDT in North Atlantic Sediments

Marine sediments should be the final repository of chlorinated hydrocarbons which escape degradation. Also, since organic decomposition in deep sea sediments is very slow (Jannasch *et al.*, 1971) they should contain a permanent record of the input and transfer pathways through the water column. The results of our sediment analyses are set out in Table 15.3 and the positions of the cores are illustrated in Figure 15.5.

There is a very crude correlation between a decreasing PCB concentration with distance from source, *e.g.*, cores #16, #14 and #10. A similar correlation, by no means smooth, exists between a decreasing PCB concentration and the overlying water depth, *e.g.*, cores #8, #7 and #6. However, compare cores #9 and #4.

Also, there is a better correlation between depth and the PCB concentration in g/kg than there is in g/m^2 (calculated by multiplying the g/cm by the depth of the core section analyzed in cm). The reasons for these discrepancies are several.

1. The sediments at a particular location in the deep sea do not necessarily represent material contributed from the surface directly above. Horizontal advection of sinking particles could carry PCB and DDT thousands of kilometers from point of entry into the sea. Nevertheless, the exponential decrease of PCB concentrations in the atmosphere over the North Atlantic with increasing distance from the United States (Harvey and Steinhauer, 1974; Bidleman and Olney, 1974) is clearly reflected in the sediment analyses.

2. In areas of strong bottom currents sediments on uneven topographies can be scoured from high points and deposited in low areas. We assume this is the explanation for core #5.

3. Coring in the deep sea, even with a good acoustic pinger and receiver, is at best a skilled art. Generally, the operation requires about 10% more wire out than there is water depth. Consequently, one does not know, in relation to the ship, where the corer penetrates, *i.e.*, on a slope, in a gully or on a ridge.

With these uncertainties in mind a few "order of magnitude" interpretations seem appropriate.

The canyons in the northwest continental shelf do not appear to be an unusually important transfer route of PCB into the deep North Atlantic. Core #9 at the head of the canyon has only twice the concentration of core #7 at the mouth. In contrast, core #10 on the New York shelf has ten

Table 15.3

PCB Concentrations in North Atlantic Sediments

Figure	WHOI Code	Position	Date	Water Depth (m)	PCB $\times 10^{-6}$ g/kg (dry wt.)	PCB $\times 10^{-6}$ g/m^2
1	K-19-4-2	See Figure 15.5	4/71	4489	0.2	2.8
2	A-II-78-2		9/73	3938	1.0	24.0
3	AOII-78-6		9/73	4590	1.3	5.5
4	A-II-78-18		10/73	5950	2.8	28.0
5	K-33-2-2		10/73	4936	0.0	--
6	K-33-2-6		10/73	5465	0.1	1.4
7	K-33-2-8		10/73	3785	1.0	12.0
8	K-33-2-9		10/73	2626	1.7	96.0
9	K-33-2-12		10/73	137	2.2	30.0
10	A-II-81-1		2/74	37	30.4	811.0
11	A-II-81-3		2/74	110	1.4	25
12	A-II-81-5		2/74	690	4.6	48
13	A-II-81-6		2/74	2325	5.0	50
14	AS-1	Buzzards Bay, Mass.	4/74	3	165.0	445
15	AS-2	Vineyard Sound, Mass.	5/74	3	189. (0-2 cm) 3.7(30-40 cm)	--
16	New Bedford Harbor		6/73	3	8400.0	--

The Carbon Cycle 215

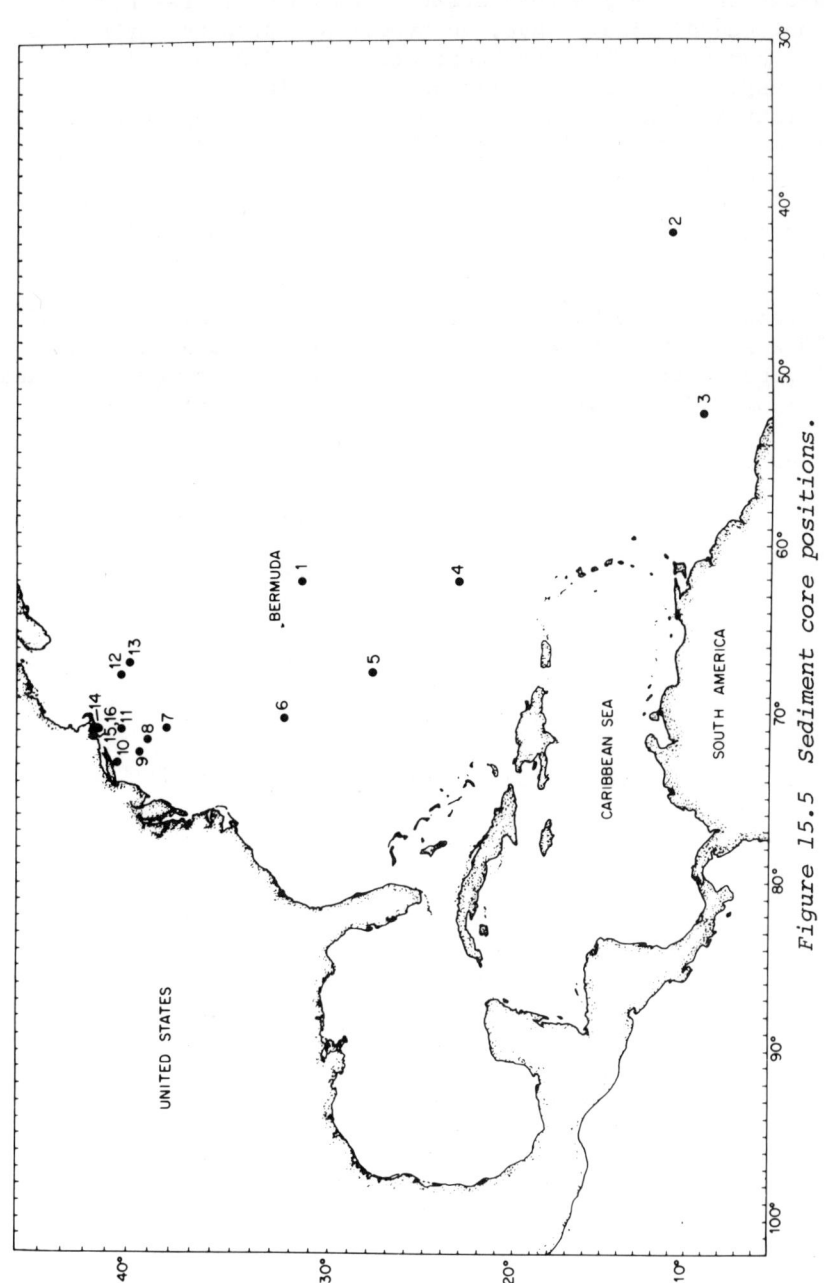

Figure 15.5 Sediment core positions.

times more PCB. Core #12 is in Lydonia Canyon and has about the same PCB concentration as core #13 taken on the continental rise. Thus, an hypothesis proposed by us that the canyons in the northwest shelf would be efficient transport vectors of pollutants directly into the deep Atlantic is not borne out. Sediments slumping down the rise (compare core #4 with core #13 in Table 15.3) are equally efficient at this transport.

A very small fraction of the PCB now in the North Atlantic is represented in the sediments. From Table 15.3 the average amount of PCB in sediments lying under water deeper than 2 km is 27×10^{-6} g/m^2. If we restrict our estimate to the North American Basin which has an area of 10×10^6 km^2, a volume of 4.2×10^7 km^3, and an average PCB concentration of 1 ng/l we develop the following fluxes:

A. PCB in North American Basin sediments:

$$27 \times 10^{-6} \frac{g}{m^2} \times 10 \times 10^6 \text{ km}^2 \times 10^6 \frac{m^2}{km^2} = 2.7 \times 10^8 \text{ g}$$

B. PCB in North American Basin water:

$$1 \times 10^{-9} \frac{g}{l} \times 4.2 \times 10^7 \text{ km}^3 \times 10^{12} \frac{l}{km^3} = 4.2 \times 10^{10} \text{ g}$$

Thus, the sediments now contain about 0.7% of the total PCB in the North American Basin.

The major production of PCB began about 1955 so we can estimate the average rate of deposition on the seabed as 1.4×10^{-6} g/m^2 yr, or about 1000 times less than the flux (or loss) from the mixed layer. The 1.4×10^{-6} g/m^2 yr of PCB found in these sediments would require only 2-3 g/m^2 yr of sedimenting particles to carry them to the bottom, a quantity easily available from organic and mineral fallout. We have obtained some different experimental verification of this rate of PCB supply to the deep sediments. During 1974 Dr. Peter Wiebe (W.H.O.I.) placed a moored sediment trap, containing a grid of sixteen 25 x 25 cm glass fiber filters, 100 m above the bottom (2150 m), for 63 days in the Tongue of the Ocean (Nassau). We analyzed four of those filters and found an average of 0.8 μg/m^2 which corresponds to a delivery rate of 4.9μg/m^2 year. This value is only 3.5 times higher than our calculated rate for the deep North American Basin.

SUMMARY OF THE BIOGEOCHEMISTRY OF PCB AND t-DDT IN THE ATLANTIC

t-DDT

t-DDT is apparently very unstable in the open marine environment and probably succumbs quickly to the combined attacks of oxygen, water, sunlight and microorganisms. A small fraction of the DDT which reaches the sea surface is incorporated into the food chain where it is successively converted to the metabolite DDE by both horizontal and vertical predatory food chains. Within the body lipids of marine organisms, t-DDT appears quite stable and body concentrations do not change markedly from the upper ocean organisms to the deep benthos.
Another small fraction of the DDT released into the environment reached the sea adsorbed on particles heavy enough to fall directly to the bottom without ever entering the pelagic food chain. This appears in the deep sediments and benthos as the parent insecticide, p,p' - DDT in concentrations and ratios usually not seen elsewhere in the marine environment. Apparently the benthic organisms have a lower capacity for the conversion of DDT to DDE. However, in our present state of knowledge we would consider t-DDT a minor hazard to the environmental quality of the Atlantic Ocean at this time.

PCB

Commercial PCB mixtures, $e.g.$, Aroclor, Clophen Kanechlor, Phenochlor, contain 50-60 chlorine homologues and isomers of which about 20-30 are resolvable by gas chromatography. The vapor pressures and water solubilities of the components of these mixtures range over two orders of magnitude. However, the gas chromatographic fingerprints produced by extracts of surface samples, $i.e.$, water, plankton, particulates, etc., are generally very good matches for the whole commercial mixtures. Thus, we must agree with Jernelov (1974) that the transfer unit of PCB to the sea must be an aerosol droplet. To quote Jernelov, "It seems quite unlikely that the 'fingerprint' of the PCBs could remain the same if the individual PCB compounds evaporated to the atmosphere, precipitated to the ocean surface, and were accumulated individually in marine organisms."
Geochemically, the PCBs are quite resistant because excellent gas chromatographic fingerprints are generally observed throughout the water column and in the sediments.

Biologically, they are susceptible to slow degradation. Organisms residing deep in the water column or on the bottom reveal PCB gas chromatographic fingerprints with very distorted features and missing peaks. On several occasions the gas chromatographic match between a PCB standard and an extract of a deep living organism is so poor that quantification cannot be justified.

Since PCBs appear so resistant to geochemical degradation, they are probably capable of being transported greater distances than t-DDT by the atmosphere and by advective processes within the sea. In the North American Basin, less than 1% of the PCB introduced has been transported directly to the sediment by sinking particles. Only a very small fraction could be metabolized by organisms since only 0.1% of the PCB is accounted for in organisms. A significant fraction of the PCB introduced to the mixed layer in mid and low latitudes may be reinjected by entrainment, codistillation or evaporation etc. into the atmosphere and eventually deposited in the colder, low pressure trough about 60°N latitude. Finally, some portion of the PCBs must be carried to intermediate depths by fragile, soluble or edible particles which on dissolution introduce PCBs into the deep water mass. The subsequent dilution in the deep water would make concentration changes impossible to detect by current methodology.

METHODS

Extraction of Seawater for PCB and t-DDT

Amberlite XAD-2 Cleanup

    1. Put XAD-2 resin (Rohm and Haas, Co.) in a 50-60 mesh sieve and run several liters of tap water through.
    2. Soxhlet extract or reflux XAD with redistilled acetonitrile for 24 hr. Change solvent and repeat for another 24 hr.
    3. Repeat extraction with benzene for 24 hr.
    4. Air dry and store under distilled water until needed.

XAD-2 Blank

    1. Pack a 2-cm diameter column with 50 cc of resin.
    2. Pour 200 ml of boiling acetonitrile through column into 1 liter of water.
    3. Extract with 2 x 75 ml of hexane.

4. Wash hexane once with water.
5. Concentrate hexane to 0.5 ml.
6. Inject 5 microliters into GC.
7. Calculate ng of interference.
8. If interference will be more than 0.4 ng per liter (depending upon how many liters you will extract), repeat benzene extraction until acceptable levels are achieved.
9. Once acceptable the resin can be used for at least 20 analyses. Naturally, it is just as convenient to clean up 2000 g as it is 50 g.

Extraction of Seawater

1. Water should be transferred from the sampling bottle into a glass carboy which has been rinsed with reagent HF.
2. Pump or gravity feed the water (20-50 liters) through a 50-cc column of resin at 200-250 ml/min.
3. Elute column with 300 ml of boiling acetonitrile at full gravity flow.
4. Proceed as in previous section.
5. If there is excessive interference on the chromatogram, pass the extract through a small amount of Florisil in a disposable pipette column and rinse column with one bed volume of 5% ether/hexane.

Note: In the open ocean the particulate matter accounts for 10% of the PCB/l. In coastal areas where the total suspended matter load becomes appreciable, the particulate PCB load can go as high as 90%. In such cases it is advisable to filter the water with 0.3-micron glass fiber filters *after* having passed through the resin. Dissolved PCB adsorption on filters is very serious.

Extraction of Marine Sediments for PCB and t-DDT

Glassware Preparation

1. Extract Soxhlet glassware and thimble for 24-40 hr with acetonitrile with one solvent change.

Extraction

1. Charge Soxhlet thimble with undried sediment and extract for 24 hr with acetonitrile.
2. Dilute acetonitrile 5:1 with water and extract with 2 x 75 ml of hexane.
3. Concentrate hexane to about 5 ml and pour through

a short column of Florisil. Wash column with one column
volume of 5% ether/hexane.
4. Add about 1 g of HCl-cleaned copper powder to hexane
(to remove S and its compounds).
5. Decant hexane and concentrate for GC.

Contribution No. 3536 from the Woods Hole Oceanographic
Institution. Supported by grant GX35212 from the Office of
International Decade of Ocean Exploration, National Science
Foundation, and the Rockefeller Foundation.

REFERENCES

Bidleman, T. F. and C. E. Olney, 1974: Chlorinated hydrocarbons in the Sargasso Sea atmosphere and surface water. *Science* 183, 516-518.

Bowen, V. T., V. E. Noshkin, H. L. Volchok and T. T. Sugihara, 1969: Strontium-90: concentrations in surface waters of the Atlantic Ocean. *Science* 164, 825-827.

Folger, D. W., 1970: Wind transport of land-derived mineral, biogenic and industrial matter over the North Atlantic. *Deep Sea Research* 17, 337-352.

Harvey, G. R., W. G. Steinhauer and J. M. Teal, 1973: Polychlorobiphenyls in North Atlantic Ocean water. *Science* 180, 643-644.

Harvey, G. R. and W. G. Steinhauer, 1974: Atmospheric transport of polychlorobiphenyls to the North Atlantic. *Atmospheric Environment* 8, 287-290.

Harvey, G. R., H. P. Miklas, V. T. Bowen and W. G. Steinhauer, 1974a: Observations on the distribution of chlorinated hydrocarbons in Atlantic Ocean organisms. *J. Mar. Research* 32, 103-118.

Harvey, G. R., W. G. Steinhauer and H. P. Miklas, 1974b: Decline of PCB concentrations in North Atlantic surface waters. *Nature* 252, 387-388.

Jannasch, H. K., K. Einhjellen, C. Wirsen, A. Farmanfarmaian, 1971: Microbial degradation of organic matter in the deep sea. *Science* 171, 672-675.

Jernelov, A., 1974: Heavy metals, metalloids and synthetic organics. In: *The Sea*, Vol. 5, E. D. Goldberg, ed., (New York: John Wiley & Sons, Inc.), pp. 799-816.

Menzel, D. W., 1974: Primary productivity, dissolved and particulate organic matter and the sites of oxidation of organic matter. *Ibid.*, pp. 659-678.

Noshkin, V. E. and V. T. Bowen, 1973: Concentrations and distributions of long-lived fallout radionuclides in open ocean sediments. In: *Radioactive Contamination of the Marine Environment* (Vienna: IAEA), pp. 671-686.

SECTION II

BIOGEOCHEMICAL CYCLING OF NITROGEN

CHAPTER 16

NITROGEN CYCLING IN TERRESTRIAL ECOSYSTEMS

E. A. PAUL

Department of Soil Science, University of
Saskatchewan, Saskatoon, Saskatchewan

INTERECOSYSTEM NITROGEN TRANSFERS

The first investigations leading to an understanding of the nitrogen cycling were conducted during the 18th and 19th centuries. These stressed the significance of inorganic transfers such as the movement and absorption of ammonia and nitrate concentrations in rain. Research on transformations, the chemical or biological change of nitrogen compounds from one form to another, were stressed in the 20th century. Transformations such as dinitrogen fixation, ammonification and denitrification often lead to inorganic transfers within and between ecosystems. Others such as uptake by microorganisms and plants usually lead to temporary immobilization within the ecosystem and some transfer of organic nitrogen. The microbiology, chemistry and significance of transformations in nature, with the exception of biological denitrification, are now fairly well understood for terrestrial ecosystems.
The present stress on ecosystem research, biogeochemistry of nutrients, nitrogen pollution, and the high cost of fertilizer nitrogen have again created a heightened interest in nitrogen transfers in nature. We have thus come full circle in the study of the nitrogen cycle, and a short survey of the early history of nitrogen research can give us a wealth of information that is still pertinent to today's problems.

The interaction of soil and air was recognized by John Evelyn when in 1676 he wrote, "The earth in years of repose recovers its vigour by the attraction of vital spirits which it receives from the air." A hundred years later in the middle of the 18th century, Lavoisier and his co-workers invented the names "azote" and "nitrogene" for the newly discovered gas. They recognized the interaction of atmospheric and plant nitrogen and analyzed the nitrogenous constituents in many ammonia compounds, saltpeter, and numerous plants (reviewed by Aulie, 1970; Lawes, Gilbert and Pugh, 1861). Three-quarters of a century elapsed before further advances in the nitrogen cycle were made.

Both the transfers and transformations of nitrogen remained a mystery until Boussingault's study of nitrogen fixation (1831-1838) proved that legumes accumulate atmospheric nitrogen. In 1844 Dumas and Boussingault published a small book entitled, *The Chemical and Physiological Balance of Organic Nature*. In this they wrote, "The atmosphere is a mysterious link that connects the animal with the vegetable and the vegetable with the animal kingdom." During the same period Liebig stressed the importance of ammonia in nitrogen transfers between the atmosphere, rain, plants and animal excreta (Justus von Liebig, 1840, *Organic Chemistry in Its Application to Agriculture and Physiology*, 1st edition, London, Taylor; also 2nd edition, 1842; 3rd edition, 1843; 4th edition, 1847).

The controversy that developed between Boussingault, Liebig and Lawes and Gilbert led to many excellent measurements of nitrogen concentrations in the atmosphere and estimates of nitrogen transfer rates. It also led to some of the most interesting scientific literature concerning the nitrogen cycle. Aulie (1970) quotes Berzelius in 1848 as saying, "Boussingault covers the same field as Liebig, but Boussingault takes the hard tiresome way of answering every question by one or more experiments. He gives his answers not so quickly but they are reliable." Liebig also demonstrated a sharp pen when the long series of field experiments by Lawes and Gilbert (Lawes and Gilbert, 1855; Lawes, Gilbert and Pugh, 1861) substantiated Boussingault concerning the puzzling ability of legumes to accumulate nitrogen and showed Liebig's ideas on mineral nitrogen movement left much to be desired. Liebig wrote "The results of Mr. Lawes have no value for his next-door neighbour, nay, they have no value for himself. . . . None of these facts are new and what is new in the opinion of Mr. Lawes is erroneous."

The very active controversy (Lawes and Gilbert, 1855) led to the measurement of ammonia and nitrate concentrations in rainfall, the ammonia content of air and the absorption of nitrogen by plants grown under controlled conditions in airstreams (summarized by Lawes, Gilbert and Pugh, 1861; Lawes and Gilbert, 1889). The early workers recognized the great difficulty in obtaining quantitative estimates of ammonia absorption by soil and noted that nitrogen evolution from the growing plants was usually greater than nitrogen uptake. A comparison of the data collected by Boussingault and Lawes and Gilbert with some more recent data collected in Japan is shown in Table 16.1. Although the

Table 16.1
Concentrations of NH_3-N in the Atmosphere and in Rainfall

Maritime air, 1968	2.8 µg m^{-3} (STP)
Hakadate air, 1967	17.5-70 µg m^{-3} (STP)
Paris air, 1852	26 µg m^{-3} (STP)
Hakadate 1967, rainfall	0.7-1.6 mg liter^{-1}
Rothamsted 1853-1857, rainfall	0.99-1.2 mg liter^{-1}
Paris 1851, rainfall	3.5 mg liter^{-1}

(From Lawes and Gilbert, 1855, Tsunogai, 1971).

data are separated by 1¼ centuries and thousands of miles, the range of ammonia concentrations in the air over a modern Japanese city spans those measured in Paris during the 1850s. In both cases, the highest concentrations occurred in the winter time. Tsunogai (1971) and Tsunogai and Ikeuchi (1968) found a low concentration of ammonia over the ocean. Similarly Boussingault had found low concentrations in rural areas without high cattle concentrations (Lawes and Gilbert, 1855).

Viets and co-workers (Hutchinson and Viets, 1969; Viets, 1974) in recent years have measured the concentrations of ammonia and its absorption under a number of conditions near feedlots. They found that under normal agricultural conditions, absorption by water surfaces brought in 3.9 kg ha^{-1} and rainfall added another 3.3 kg ha^{-1} yr^{-1}. Absorption by water surfaces 2 km and 0.4 km away from a 90,000 head feedlot resulted in the absorption of 34 and 73 kg ha^{-1}, respectively. However, the increase due to rainfall in both cases was only 5 kg ha^{-1} showing that rainfall, especially in an arid climate, adds only a small percentage of that absorbed by a water surface if the ammonia concentrations are high. Volatile amines especially near cattle feedlots can contribute a significant portion of the volatile bases absorbed by acid traps (Elliot, Sherman and Viets, 1971).

The high potential for interpool transfers is shown in Figure 16.1. Volatilization of ammonia and denitrification are major processes for the loss of nitrogen from terrestrial systems. This figure includes data from Martin (1970), Hutchinson (1954) and Stevenson (1965) and shows that the terrestrial input by precipitation could equal that of all biological and industrial fixation combined. These estimates are similar to those of Robinson and Robbins (1970), but higher than those of Delwiche (1970) and Porter (1975).

NITROGEN TRANSFERS WITHIN ECOSYSTEMS

The ecosystem studies conducted within the tundra, forest, desert and grassland biomes have made it possible to quantitatively estimate some of the nitrogen transfers within the ecosystems and between the soil and the atmosphere. The extent of biological fixation of dinitrogen is one of the most variable factors in the nitrogen cycle depending on, among other things, the carbon inputs and on the flux of nitrogen through the system.

Grassland soil, except under certain conditions where grasses of high photosynthetic capacities are growing (Day, Neves and Dobereiner, 1975), tend to show low fixation levels as typified by the grassland soils of Saskatchewan in Table 16.2 These soils are in a steady state with very

Table 16.2
Nitrogen Flux Attributable to Biological Fixation

	$g\ m^{-2}\ yr^{-1a}$	Reference
Grass (Sask)	0.1-0.2	Vlassak et al. (1973)
Wheat (Sask)	0.05-0.01	Vlassak et al. (1973)
Strip Mine Wastes (Sask)	1.5-2.0	Anderson (unpubl.)
V. faba (Sask)	15-25	Paul (unpubl.)
Desert (Utah)	1.6	Rychert & Skujins (1974)
Tundra, Meadow, (Devon Isle)	0.1-0.4	Stutz & Bliss (1973)
Tundra, Ridge, (Devon Isle)	0.007-0.03	Stutz & Bliss (1973)
Paddy(Philippines)	8.0	Yoshida & Ancajas (1973)
Old Field, (Rothamsted)	5.5	Harris & Dart (1973)
Marine Angiosperms	10-50	Patriquin & Knowles (1972)

$^a g\ m^{-2} \times 10 = kg\ ha^{-1}$

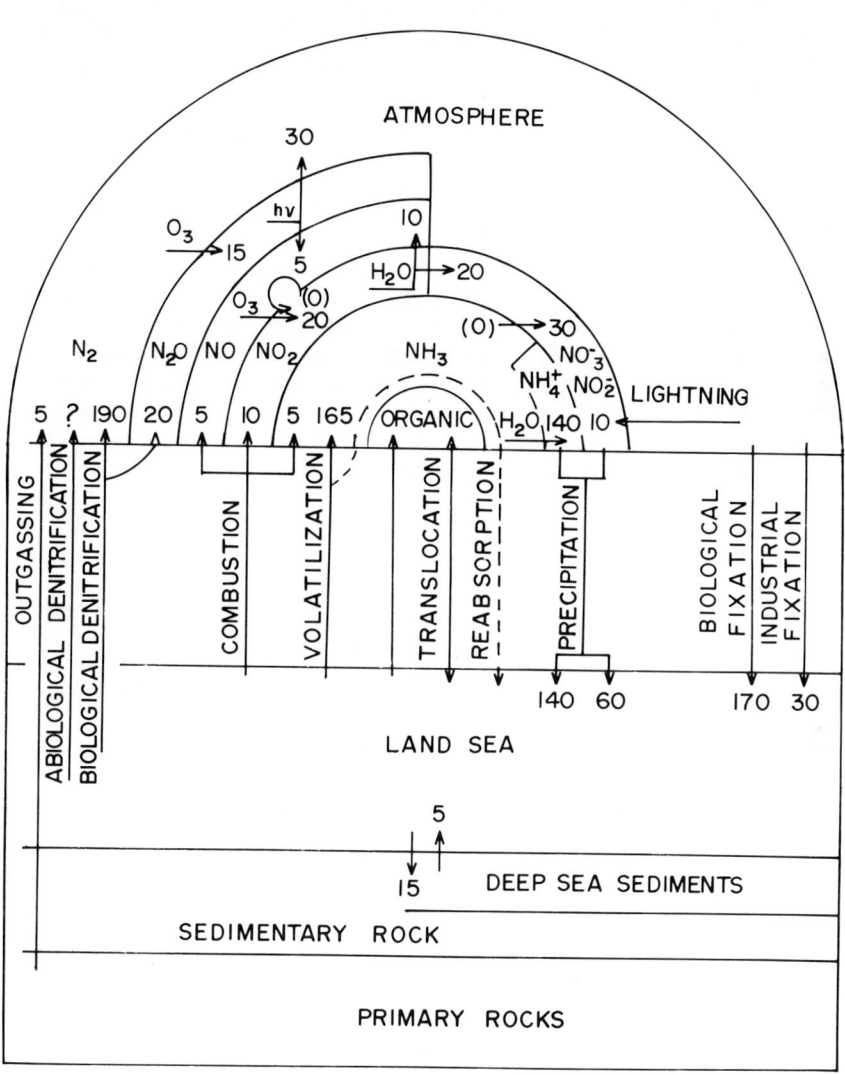

Figure 16.1 Nitrogen interpool transfers expressed in 10^6 metric tons per year (Burns and Hardy, 1975).

little mineral nitrogen. Wheat grains in Saskatchewan are
grown on a wheat-summerfallow system where nitrate is usually in excess. They show very low levels of fixation.
Similar climatic conditions led to 1.5 to 2 g m^{-2} of nitrogen input on the strip mine wastes which are accumulating
organic matter. The legume *Vicia faba* grown under Saskatchewan conditions fixes about 75% of its high nitrogen
requirements with the other 25% coming from the transformations of soil organic matter. This level of fixation is
about 1/3 of the highest levels of fixation shown where
legumes grow year-round (Date, 1973).

The deserts of Utah have a very high nitrogen input
attributable to fixation by blue-green algae. Later in
this paper this high level of nitrogen fixation will be
related to the large losses also occurring under these conditions. Water-saturated conditions such as paddy fields
and the marine angiosperms have shown high levels of nitrogen fixation in a number of situations. These high levels
of fixation also must be related to large withdrawals of
nitrogen either by cropping or through removal of the plant
materials by water movement.

The nitrogen fluxes for Arctic tundra (Figure 16.2)
represent a system with very slow turnover rates. The low

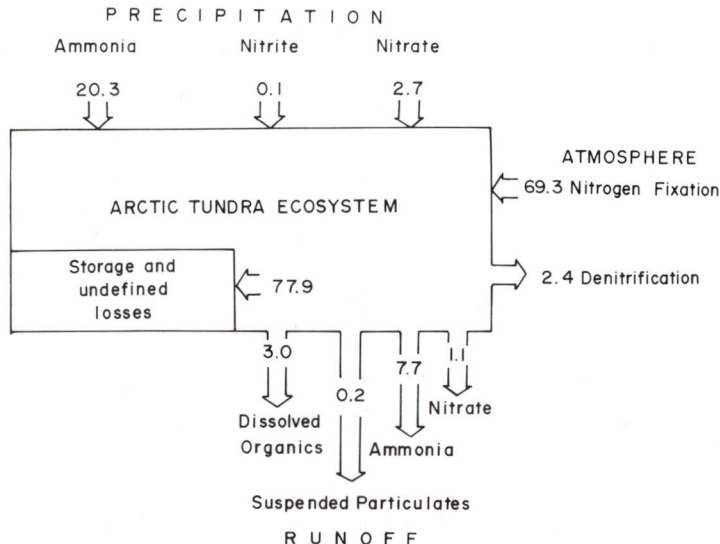

Figure 16.2 Nitrogen flux rates in Arctic tundra (mg m^{-2} per year) (Barsdate and Alexander, 1975).

inputs of nitrogen, primarily by algal fixation, and ammonia absorption are not offset by transfers out of the system either as dissolved organics, ammonia volatilization or denitrification. This leads to a storage of nitrogen with the accumulation of peat under permafrost conditions. The magnitude of nitrogen transfers in a Utah desert under opposite climatic conditions is much greater (Skujins and West, 1975). This system is also primarily dependent upon the input of nitrogen through blue-green algal fixation systems (Table 16.3). The high levels of fixation are offset by very large denitrification.

Table 16.3
Nitrogen Transfers in Utah Desert (Skujins and West, 1975)

	$g\ m^{-2}\ yr^{-1}$
Fixation	1.6
Rainfall	0.4
Denitrification	1.9
NH_3 volatilization	0.1
Plant Uptake	1.2

Nitrogen cycling in a mature oak-hickory forest is expressed in terms of standing crop and estimated turnover time in Table 16.4. The data for standing crop of nitrogen show that the concentration of nitrogen in the biomass greatly exceeds that in the litter. The initial estimates for mycorrhizae show that the fungi directly associated with the root system contain more nitrogen than the rest of the soil microflora. All terrestrial plants, with the exception of a few groups such as the *Cruciferae* and *Chenopodiaceae*, contain either ectotrophic or endotrophic mycorrhizae (Gerdemann, 1971). In the future, detailed considerations of nitrogen cycling will have to take both the biomass and acitvity of these organisms into account.

All components of the system show relatively low turnover times except for the soil fauna. The 0.15 turnover time for the microflora corresponds to that previously estimated for grasslands (Clark and Paul, 1970). The concentration of NO_3^- is higher than would be expected in a climax vegetation (Rice and Parncholy, 1973). However, the modeling of the system has resulted in the use of fairly low turnover rates for the inorganic soil nitrogen constituents.

Table 16.4
Parameters for Nitrogen Cycling in Mature
Oak-Hickory Forest, North Carolina
(Mitchell, Waide and Todd, 1974)

	Standing Crop ($g\ m^{-2}$)	Turnover Time (yr)
Reproductives and Leaves	9.8	1.1
Branches	12	27
Stems	19	654^a
Roots	15	3.6
Litter	14	2.9
Soil Fauna	0.2	0.089
Microflora	7.8	0.15
Mycorrhizae	9.5	1.0
NO_3^--N	3.4	0.19
NH_4^+-N	5.3	0.19
Soil Organic Matter	390	150
Total	483	

^{a}Probably overestimate: Half-life = $T_{1/2}$ = 0.693 x (Turnover time)

NITROGEN CYCLING THROUGH THE ORGANIC MATTER OF GRASSLANDS

The nitrogen of the aboveground green vegetation of a semi-arid grassland is only 1/20 that of the previously discussed oak forest (Table 16.5). The amounts in the microorganisms however are similar. The large amount of nitrogen tied up in the soil fungal flora is noteworthy. This together with the bacteria constitute a major storehouse of nitrogen which is five times as great as the nitrogen tied up in the green vegetation at any one time. Although the roots constitute a large store of nitrogen, because of much lower turnover rates, the annual underground production utilizes only 40% as much nitrogen as the shoot production (Table 16.6). Clark (1974) has shown that grasslands have a great capability for maintaining nitrogen within their vegetative system. Fertilizer ^{15}N was rapidly taken up by the plants. Later in the growing season, before senescence of the green vegetation, the nitrogen in much of the aboveground vegetation was transferred below ground. This nitrogen was retransported to the tops with the initiation of new regrowth. After two years in the field, 80% of the nitrogen originally incorporated into plant parts was still in the vegetation, with slightly

Table 16.5
Nitrogen Components of Grassland (Matador, Saskatchewan)

Material	$g\ m^{-2}$
Green Vegetation	2.2
Dead Litter	7.0
Roots	13.5
Microorganisms on Plant Parts	0.2
Fungi	10.4
Bacteria	2.4
Fauna	0.3
Soil Organic Matter	540.0

[a] Does not contain data for endotrophic mycorrhiza.

Table 16.6
Annual Matador Nitrogen Flow Rates

	$g\ m^{-2}\ yr^{-1}$	$kg\ ha^{-1}\ yr^{-1}$
N fixation	0.2	2
Rainfall	0.2	2
Shoot Production	4.9 (127[a])	49
Underground Production	2.0 (90[a])	20
Max. Standing Shoots	2.2 (56[a])	22

[a] Carbon equivalents

less than half being in the living vegetation and the remainder in identifiable dead shoots and senescent and detrital roots.

Grasslands under native vegetation tend to have very low mineral nitrate levels. The nitrogen flux within the plants and the ammonia-organic nitrogen flux through the plant and microbial biomass results in low losses of nitrogen from the system. Grasslands at or approaching their steady state may respond to added fertilizer nitrogen. Their maintenance, however, is not dependent upon external nitrogen supplies, and fixation levels under these conditions are low.

In all the ecosystems studied with the exception of the desert, soil organic matter is a major storehouse of nitrogen. Measurements of nitrogen cycling must take the soil organic matter transfers into account. A number of methods have been used to fractionate soil organic matter such that interpretation of turnover rates can be made. The distribution of amino acid and amino sugar nitrogen and the hydrolyzable nitrogen for the Saskatchewan Matador soil (Table 16.7) is very similar to that of other soil organic matter (Moore and Russell, 1970; Sowden, 1966; Given and Dickinson, 1975).

Table 16.7
Nitrogen Content of Soil Organic Fractions
(McGill, Paul and Sorensen, 1974)

	Soil N	Labeled N Day 4	Labeled N Day 71	$T_{1/2}$ of Labeled N[a]
	$\mu g\ g^{-1}$			days
Amino Acids	725	59	44	2700
Amino Sugars	170	20	9	224
NH_4 Released on Hydrolysis	525	20	22	940
Unidentified Hydrolyzable	821	86	78	
Total Hydrolyzable	2240	185	153	
Not Hydrolyzable in 6 N HCl	260	1.4	1.5	
Total Organic N	2500	186.4	154.5	
N added		195		

[a] $T_{1/2}$ for 90–268 day period.

The use of ^{15}N and ^{14}C to label the soil microbial population made feasible measurements of the flux of nitrogen through the microbial biomass and the various soil organic matter constituents separated by fractionation techniques (McGill et al., 1972; McGill, Paul and Sorensen, 1974). Table 16.7 shows that the amino acids accounted for 29% and the unidentified hydrolyzable components for 33% of the original organic nitrogen. On day 4, when the microbial population had reached a peak, the amino acids accounted for 32% of the immobilized ^{15}N, and the unidentified hydrolyzable material accounted for 46%.

By day 71, the system was approaching a steady state. At this time the amino acids plus amino sugars comprised 42% of the organic ^{15}N. Their decay contributed 54% of the mineralized nitrogen. The unidentified hydrolyzable fraction increased in percentage to 50% of the total organic ^{15}N. Its turnover accounted for 24% of the mineralized ^{15}N. The stability and internal cycling of nitrogen also are shown by the half-lives of the soil organic constituents calculated for the incubation period between 90 and 268 days. The amino sugars with a half-life of 224 days showed the greatest turnover and thus the largest net release of nitrogen. The amino acids were much more stable..

Tracers incorporated into the microbial population under cultivated conditions in the field showed similar high stabilization or recycling of the microbiologically immobilized carbon and nitrogen (Shields et al., 1973). In this clay soil, treatment with chloroform or freezing and thawing were required before much of the stabilized nitrogen could be mineralized by the microbial populations (Shields, Paul and Lowe, 1974). Both these treatments are known to kill the population and possibly also release metabolites stabilized by the soil system.

Interpretative modeling of nitrogen flow through the soil subsystem requires measurements of the degradation rates of plant residues. Some degradation data are shown in Figure 16.3 where the loss of added straw ^{15}N and ^{14}C with time is shown for the cultivated Sceptre soil at Matador, Saskatchewan. The time period shown on the abscissa has eliminated the period during which the soil was frozen. Dryland agriculture maintains the straw residue on the surface so that as little decomposition occurs as possible. Although 50% of the straw nitrogen was still identifiable by the end of the first growing season, three years in the field had resulted in the microbial degradation of the identifiable material or its transfer to organic matter.

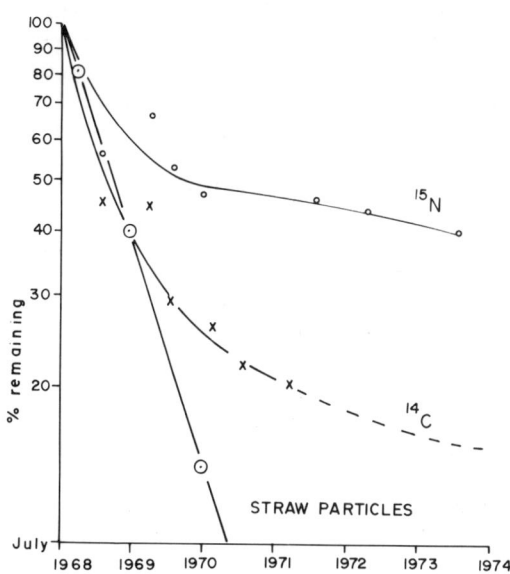

Figure 16.3 Loss of ^{15}N - ^{14}C labeled plant residue from a cultivated Chernozemic soil (Matador).

At the end of the first year, 40% of the original straw nitrogen had been lost from the soil system either through plant uptake, leaching, volatilization or denitrification. Continuation of the field experiments for a 6-year period showed that a total of 60% of the nitrogen had been removed from the system with the remainder being found in the soil organic fraction. The half-life of the nitrogen remaining in the soil organic matter during this period represented 22 months in the field.

The data for ^{14}C give a more appropriate estimate for mineralization of plant residues than does that for nitrogen. By 1974, 80% of the original ^{14}C had been evolved, whereas 40% of the nitrogen still remained in this system. This difference was attributed to two factors: 1) an initital greater C loss as the C/N ratio dropped from 50:1 to that of the stabilized soil organic constituents (10:1); 2) reutilization and immobilization of a portion of the mineralized ^{15}N but loss of all the ^{14}C which appeared as CO_2 during degradation.

The straw residue experiment conducted under field conditions was fractionated by a physical dispersion system. Table 16.8 shows the distribution of the soil nitrogen in the various fractions after one summer in the field. A simple nitrogen transfer model based on the nitrogen originally in the straw particles and that in the labeled organic matter was used to predict the nitrogen that would be mineralized from the various fractions (McGill, Paul and Sorensen, 1974). Equations for a series of first-order reactions were based on the ^{15}N in the various fractions during sampling periods. The calculations showed that a sodium pyrophosphate, acid-soluble fraction (fulvic acids) comprised 12% of the original straw nitrogen but contributed 30% of the nitrogen mineralized during the four-year period (Table 16.8). The organic matter associated with the mineral constituents smaller than 0.04 µm contributed an equal amount of nitrogen. However, this fraction comprised three times as great a portion of the total soil nitrogen and twice as much labeled nitrogen. Its turnover rate as shown by the half-lives in Table 16.8 was, therefore, much lower. The sodium pyrophosphate acid-insoluble fraction (humic acids) had slow turnover rates and contributed only small amounts of the nitrogen mineralized. As the transformations proceeded with time, mineralization from the greater than 0.2 µm fractions accounted for the greater portion of ^{15}N moving through the system.

The turnover of the large concentrations of stable soil organic matter is difficult to study with normal tracer

Table 16.8
Turnover of Straw Residue N in Soil Fractions during
Four Years in the Field (Matador)

Fractions	Soil N ($\mu g\ g^{-1}$)	^{15}N in Fraction (%)	Half-life	Predicted ^{15}N Mineralized (%)
Fulvic acids 1967	250	11.9	0.7	9.9
" 1968-71		6.5	0.9	16.9
Fine Colloids 1967	760	18.5	1.2	8.6
(<0.04 μm) 1968-71		13.6	2.6	18.2
Humic acids 1967	187	5.1	3.0	1.2
" 1968-71		4.3	4.6	2.5
0.04 - 0.2 μm 1967	292	4.1	1.3	1.9
1968		3.1	7.0	4.0
>0.2 μm 1967	934	22.8	1.4	4.0
1968		17.8	3.6	12.3
Total 1967	2423	61.4	—	25.8
1968-71		45.3	—	56.7

Half-lives corrected to annual basis ($\mu g\ g^{-1} = \frac{\%}{100} \times 37.1$)

techniques. Carbon and nitrogen analysis, and fractionation
of the soil samples from long-term fertility plots, or
from comparative virgin and cultivated soils in conjunction with carbon dating techniques, can prove most useful.
Carbon dating of the resistant fractions has been utilized
in conjunction with the biochemical fractionation technique
which separates materials such as amino acids and amino
sugars. It has also been used in conjunction with physical-chemical techniques which separate sodium pyrophosphate-soluble fractions and dispersed clay organic matter complexes.

The effect of 60 to 75 years cultivation on the carbon
and nitrogen contents of two Saskatchewan soils is shown
in Table 16.9. The arid Brown Chernozemic soil has not
quite lost 50% of its C, whereas the Black Chernozemic has
lost more than 50%. Table 16.9 indicates that the Chernozemic Black soils have lost 440 g m^{-2} of N. This is equivalent
to 4000 lb/acre or 2 tons N. An average of 1.5 tons have

Table 16.9
C and N Contents and Future Equilibrium Levels of Two
Saskatchewan Soil Types (g m^{-2}) to 30 cm

	Brown		Black	
	C	N	C	N
Original Present	6,600	550	13,200	1,100
Present (1973)	3,850	385	6,600	660
Equilibrium Level	1,980	220	2,970	330
Potential Available	–	165	–	330

kg ha^{-1} = g m^{-2} x 10

been lost from each of the 70 million acres of land cultivated in Western Canada since man first turned the prairie
sod. This total of 105 million tons is 100 times greater
than the present annual capacity of Canada's fertilizer
industry.

Calculations showed that 20% of the nitrogen that was
lost was exported in grain or animal protein. Another 20%
can be found as NO_3^- N below the root zone. The rest is
unaccounted for. The losses must be attributed to denitrification, erosion, leaching or ammonia volatilization.

The equilibrium level of the soil organic matter was
calculated from carbon dating of the total soil organic
matter and of the acid hydrolyzable carbon. This made it
possible to determine the turnover rates of an active and
a resistant fraction (Martel and Paul, 1974). These data
were then used together with known addition and turnover
rates of straw residue nitrogen to calculate the future

equilibrium level of the soil organic matter using the equations of Bartholomew and Kirkham (1960).

The data in Table 16.9 indicate that the present summerfallow-wheat rotation is leading to a continued loss of N. Contribution from the native soil reserves has been estimated to be 0.9 g m^{-2} for the semi-arid Brown soil and 1.6 g m^{-2} annually for the Black soil (Table 16.10). In a wheat-summerfallow system where nitrogen mineralization is encouraged and a crop is grown every two years, this N together with that coming in from rainfall and nitrogen fixation plus the turnover of the residues and soil biomass contributes enough nitrogen for plant growth. The values for the turnover of residues and soil biomass

Table 16.10
N Released in Cropped Soils (g m^{-2})
under a Wheat-Summerfallow System

	Brown	Black
N Release Total	3.3	6.6
Rainfall	0.21	0.33
N Fixation	0.21	0.22
Turnover of Residues and Soil Biomass	2.0	4.4
From Native Soil Reserves	0.9	1.6

have been calculated from the turnover rates of straw residues incorporated under field conditions and from the microbiologically immobilized nitrogen which is later mineralized. Thus, this table is a synthesis of all the data given for Saskatchewan agricultural and grassland soils. The sum of the various components in this table equals the long term nitrogen mineralization capacity known to exist for these soils under field conditions.

CONCLUSIONS

This paper has attempted to relate some of the major nitrogen transfers in tundra, desert, forest and grassland soil conditions. The data have shown that the nitrogen fluxes are very dependent on the type of ecosystem. The tundra has very low loss rates and very low inputs of nitrogen into the system. The desert also with low primary productivity shows much greater fluxes both for fixation and loss.

The transfers of nitrogen very often follow a transformation induced by microbial growth. These transformations in the case of nitrification, utilize the energy within the nitrogenous molecule. In all other conditions, the transformations are dependent on the associated C supply. The flow of nitrogen through ecosystems, therefore, is very dependent on the flow of carbon. The soil organic matter, which constitutes the largest reservoir of potentially available nitrogen, has a C/N ratio of 10:1 in many parts of the world. The stabilization within organic matter, therefore, also is dependent on the soil carbon content and indirectly dependent on carbon flux.

The recognition of the large fluxes of nitrogen between the soil subsystem, the atmosphere and ground water makes it imperative that measurements of ammonia volatilization, denitrification and leaching be conducted in detail at a number of sites. Although estimates of the movement of terrestrial nitrogen via the atmosphere and ground water to marine sites have been made, much of the data are based on only one or two measurements. Most of the techniques for the required measurements are available. However, as in ecosystem studies, measurements of transfer rates will require a great deal of cooperative scientific research in the various disciplines interested in this problem.

REFERENCES

Aulie, R. P., 1970: Boussingault and the nitrogen cycle. *Proc. Amer. Phil. Soc.* **114**, 411-422.

Bartholomew, W. V. and D. Kirkham, 1960: Mathematical descriptions and interpretation of culture induced soil nitrogen changes. *Trans. 7th Int. Cong. Soil Sci.* **2**, 471-477.

Burns, R. C. and R. W. F. Hardy, 1975: *Nitrogen Fixation by Bacteria and Higher Plants.* (New York: Springer-Verlag) in press.

Clark, F. E., 1975: Partitioning of added isotope nitrogen in a blue grama grassland. Symp. of Below Ground Ecosystems: A Synthesis of Plant Associated Processes, Ft. Collins. In press.

Clark, F. E. and E. A. Paul, 1970: The microflora of grassland. *Adv. Agron.* **22**, 375-435.

Date, R. A., 1973: Nitrogen: A major limitation in the productivity of natural communities of crops and pastures in the Pacific area. *Soil Biol. Biochem.* **5**, 5-18.

Day, J. N., M. C. P. Neves and J. Dobereiner, 1975: Nitrogenase activity on the roots of tropical forage grasses. *Soil Biol. Biochem.* **7**, 107-112.

Delwiche, C., 1970: The nitrogen cycle. *Scien. Amer.* 223, 136-146.

Elliot, L. F., G. E. Sherman and F. G. Viets, Jr., 1971: Volatilization of nitrogen containing compounds from beef cattle areas. *Soil Sci. Soc. Amer. Proc.* 35, 752-755.

Gerdemann, J. W., 1971: Mycorrhizae. In: *The Plant Root and Its Environment.* E. W. Carson, ed., (Charlottesville: University Press of Virginia), pp. 205-218.

Given, P. H. and C. H. Dickinson, 1975: Biochemistry and microbiology of peats. In: *Soil Biochemistry,* Vol. 3, E. A. Paul and A. D. McLaren, eds., (New York: Marcel Dekker, Inc.), pp. 124-212.

Harris, D. and P. J. Dart, 1973: Nitrogenase activity in the rhizosphere of *Stachys sylvatica* and some Dicotyledenous plants. *Soil Biol. Biochem.* 5, 277-279.

Hutchinson, G. E., 1954: The biochemistry of the terrestrial atmosphere. In: *The Earth as a Planet,* Vol. 2, G. P. Kuiper, ed., pp. 371-433.

Hutchinson, G. L. and F. G. Viets, Jr., 1969: Nitrogen enrichment of surface water by absorption of ammonia volatilized from cattle feedlots. *Sci.* 166, 514-515.

Lawes, J. B. and J. H. Gilbert, 1855: On the amounts of, and methods of estimating ammonia and nitric acid in rain water. Report of the British Assoc. for the Advancement of Science for 1854. pp. 1-15.

Lawes, J. B. and J. H. Gilbert, 1889: On the present position of the sources of the nitrogen of vegetation with some new results and preliminary notice of new lines of investigation. *Phil. Soc. Trans.* 180, 1-107.

Lawes, J. B., J. H. Gilbert and E. Pugh, 1861: The source of the nitrogen of vegetation with special reference to the question whether plants assimilate free or uncombined nitrogen. *Phil. Trans.,* Part 2, pp. 433-468.

Martel, Y. and E. A. Paul, 1974: Effects of cultivation on the organic matter of grassland soils as determined by fractionation and radiocarbon dating. *Can. J. Soil Sci.* 54, 419-426.

Martin, E. F., 1970: The nitrogen cycle. *Marine Chemistry,* Vol. 2 (New York: Marcel Dekker, Inc.), pp. 225-266.

McGill, W. B., E. A. Paul, J. A. Shields and W. E. Lowe, 1973: Turnover of microbial populations and their metabolites in soil. *Bull. Ecol. Res. Comm.* (Stockholm) 17, 293-301.

McGill, W. B., E. A. Paul and H. L. Sorensen, 1974: The role of microbial metabolites in the dynamics of soil nitrogen. Matador Project Tech. Rpt. No. 46, April, 1974.

Mitchell, J. E., J. B. Waide and R. L. Todd, 1974: A preliminary compartment model of the nitrogen cycle in a deciduous forest ecosystem. *Proc. Symp. on Mineral Cycling in Southeastern Ecosystems*, Augusta, Ga. (In press).

Moore, A. W. and J. S. Russell, 1970: Changes in chemical fractions of nitrogen during incubation of soils. *Aust. J. Soil Res.* $\underline{8}$, 21-30.

Patriquin, B. and R. Knowles, 1972: Nitrogen fixation in the rhizosphere of marine angiosperms. *Marine Biol.* $\underline{16}$, 49-58.

Paul, E. A., 1970: Plant components and soil organic matter. *Adv. Phytochem.* $\underline{3}$, 59-104.

Porter, L.K., 1975: Nitrogen transfer in ecosystems. In: *Soil Biochemistry*, Vol. 4, E. A. Paul and A. D. McLaren, eds., (New York: Marcel Dekker, Inc.), pp. 1-31.

Rice, E. L. and S. K. Parncholy, 1973: Inhibition of nitrification by climax ecosystems. *Amer. J. Bot.* $\underline{59}$, 1033.

Robinson, E. and R. C. Robbins, 1970: Gaseous nitrogen compound pollutants from urban and natural sources. *J. Air Poll. Cont. Assoc.* $\underline{20}$, 303-306.

Rychert, R. C. and J. Skujins, 1974: Nitrogen fixation by blue-green algae lichen crusts in the Great Basin Desert. *Soil Sci. Soc. Amer. Proc.* $\underline{38}$, 768-771.

Shields, J. A., E. A. Paul and W. E. Lowe, 1974: Factors influencing the stability of labeled microbial materials in soils. *Soil Biol. Biochem.* $\underline{6}$, 31-37.

Shields, J. A., E. A. Paul, W. E. Lowe and D. Parkinson, 1973: Turnover of microbial tissue in soil under field conditions. *Soil Biol. Biochem.* $\underline{5}$, 753-764.

Skujins, J. and N. West, 1975: Nitrogen inputs and losses from a desert ecosystem. *Proc. of the 141st Meeting of the American Association of Science*. (In press).

Sowden, F. J., 1966: Nature of the amino compounds of soil. I. Isolation and fractionation. *Soil Sci.* $\underline{102}$, 202-207.

Stevenson, F. J., 1965: Origin and distribution of nitrogen in soil. In: *Soil Nitrogen*, W. V. Bartholomew and F. E. Clark, eds., (Madison: American Society of Agronomy), pp. 1-42.

Stutz, R. C. and L. C. Bliss, 1973: Acetylene reduction assay for nitrogen fixation under field conditions in remote areas. *Plant and Soil* 38, 209-213.

Tsunogai, S., 1971: Ammonia in the oceanic atmosphere and the cycle of nitrogen compounds through the atmosphere and the hydrosphere. *Geochem. J.* 5, 57-67.

Tsunogai, S. and K. Ikeuchi, 1968: Ammonia in the atmosphere. *Geochem. J.* 2, 157-166.

Varsdate, R. J. and V. Alexander, 1975: The nitrogen balance of Arctic tundra. *J. Environ. Qual.* 4, 111-117.

Viets, F. G., Jr., 1971: The mounting problem of cattle feedlot pollution. *Agr. Sci. Rev.* 9, 1-8.

Viets, F. G., Jr., 1974: Fate of nitrogen under intensive animal feeding. *Fed. Proc.* 33, 1178-1182.

Vlassak, K., E. A. Paul and R. E. Harris, 1973: Assessment of biological nitrogen fixation in grassland associated sites. *Plant and Soil* 38, 637-649.

Yoshida, T. and R. Ancajas, 1973: The fixation of atmospheric nitrogen in the rice rhizosphere. *Soil Biol. Biochem.* 5, 153-155.

CHAPTER 17

SOIL NITROGEN TRANSFORMATIONS:
A MODELING STUDY

H. E. DONER
A. D. MCLAREN

Department of Soils and Plant Nutrition, University of
California, Berkeley, California 94720

INTRODUCTION

Pathways for nitrogen transformations in soil have been
studied extensively for many years (Schoenbein, 1868; Meusel,
1875; Winogradsky, 1949; Quastel and Scholefield, 1951;
Macura, 1966). Rates of chemical transformation of nitrogen
from one compound to another are changed significantly by
both soil biological and physical factors. Enzymes and
soil microorganisms catalyze most transformation reactions
(Skujins, 1967) and the microorganisms use nitrogen for maintenance and growth. Also, such factors as water content of
soil, temperature, pH and oxygen gas partial pressure affect
both reaction rates and extents of the reactions.
 The results of several reports on soil nitrogen by a
group at the University of California, Berkeley, are here
summarized. A mathematical approach to nitrogen transformations during solution flow in soil was developed. The studies
were conducted in both laboratory soil columns and in
field plots to test and refine the theory. Studies were
restricted to the reactions urea $\rightarrow NH_4^+ \rightarrow NO_2^- \rightarrow NO_3^- \rightarrow N_2O$
plus N_2 in an agriculturally important soil.

THEORY

Basic to an account of some of the important biological and physical factors for nitrogen transformation are the following equations. The microbiological reaction rate is

$$-\frac{d[N_i]}{dt} = \frac{Adm}{dt} + \alpha\,m + \beta\,m \tag{1}$$

where N_i is a general notation for concentrations of nitrogen in any form, t is time, A is a proportionality constant related to N utilized per unit weight of biomass synthesized, m is biomass, α is N utilized per biomass per unit time for maintenance, and β is N utilized per biomass in wasted metabolism (McLaren, 1970).

The material balance equation for one dimensional flow is (Cho, 1971; McLaren, 1973a; Misra et al., 1974):

$$\left(\frac{\partial[N_i]}{\partial t}\right)_x = D\left(\frac{\partial^2[N_i]}{\partial X^2}\right)_t - f\left(\frac{\partial[N_i]}{\partial X}\right)_t + \phi \tag{2}$$

where D is a hydrodynamic dispersion coefficient, X is distance, ϕ is a reaction term equal to $d[N_i]/dt$, equation (1), and f is the soil solution pore velocity. Equation (2) gives the concentration of nitrogen in solution at any given soil depth if ion exchange reactions are ignored as a first approximation (Kirda et al., 1974).

Nitrification ($NH_4^+ \rightarrow NO_2^- \rightarrow NO_3^-$) under steady-state conditions was chosen first because it is relatively simple (McLaren, 1969). In the steady state, concentration at depth X does not change with time and

$$\left(\frac{\partial[N_i]}{\partial t}\right)_x = 0.$$

Also if the microbial population is constant, $dm/dt = 0$. This can be true when m is maximal, for example, as in Figure 17.1. If D is neglected, which is justified by experiment under our conditions, we are left with

$$\frac{d[N_i]}{dX} = \frac{\phi}{f} \qquad (3)$$

During the steady state with m a constant, $\bar{m}$, ϕ is given by [cf. equation (1)]

$$\phi = -(\alpha + \beta)\bar{m} \qquad (4)$$

and the quantity,

$$(\alpha + \beta) = \frac{(\alpha + \beta)_\infty [N_i]}{K_m + [N_i]} \qquad (5)$$

K_m is a saturation constant and $(\alpha + \beta)_\infty$ is a maximum. Upon combining (3), (4) and (5) and integrating (Ardakani et al., 1973) we have our working equation

$$K_m \ln \frac{[N_i]}{[N_i]_o} + [N_i] - [N_i]_o = -K_i X \qquad (6)$$

$K_i = (\alpha + \beta)_\infty \bar{m}_i f^{-1}$ is the apparent rate constant for the reaction in a flowing system involving a biochemical transformation dependent on a microbial biomass $\bar{m}_i$ specific for a given reaction of some compound of nitrogen. In order to give a specific meaning to any value of K_i, the biomass of microbes, $\bar{m}$, involved in the reaction and flow rate, f, must be determined (McLaren, 1973b). K, the normalized rate constant is equal to $(\alpha + \beta)_\infty$.

LABORATORY AND FIELD EXPERIMENTAL RESULTS

Nitrification

Laboratory soil columns and field plots were studied to determine apparent rate constants for NH_4^+ and NO_2^- oxidation (Ardakani et al., 1973, 1974a, 1974b). To evaluate K_i it was necessary to establish steady-state conditions and to measure numbers of NH_4^+ and NO_2^- oxidizing microorganisms, $\bar{m}$; soil solution flow rate, f; and NH_4^+, NO_2^-, and NO_3^- concentrations at various soil depths. From the literature and our experimental work, values for K_m were found to be about 8 ppm NH_4^+ -N and 23 ppm NO_2^- -N for

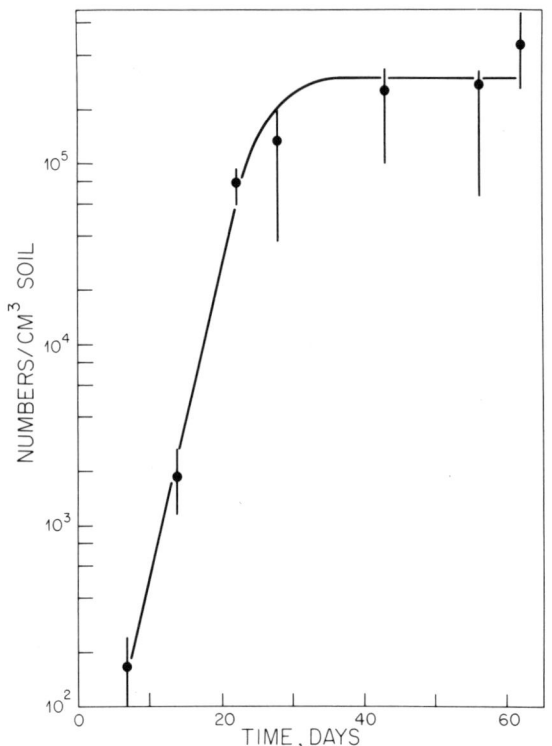

Figure 17.1 Numbers of nitrite oxidizers in a mixture of soil plus sand in a column as a function of time of infiltration of nitrite solution. Growth rate constant is 0.41 and the solid line gives a plot of the logistic equation (m_{max} is proportional to $\bar{m}$ of biomass)

$$m = \frac{m_{max}}{1 + \left(\dfrac{m_{max} - m_o}{m_o}\right) e^{-\gamma t}}$$

Here m_o is the initial population and m_{max} is the maximum population. Note that $dm/dt = 0$ after about 40 days of infiltration. (Ardakani et al., 1973).

NH_4^+ and NO_2^- oxidation, respectively. A comparison of $(\alpha + \beta)_\infty$ constants from laboratory and field studies are in good agreement (Table 17.1). These data provide confidence that the parameters measured, *i.e.*, microbial numbers, ionic nitrogen concentrations, and flow rate, are important in evaluating nitrification. The agreements also seem to justify the two approximations specified above.

Table 17.1
Maintenance Plus Waste Constants for Oxidation of NH_4^+ and NO_2^- in Laboratory and Field Studies

Reaction	$(\alpha + \beta)_\infty$	
	Laboratory	Field
$NH_4^+ \rightarrow NO_2^-$	6	3
$NO_2^- \rightarrow NO_3^-$	1	0.6

units in ppm N cm^3 $bacterium^{-1}$ h^{-1} 10^{-3}

From published data and equation (6) we can draw field profiles of N_i. As an example, we calculate a K of 4×10^{-3} (ppm N cm^3 $bacterium^{-1}$ h^{-1}) for NH_4^+ oxidation, a K of 0.8×10^{-3} ppm N cm^3 $bacterium^{-1}h^{-1}$ for NO_2^- oxidation, with a flow rate of 1 cm h^{-1}, 10^4 NH_4^+ oxidizers cm^{-3}, 5×10^5 NO_2^- oxidizers cm^{-3} and an initial concentration of 100 ppm NH_4^+ in the soil influent.

Almost complete nitrification should occur in the field within 3 cm of soil surface in steady state (Figure 17.2). Predictions indicate only small amounts of NO_2^-. The results conform to field observations (Ardakani, Schulz and McLaren, 1974).

Urea Hydrolysis

Urea hydrolysis in the soil obeys first order kinetics (Ardakani *et al.*, 1975). Apparently the K_m is very large compared to the urea concentration used (100 ppm-N). The apparent rate constant was approximately 1 h^{-1} for the reaction $(NH_2)_2 CO \rightarrow NH_4^+$. Profiles for transformation of urea to NO_3^- were as expected from the model for a soil column; equations similar to equation (6) were applicable for each intermediate. But from this experiment it was clear that microbial populations in soil columns cannot be maintained constant for long. Both microbial counts and solution composition changed with time and depth. Following

homogenization of the soil in the column by mixing and
repacking in the column, urea hydrolysis and nitrification
conformed to steady-state kinetics for a time, however.

Denitrification

Denitrification studies in laboratory soil columns and
field plots have also been conducted (Doner, 1975;
Doner et al., 1974, 1975;   Volz et al., 1975). But in
contrast to the nitrification regime, oxygen was reduced
by flooding the soil surface. Microbial counts of denitrifiers, nitrate reducers, total cell biomass and chemical
measurements for NO_3^- and NO_2^- in soil solution were conducted
Denitrification was greatest during the first several
hours after wetting dry soil (Doner et al., 1974; Volz et al.,
1975). During this time, the denitrifier population increased
from 10^2 to 10^4 cells (g of soil)$^{-1}$ (Doner et al., 1975),
and organic matter losses were observed concurrently (Doner,
1975).   Since denitrification depends on an energy source,
notably soil organic matter (O.M.), lower rates of denitrification observed with continual soil leaching was doubtless
due to O.M. removal.
    Doner et al. (1974), reported apparent zero order kine-
tics in steady state with an apparent rate constant of
$0.06\,\mu g\,N\,h^{-1}$ (g soil)$^{-1}$ or $0.3\,\mu g\,N\,ml^{-1}\,h^{-1}$, for the reaction
$NO_3^- \rightarrow (N_2 + N_2O)$. No microbial counts were taken in this
experiment. Field studies with soil of the same type
gave apparent zero order constants of between 0.01 and 0.05
$\mu g\,N\,h^{-1}$ (g soil)$^{-1}$ for the upper 16 cm (Volz et al., 1975).
The number of denitrifiers ranged between 350 and 2.9×10^4 bacterium (g soil)$^{-1}$ with an average of about 5×10^3
bacterium (g soil)$^{-1}$ or 8×10^3 bacterium cm^{-3}. With a
zero order rate constant of $0.03\,\mu g\,N\,h^{-1}$ (g soil)$^{-1}$ and an
average denitrifier count, an apparent rate constant for
maintenance plus waste is $0.6 \times 10^{-5}\,\mu g\,N\,h^{-1}$ bacterium^{-1}
(5×10^{-5} ppm N cm^3 h^{-1} bacterium^{-1}). Nitrification appears
to be approximately 10^2 faster than denitrification with
optimum conditions for each.
    By equation (6), for a zero order reaction only

$$[NO_3^-]_o - [NO_3^-] = K_i X \qquad (7)$$

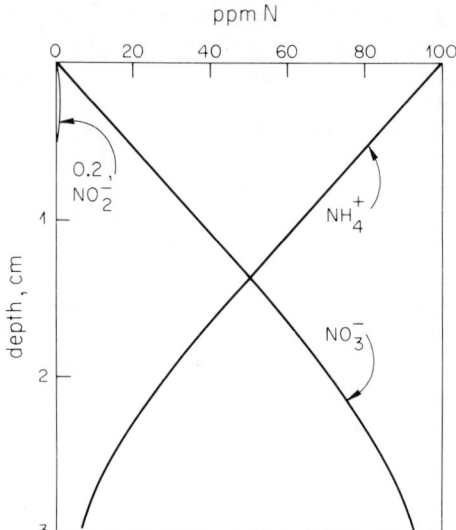

Figure 17.2 Nitrification with soil depth as calculated from experimental data using equation (6).

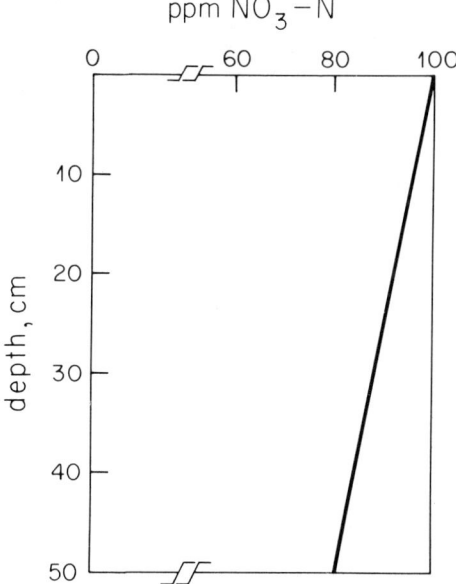

Figure 17.3 Denitrification with soil depth as calculated from experimental data using equation (7).

Assuming $\bar{m} = 8 \times 10^3$ bacterium cm^{-3}, and $f = 1$ cm h^{-1}, or $K = 5 \times 10^{-5}$ ppm N cm^3 h^{-1} bacterium^{-1}, and $[NO_3^-]_0 = 100$ ppm NO_3^- -N, we can plot an NO_3^- profile (Figure 17.3). Under these conditions it would take a soil column 2.5 m long to remove all NO_3^-.

It is generally agreed that biological denitrification is a much more important path for NO_3^- loss than are chemical modes (Broadbent and Clark, 1965). Since, however, a general method for counting denitrifiers remains somewhat controversial (Focht and Joseph, 1973; Belser, 1974), a degree of uncertainty exists for a biomass value. Evaluation of soil organic matter availability presents another problem in formulating a model for denitrification. Doner et al. (1974) suggested a modification to equation (5), which takes into account organic matter oxidation, namely

$$\phi = \frac{d[NO_3^-]}{dt} = -\frac{(\alpha + \beta)\infty \bar{m} [NO_3^-] [O.M]}{K_m + [NO_3^-]} \quad (8)$$

where O.M. = available organic matter and is taken as large compared with the nitrate supplied. The value of K_m is dependent on O.M. in this case. Small amounts of available organic matter utilized makes the rate loss of NO_3^- appear zero order. The assumptions seem warranted by experiment (Doner et al., 1974).

Let us consider a hypothetical situation in which 100 ppm NH_4^+ is added to soil in a continuous manner. Ammonium is oxidized to NO_3^- within a few centimeters of the surface. If an anoxic layer is encountered some denitrification will ensue, but some nitrate will persist in the soil and, potentially, can reach ground waters.

Under field crops plant roots can drastically reduce NO_3^- hazards. Our model must take root adsorption into account to be complete, and this is being investigated. Denitrification in soils with plants will be more rapid than our model predicts due to larger amounts of "available" organic matter from root exudation (David Barber, private communication).

A curious fact has been revealed from our nitrogen transformation studies. In the field, upon addition of NO_3^- (not necessarily anaerobic) growth of *nitrite*-oxidizing organisms was observed (Volz et al., 1975b; Belser, 1974). The only source of NO_2^- was from NO_3^- reduction. Apparently

NO_3^- reduction occurred in anoxic microsites to form NO_2^-. Nitrite was then transported or diffused to more oxic sites and was oxidized to NO_3^-. The net result is a recycling of NO_3^- to NO_2^- and back to NO_3^-.

Energetics of Microorganisms

It was pointed out elsewhere (McLaren, 1973b) that one needs to measure rates of conversion of substrates per organism in biogeochemical systems in order to obtain realistic mathematical models. This concept may be expanded somewhat.
For the rate of consumption of a substrate S by a microorganism we can write

$$-dS/dt = A \frac{dm}{dt} + \alpha m + \beta m = \frac{1}{Y} dm/dt \qquad (9)$$

and $dm/dt = \gamma m$. The molar yield coefficient Y (Stouthhamer, 1969) is equal to $\gamma/(A\gamma + \alpha + \beta)$. Y is the grams dry weight/mole of substrate used and if the ATP yield (moles ATP produced/mole substrate used) is known we can calculate the amount of dry weight of microorganisms produced per mole of ATP formed, Y_{ATP}. From equation (9) since the doubling time T is given by $0.69/\gamma$, we also may write

$$-\frac{dS}{m_0} = \frac{0.69}{YT} dt \qquad (10)$$

or

$$\frac{\Delta S}{m_0} T = \frac{0.69}{Y_{ATP}} T = \text{"action per microbe"}. \qquad (11)$$

Substrate consumption, ΔS, is the substrate consumed during the time that the population increases exponentially from m_0 to $2m_0$. Since ΔS is expressible in terms of ATP formed, which in turn is related to a change in free energy during growth, the quantity $0.69 T/Y_{ATP}$ may be thought of as *action*, *i.e., energy x times* of doubling per organism (Grimsehl, 1932). In spite of great differences in yields and growth doubling times, products of these quantities are surprisingly similar for several bacteria (Table 17.2). This suggests that a less efficient organism takes a longer time to grow (the bacterial cells of each species listed in the table are of about the same size). Note that the *action* for a much larger organism, the yeast, falls among values for the other fermenters.

Table 17.2
Some Yield Coefficients and Growth Rates from the
Literature (Spector, 1956; Brock, 1966;
Stouthamer, 1969; Forrest, 1969).

Organism	Y_{ATP}	T min.	$T \cdot Y_{ATP}$	Action (moles ATP x min/g)
Aerobacter aerogenes	11	17	190	1.1
Escherichia coli	11	16	180	1.0
Nitrobacter winogradski	0.17	420	70	1704
Pseudomonas fluorescens	2.4	40	96	11.5
Streptococcus fecalis	11	27	300	1.7
Streptococcus lactis	10	26	260	1.8
Saccharomyces cerevisiae	10	120	1200	8.3

Nitrobacter is of course quite inefficient, and perhaps utilizes only 5-10% of available nitrite for growth *per se* in pure culture (Dessers et al., 1970). Even longer growth times for this organism are observed in soil (Gray and Williams, 1971). Evidently the physical-chemical environment can reduce the efficiency of growth, sometimes by diffusion limitations of nutrient availability (Saunders and Bazin, 1973).

Comments on Earlier Work

In their classic papers on nitrification Lees and Quastel (1946a, 1946b) concluded that soil nitrification is due entirely to the activity of cells that are multiplying and increases with the amount of soil on which NH_4^+ is absorbed. In their second paper they concluded, however, that soils saturated with nitrifiers can nitrify at maximal rates and that only adsorbed NH_4^+ is oxidized. In our work we have shown that soils oxidize NH_4^+ at rates that are proportional to the number of nitrifying organisms even though net multiplication has ceased (cf. Schmidt, 1974; McLaren, in press) and that displacement of adsorbed NH_4^+ by flowing solution KCl enhances nitrification in a soil column (S. Ardakani and McLaren, unpublished).

Lees and Quastel stated that nitrification rates increase if sterile soil is added to the system, in amounts proportional to the increase in exchange capacity, and added sand is without effect. They did not count nitrifying organisms, however. Had they done so they probably would have found that numbers increased significantly after adding soil but not so with added sand. Since rates are proportional to the numbers of nitrifiers in soil (Ardakani et al., 1974), addition of sterile soil increased the surface area available for growth irrespective of the augmented exchange capacity. Exchange capacity is not paramount (Faurie, et al.1975).

In still another experiment Lees and Quastel added the same amount of NH_4^+ to two perfusion flasks but used two different volumes of water. This was supposed to give different fractions of adsorbed NH_4^+ in each. Since, however, the concentrations of NH_4^+ in both were much greater than K_m (giving maximum, zero order kinetics) this experiment also reveals nothing about the site of nitrificaiton in soil.

Let us consider the matter from a topological viewpoint. Nitrification takes place readily on glass beads with low exchange capacity (Bazin and Saunders, 1973). If the particulate surfaces are negatively charged, however, they can adsorb ammonium by ion exchange. On the other hand, the microbial cell has a wall within which is a protoplast surrounded by a membrane containing the nitrogen-oxidizing enzyme system. To reach the membrane ammonium ions must diffuse through the wall and once within the wall it matters not from whence they came, be it from the solution phase or from exchange sites. One side of an adsorbed cell is in contact with the exchange sites and the other with solution, so the only effects of a soil particulate surface is to concentrate NH_4^+, H^+ and other cations on one side of the cell. The influence of surface pH is discussed elsewhere (Hattori, 1973).

In batch soil cultures diffusiton control of ions must be rate-limiting. In perfusion, $i.e.$, flowing systems, mass transport also influences the N-oxidation rates, and in fact these rates are slower the faster the flow rate, even in steady states with constant amounts of adsorbed NH_4^+.

Finally, an ion exchange phenomenon has never been suggested for nitrite oxidation by *Nitrobacter* in soil. Correspondingly, ion exchange with a preferential oxidation of adsorbed NH_4^+ by *Nitrosomonas* seems unrealistic.

In summary, our studies have shown that a useful mathematical model can assist in understanding the mechanism of soil nitrogen transformations. Both biological and physical parameters must be evaluated. Rates of nitrification and denitrification are affected by the population densities of ammonium and nitrate oxidizers and of denitrifiers. In a quantitative sense, the role of soil organic matter in the activity of denitrifiers has yet to be clarified. On this point work is in progress.

ACKNOWLEDGMENT

Appreciation is extended to RANN-NSF for partial support of this work.

REFERENCES

Ardakani, M. S., J. T. Rehbock and A. D. McLaren, 1973: Oxidation of nitrite to nitrate in a soil column. *Soil Sci. Soc. Amer. Proc.* 37, 53-56.

Ardakani, M. S., J. T. Rehbock and A. D. McLaren, 1974a: Oxidation of ammonium to a nitrate in a soil column. *Soil Sci. Soc. Amer. Proc.* 38, 96-99.

Ardakani, M. S., R. K. Schulz and A. D. McLaren, 1974b: A kinetic study of ammonium and nitrite oxidation in a soil field plot. *Soil Sci. Soc. Amer. Proc.* 38, 273-277.

Ardakani, M. S., M. G. Volz and A. D. McLaren, 1975: Steady state reactions of urea nitrogen in soil. *Can. J. Soil Sci.* 55, 83-91.

Bazin, M. J. and P. S. Saunders, 1973: Dynamics of nitrification in continuous flow system. *Soil Biol. Biochem.* 5, 531-543.

Belser, L. W., 1974: Ecology of nitrite-oxidizing bacteria: nitrate reduction as a source of nitrite for nitrite-oxidizers growth. Ph.D. dissertation, University of California, Berkeley.

Broadbent, F. E. and F. Clark, 1965: Denitrification. In: *Soil Nitrogen,* W. V. Bartholomew and F. E. Clark, eds, (Madison: Amer. Soc. Agronomy Inc., Publ.), pp. 344-359.

Brock, T. D., 1966: *Microbial Ecology.* (Englewood Cliffs: Prentice Hall), p. 95.

Cho, C. M., 1971: Convective transport of ammonium with nitrification in soil. *Can. J. Soil Sci.* 51, 339-350.

Dessers, A., C. Chang and H. Laudelout, 1970: Calorimetric determination of free energy efficiency in *Nitrobacter winogradskyi. J. Gen. Microbiology.* 64, 71-76.

Doner, H. E., M. G. Volz and A. D. McLaren, 1974: Column studies of denitrification in soil. *Soil Biol. Biochem.* 6, 341-346.

Doner, H. E., 1975: Disappearance of nitrate under transient conditions in columns of soil. *Soil Biol. Biochem.* 7, 257-259.

Doner, H. E., M. G. Volz, L. W. Belser and Jan-Per Løken, 1975: Short term nitrate losses and associated microbial populations in soil columns. *Soil Biol. Biochem.* 7, 261-263.

Faurie, G, A. Josserand and R. Bardin, 1975: Influence of day minerals on ammonium retention and nitrification. *Rev. Ecol. Biol. Sol.* 12, 201-210.

Focht, D. D. and H. Joseph, 1973: An improved method for enumeration of denitrifying bacteria. *Soil Sci. Soc. Amer. Proc.* 37, 698-699.

Forrest, W. W., 1969: Energetic aspects of microbial growth. In: *Microbial Growth*, P. M. Meadow and S. J. Pirt, eds., (Cambridge: University Press), p. 65-86.

Gray, J. R. G. and S. T. Williams, 1971: *Soil Micro-organisms*. (Edinburgh: Oliver and Boyd).

Grimsehl, E., 1932: *A Textbook of Physics*. Vol. 1, (London: Blackie and Son Ltd.), p. 91.

Hattori, T., 1973: *Microbial Life in the Soil*. (New York Marcel Dekker).

Kirda, C., J. L. Starr, C. Misra, J. W. Biggar and D. R. Nielsen, 1974: Nitrification and denitrification during miscible displacement in unsaturated soil. *Soil Sci. Soc. Amer. Proc.* 38, 772-776.

Lees, H. and J. H. Quastel, 1946a: Biochemistry of nitrification in soil I. *Biochem. J.* 40, 803-815.

Lees, H. and J. H. Quastel, 1946b: Biochemistry of nitrification in soil II. *Biochem. J.* 40, 815-823.

McLaren, A. D., 1969: Steady state studies of nitrification in soil: Theoretical considerations. *Soil Sci. Soc. Amer. Proc.* 33, 273-276.

McLaren, A. D., 1970: Temporal and vectorial reactions of nitrogen in soil: A review. *Can. J. Soil Sci.* 50, 97-109.

McLaren, A. D., 1973a: Nitrification in soil; systems approaching a steady state: A correction. *Soil Sci. Soc. Amer. Proc.* 37, 336-337.

McLaren, A. D., 1973b: A need for counting microorganisms in soil mineral cycles. *Environmental Letters* 5, 143-154.

McLaren, A. D. : Comments on an autoecological study of microorganisms. *Soil Sci.* In press.

Macura, J., 1966: Application of continuous flow methods in soil microbiology. In: *Theoretical and Methodological Basis of Continuous Culture of Microorganisms*, I. Mälek and Z. Fenel, eds., (New York: Academic Press), p. 462-492.

Meusel, E., 1875: Nutritbildung durch bakterien. *Ber. deut. Chem. Gesell.* **8**, 1214-1653.

Misra, C., D. R. Nielsen and J. W. Biggar, 1974: Soil nitrogen transformations during leaching: I. Theoretical considerations. *Soil Sci. Soc. Amer. Proc.* **38**, 294-299.

Quastel, J. H. and S G. Scholefield, 1951: Biochemistry of nitrification in soil. *Bact. Rev.* **15**, 1-53.

Saunders, P. T. and M. J. Bazin, 1973: Attachment of microorganisms in a packed column: metabolite diffusion through the microbial film as a limiting factor. *J. Appl. Chem. Biotechnol.* **23**, 847-853.

Schmidt, E. L., 1974: Quantitative autecological study of microorganisms in soil by immunofluorescence. *Soil Sci.* **118**, 141-149.

Schoenbein, C. Z., 1868: Uber die umwandlung der nitrate in nitrite durch conderven und audere organishe gebilde. *J. Prakt. Chem.* **105**, 208-214.

Skujins, J. J., 1967: Enzymes in soil. In: *Soil Biochemistry*, A. D. McLaren and G. H. Peterson, eds., (New York: Marcel Dekker), pp. 371-414.

Spector, W. S., ed., 1956: *Handbook of Biological Data*. Div. of Biol. and Agricul. NAS, NRC, Wright Air Development center. WADC Tech. Rpt. 56-273; ASTIA Document no. AD 110501. p. 97.

Stouthamer, A. D., 1969: Determination and significance of molar growth yields. In: *Methods in Microbiology*, J. R. Norris and D. W. Ribbons, eds., (New York: Academic Press), pp. 629-663.

Volz, M. G., L. W. Belser, M. S. Ardakani and A. D. McLaren, 1975a: Nitrate reduction and associated microbial populations in a ponded Hanford sandy loam. *J. Environ. Qual.* **4**, 99-102.

Volz, M. G., L. W. Belser, M. S. Ardakani and A. D. McLaren, 1975b: Nitrate reduction and nitrite utilization by nitrifiers in an unsaturated Hanford sandy loam. *J. Environ. Qual.* **4**, 179-182.

Winogradsky, S., 1949: *Microbiologie due sol*. (Paris: Masson et Cie), 861 p.

CHAPTER 18

THE INFLUENCE OF ENVIRONMENTAL FACTORS
ON NITRIFICATION

J. DE LEVAL
J. REMACLE

Département de Botanique, Université de Liège
Sart Tilman, B-4000 Liège, Belgique

INTRODUCTION

This report deals with the study of the rate of nitrification and is aimed at providing the necessary $N-NH_4$ oxidation constants which may be incorporated into the mathematical model of pollution in rivers.

With this objective in mind, the productivity of an aquatic strain of nitrifiers was investigated in relation to the temperature, the acidity and the substrate concentration. In this paper, only the **nitratation** experiments are considered.

MATERIALS AND METHODS

Many strains have been isolated from a polluted river by the technique of Schmidt, Molina and Chiang (1972). After purification productivities were checked against that of a strain received from Prof. Laudelout's laboratory (Louvain, Belgium).

Strain "7" was selected for its good productivity, as evaluated by chemostat culture after the microbial population had reached the steady state. At the steady state, the characteristics of the culture were very stable as shown by preliminary tests (Tables 18.1 and 18.2) performed with an aquatic strain of nitrifier. The maximum deviation from the mean productivity equals ±4%.

Table 18.1
Productivity of Nitrobacter sp. under Steady Conditions

Number of Days after Reaching the Steady State	Productivity ($\mu g \cdot l^{-1} hr^{-1}$)	
	A	B
2	341	366
3	207	291
4	308	292
4.5	305	343
5	360	361
5.5	378	392
6	391	396
6.5	437	386
7	420	415
7.5	410	360
8	435	353
8.5	396	418
9	451	385
10	445	375
Mean	412 ± 18	384 ± 13

A: Conductivity= 434 μmhos.
B: Conductivity= 1663 μmhos.

Table 18.2
Productivity in Relation to Illumination ($\mu g \cdot l^{-1} hr^{-1}$)

Number of Days after Reaching the Steady State	Culture	
	In the Dark	In Diffuse Daylight
3	468	457
4	467	451
5	433	463
6	456	438
7	480	477
8	466	463
9	463	457
Mean	461 ± 12	458 ± 11

Table 18.1 also shows that the productivity is the same at two conductivities, but further experiments should be run to confirm these results. No difference is noted in cultures exposed to diffuse daylight or darkness (Table 18.2).

The nitrite concentrations were estimated by the colorimetric method (see Standard Methods, 1965). The nitrate concentrations were determined by the chromotropic acid method, except that sulfamic acid was substituted for urea sulfite.

RESULTS

Determination of the Microbial
Parameters in Monod's Growth Model

By evaluating the nitrate production of *Nitrobacter* strain 7 in batch and in continuous cultures, μ, K_s and K_i were estimated to be equal to 0.06 h^{-1}, 3.2 mg N-NO$_3$ l^{-1} 209.0 mg N-NO$_3$ l^{-1}, respectively. (Temperature: 23°C, pH 7.5). K_s and K_i were calculated from Haldane's equation (see below).

Productivity in Relation to the
Chemical Conditions of the Medium.

Optimal Temperature and pH

The results are given in Figure 18.1. As can be seen, the optimal temperature is 23°C. It is possible to fit a polynomial curve of form $y = a + bx + cx^2$ to the results, where y = N-NO$_3$ produced, $\mu g \ l^{-1} \ h^{-1}$; x = temperature in °C. With our data, the resultant expression is, $y = 73.14 + 16.94x - 0.33 \ x^2$.

The optimal pH value is 7.5 (Figure 18.1). The productivity *vs* the pH can also be described by a polynomial equation:

$$y = a + bx + cx^2 + dx^3 + fx^4$$

where y = N-NO$_3$ produced ($\mu g \cdot l^{-1} \ hr^{-1}$)
      x = pH

The results of the experiments lead to the equation:

$$y = 77{,}930 + 44{,}540 \ x + 9{,}318 \ x^2 - 844 \ x^3 + 28.03 \ x^4$$

*Nitrification in Relation to
the N-NO$_2$ Concentration*

Since the optimal temperature and pH are determined, it is also of interest to evaluate the nitrification in relation to the substrate concentration. For this purpose, *Nitrobacter* cultures (strain 7) are fed with saline solution which contains 0.2 to 400 ppm N-NO$_2$.

The temperature and the pH are fixed at 23°C and 7.5, respectively; the dilution rate is maintained around 0.05 h^{-1} throughout the experiments. The results are given in Table 18.3. The maximum productivity is reached when the culture receives a flow of solution which contains 100 ppm

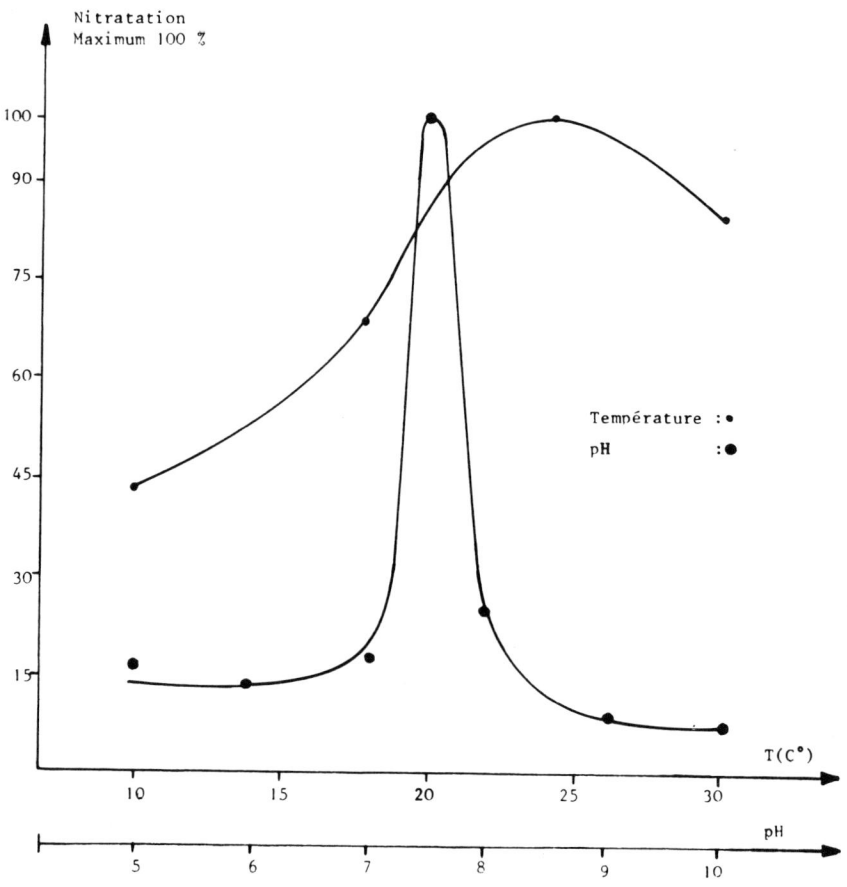

Figure 18.1 Nitratation as a function of temperature and pH.

$N-NO_2$, *i.e.*, when the culture contains 26 ppm $N-NO_2$. Below 20 ppm $N-NO_2$ in the feeding medium, all the nitrite is oxidized and the efficiency is 100%.

When the concentration in the feeding medium is above 20 ppm, the nitrite becomes toxic and the efficiency goes down from 95% to 4.5%. Thus, as can be seen in Figure 18.2, the optimal concentration of $N-NO_2$ is not the same when the productivity and the efficiency are considered.

The relation between the productivity and the substrate concentration can be examined by varying either the substrate concentration in the feeding medium or the substrate concentration in the culture. In the first case, the graph (see Figure 18.3) of productivity (µg $N-NO_3$ l^- hr^- vs log (g $NaNO_3$ l^{-1}) can be described by a Gaussian curve:

Table 18.3
Productivity and Efficiency of *Nitrobacter* Strain 7

$N-NO_2$ Concentration, $mg.l^{-1}$		Productivity $N-NO_3$, mg/l hr	Efficiency $\dfrac{N-NO_3}{N-NO_2}$ x 100
In the Feeding Medium	In the Culture		
0.2	≃0	0.014	100
2	≃0	0.139	100
10	≃0	0.601	100
20	1	1.109	95
100	26	4.130	74
200	154.8	3.665	22.6
300	278.4	1.719	7.2
400	382.0	1.293	4.5

$$y = 3400\ e^{-(2.6\ x^2 + 1.35\ x)}$$

where y = productivity ($\mu g\ N-NO_3\ l^{-1}hr^{-1}$)
x = Na concentration ($\log g\ NaNO_2\ l^{-1}$) in the feeding medium.

This equation is only verified when the $NaNO_2$ concentrations are higher than $0.01\ g\ NaNO_2\ l^{-1}$. The theoretical curve fits well with the experimental results (Figure 18.3) in this concentration range.

In the second case, the relation between productivity and the $N-NO_2$ concentration in the culture may be described by Haldane's equation,

$$P = \dfrac{\hat{P}\ s}{(K_s + s)\ (1 + \dfrac{s}{K_i})}$$

i.e.,

$$P = \dfrac{4.13\ s}{(3.2 + s)\ (1 + \dfrac{s}{209.9})}$$

P = productivity: $mg\ N-NO_3\ l^{-1}hr^{-1}$
$\hat{P}$ = maximum productivity
s = substrate concentration in the culture, $mg\ N-NO_2\ l^{-1}$
K_s = saturation constant, $mg\ N-NO_2\ l^{-1}$
K_i = inhibition constant, $mg\ N-NO_2\ l^{-1}$

This equation fits rather well the observed values (Table 18.4).

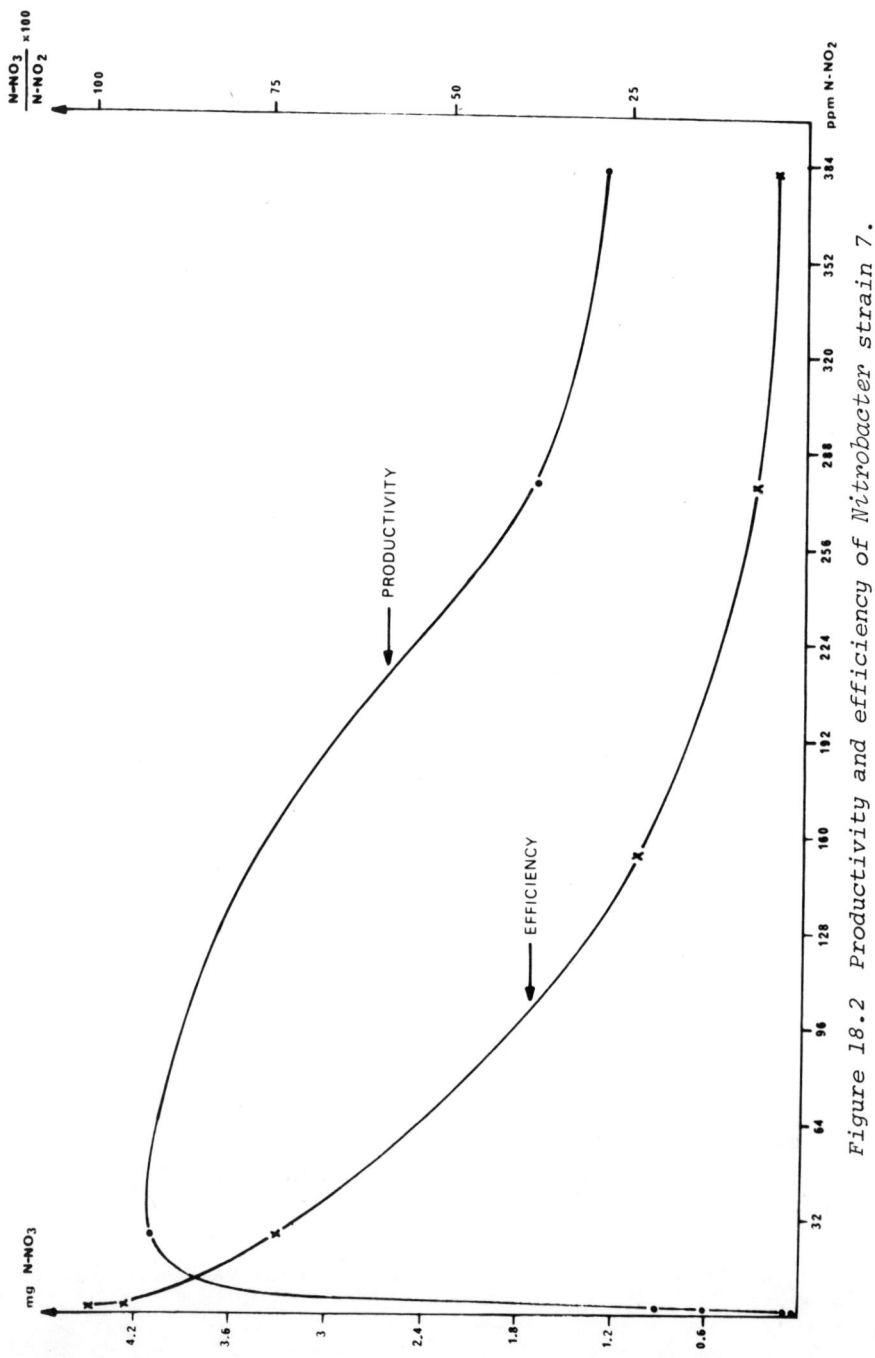

Figure 18.2 Productivity and efficiency of Nitrobacter strain 7.

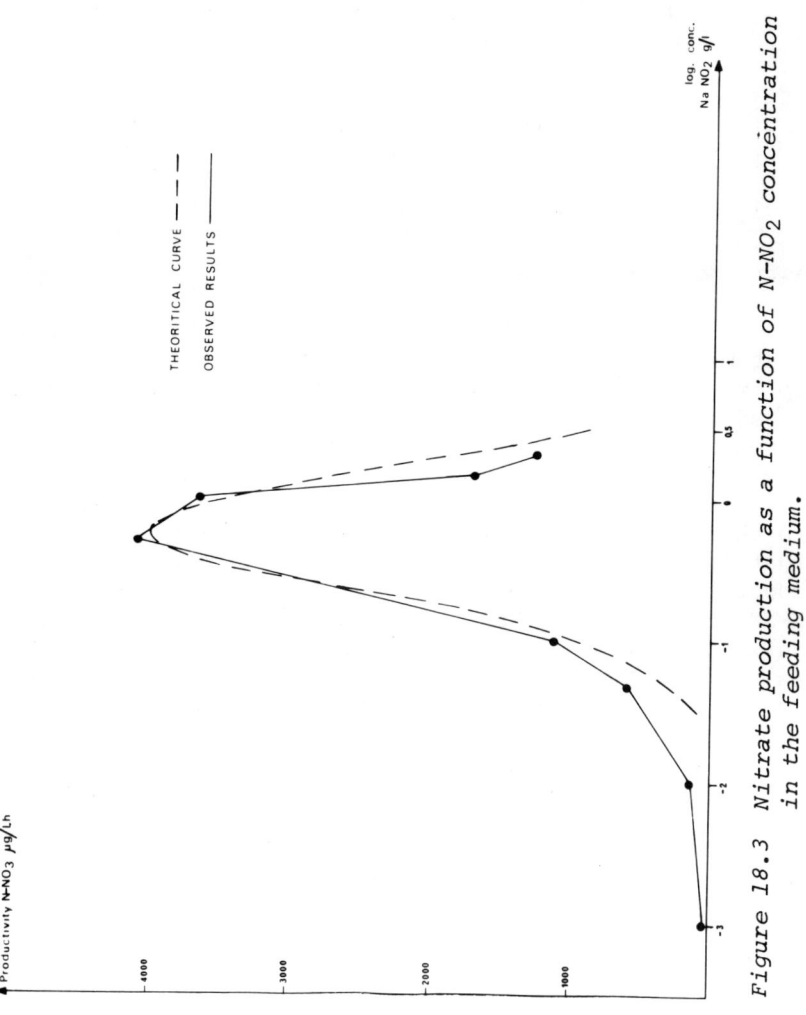

Figure 18.3 Nitrate production as a function of $N-NO_2$ concentration in the feeding medium.

Table 18.4
Inhibition of Nitratation by N-NO

N-NO$_2$ Concentration (mg l^{-1})	Productivity N-NO$_3$ (mg l^{-1})	
	Observed Values	Calculated Values
≈0	0.01	–
≈0	0.14	–
≈0	0.60	–
1	1.11	1.23
26	4.13	4.13
155	3.66	2.94
278	1.72	2.22
382	1.30	1.84

Nitrification in Mixed Culture

In order to make the results pertinent to the river systems, mixed cultures of *Nitrosomonas sp.* and *Nitrobacter sp.* were grown in two series of chemostats where the pH is adjusted to 7.5 and 8.5, respectively, and the dilution rates are fixed at ±0.05 h^{-1} and ± 0.0125 h^{-1}. The results are summarized in Table 18.5.

First, no nitrate is detected at pH 7.5 owing probably to the very low production of N-NO$_2$. Secondly it seems that *Nitrobacter* strain 7 is very sensitive to the N-NH$_4$ concentration because the efficiency is lowered when the N-NH$_4$ concentration is raised. The influence of the variations in N-NH$_4$ concentrations is not clear, however, as the quantity of N-NO$_2$ produced in the culture must also be considered. If, for example, the efficiency falls from 96.5% to 73.8% (D = ± 0.05 h^{-1}) and from 96.9% to 45.4% (D = ± 0.0125 h^{-1}) when the N-NH$_4$ concentration rises from 100 to 200 ppm, the amounts of N-NO$_2$ will be quite different in the two cases. Thus, at D = ± 0.5 h^{-1} the N-NO$_2$ concentrations are 517.7 µg l^{-1} and 67.7 µg l^{-1} and the same range in concentration is also observed at D = ± 0.0125 h^{-1} (567.2 µg l^{-1} and 44 µg l^{-1}). However, we can compare the efficiency in systems with similar N-NO$_2$ concentrations.

At D = ± 0.05 h^{-1}, the efficiencies are positively correlated with the N-NO$_2$ concentrations and it may be concluded that N-NH$_4$ does not affect the nitrification. But in the case of D = ± 0.0125 h^{-1}, the N-NH$_4$ concentration seems to influence the nitrification rate because the efficiency goes down when the N-NH$_4$ and N-NO$_2$ concentrations are increased. For example, 20 ppm N-NH$_4$ produces 33.6 µg N-NO$_2$ l^{-1} which is converted to N-NO$_3$ with an efficiency of 59.5% whereas 200 ppm N-NH$_4$ produces 44 µg N-NO$_2$ l^{-1} which

Table 18.5

Nitrobacter Strain 7 Efficiency (pH Culture: 8.5)

N-NH$_4$ Conc. (μg l^{-1})	Dilution rate: 0.05 h^{-1}			Dilution rate: 0.0125 h^{-1}		
	N-NO$_2$ Conc. (μg l^{-1})	N-NO$_3$ Conc. (μg l^{-1})	Efficiency $\dfrac{\text{N-NO}_3}{\text{N-NO}_2} \times 100$	N-NO$_2$ Conc. (μg l^{-1})	N-NO$_3$ Conc. (μg l^{-1})	Efficiency $\dfrac{\text{N-NO}_3}{\text{N-NO}_2} \times 100$
2	96.3	80	83	15.7	0	0
20	106.3	90	84.6	33.6	20	59.5
100	517.7	500	96.5	567.2	550	96.9
200	67.7	50	73.8	44	20	45.4

is oxidized to N-NO$_2$ with a yield of 45.4%. Besides it must be noted that the residence time (1/D) is much longer at D = ±0.0125 h^{-1} than at D = ±0.05 h^{-1}.

DISCUSSION

The diffuse daylight does not modify the rate of nitrification; heavy light may only inhibit the growth of the nitrifiers by killing the cells (Painter, 1970). Further experiments are needed to determine the influence of conductivity upon the nitrification.

The optimal temperature is different from the values currently mentioned in the literature (Painter, 1970; Goering, 1972). Strain 7 is more mesophilic than the other strains which have been tested (Painter, 1970). In pure culture, *Nitrobacter* strain 7 is very dependent upon the pH, but in mixed culture we have observed that the concentrations of N-NO$_2$ affect the influence of pH on the productivity.

The bacterial constants of Monod's equation, $\hat{\mu}$ = 0.06 h^{-1}, K_s = 3.2 mg l^{-1} are consistent with the previous results (Painter, 1970; Goering, 1972). In batch cultures, we have also determined a K_s value to be 5.5 mg l^{-1}. As for K_i, strain 7 gives a lower value than in earlier reports (Boon and Laudelout, 1962; Edwards, 1970).

It should be noted that the equations which describe the response of nitrifiers to the culture parameters are useful for elaborating a mathematical model of river pollution. This topic will be taken up in future reports.

ACKNOWLEDGMENTS

This research was supported by "Le 1er Programme, environment, de la commission interministérielle de la politique scientifique." Brussels, Belgium.

Our thanks go to Mrs. C. Beaufays and Dr. C. P. Dang for the mathematical treatment of the results.

REFERENCES

Boon, B. and H. Laudelout, 1962: Kinetics of nitrate oxidation by *Nitrobacter winogradskii*. *Biochem. J.* 85, 440-447.

Edwards, V. H., 1970: The influence of high substrate concentrations on microbial kinetics. *Biotechnol. Bioeng.* 12, 670-712.

Goering, J. J., 1972: The role of nitrogen in eutrophic processes. In: *Water Pollution Microbiology*, R. Mitchell, ed. (New York: John Wiley & Sons), pp. 43-68.

Kiff, R. J., 1972: The ecology of nitrification/denitrification systems in activated sludge. *J. Water Poll. Control* 71, 475-484.

Painter, H. A., 1970: A review of literature on inorganic nitrogen metabolism in microorganisms. *Water Res.* 4, 393-450.

Schmidt, E. L., J. A. E. Molina, and C. Chiang, 1973: Isolation of chemo-autotrophic nitrifiers from Moroccan soils. In: *Modern Methods in the Study of Microbial Ecology*, Th. Rosswall, ed., Bull. No. 17 NFR, Sweden, pp. 166-167.

CHAPTER 19

THE FATE OF NITROGEN IN SOIL: LOSSES BY
DENITRIFICATION AND LEACHING

R. J. DOWDELL

Agricultural Research Council, Letcombe Laboratory
Wantage, Oxfordshire, England

INTRODUCTION

The importance of nitrogen in sustaining high agricultural productivity is reflected in the increasing worldwide use of nitrogen fertilizers. The growing concern over the impact of modern agriculture on the environment, particularly water supplies, together with the high cost, in both monetary and energy terms, of nitrogen fertilizers makes it imperative that such materials be used efficiently. Therefore an understanding of the ways in which nitrogen may be lost from soil and an assessment of the magnitude of these losses is of great importance.
The primary mechanisms whereby nitrogen may be lost from soil are denitrification and leaching. These two processes can be considered to be related to some extent since they both occur when the soil water content is high, with denitrification tending to reduce the nitrate content of the drainage water. This chapter reports observations on the occurrence of denitrification in English field soils and the preliminary results of a lysimeter investigation of the magnitude of leaching and denitrification losses of fertilizer nitrogen.
The reduction of nitrate to nitrogen gas by denitrification is widely accepted to include the production of nitrous oxide as an intermediate (Wijler and Delwiche, 1954; Cady and Bartholomew, 1960; Cooper and Smith, 1963).

Hence the presence of nitrous oxide in the soil atmosphere can provide a useful index of denitrification activity. The first field observation of the natural production of nitrous oxide in soil appears to have been made by Burford and Millington (1968); subsequent reports have been made by Dowdell et al. (1972), Burford and Stefanson (1973), Dowdell and Smith (1974).

Before any estimate can be made of the total nitrogen loss by denitrification from determinations of nitrous oxide concentration, data are required for the rate of gaseous diffusion between the point at which the gas is measured and the soil surface, the air-filled pore space of the soil and the ratio of nitrogen to nitrous oxide liberated in the denitrification process. In the absence of this information, measurement of both gaseous and leaching losses of nitrogen can be obtained by use of suitably constructed lysimeters. Since we are primarily concerned with the fate of applied nitrogen in this experiment, it is essential to distinguish the fertilizer nitrogen from that native to the plants and soil; this is achieved by the use of nitrogen-15 labelled fertilizers.

OCCURRENCE OF NITROUS OXIDE

Experimental Methods

Studies of the soil atmosphere were made at two sites in southern England on clay loam soils during the winter of 1971-72; details of experimental design are given by Smith and Dowdell (1974).

The soil atmosphere was sampled at three depths (15, 30 and 60 cm) by means of probes consisting of porous bronze chambers (50 mm diameter, 66 mm long); fuller details are given by Dowdell et al. (1972). The nitrous oxide content of the soil atmosphere, or water if the probe was below the water table, was analyzed by gas chromatography (Smith and Dowdell, 1973).

Results and Discussion

Nitrous oxide was found in samples of the soil atmosphere and water at both experimental sites throughout the sampling period. The mean values of all probes at each sampling depth for one of the sites is shown in Figure 19.1. Small amounts of nitrous oxide (40 ppm) were found in January and February, but declined during March; in April to July, however high mean values (60-170 ppm) especially at 60 cm were found. The rise in concentration progressed through the soil profile, and coincided with the increase in soil temperatures during these months.

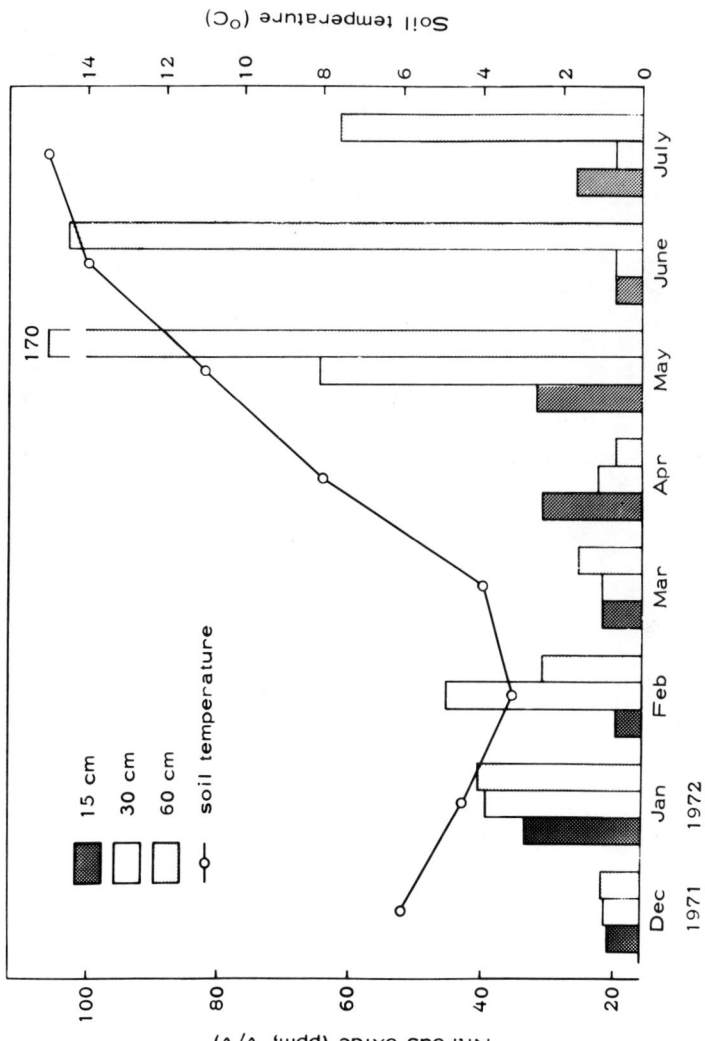

Figure 19.1 Mean concentrations of nitrous oxide in an Oxford clay soil at three depths and the mean soil temperature at 30 cm. (Dowdell and Smith, 1974)

The concentration of nitrous oxide observed varied widely with some individual samples containing 1500-6500 ppm. On any given sampling occasion nitrous oxide was present in only 5-15% of the samples, except for two occasions when the frequency exceeded 33%.

In these experiments it was observed that nitrous oxide occurred transiently in the soil atmosphere. It was not found for more than three consecutive weeks at any sampling point and it was rare for the gas to be detected at that point again during the experiment. This can be explained by studying the relationship between nitrous oxide and nitrate-N both dissolved in the soil water (Figure 19.2). While nitrate was present in the soil water, nitrous oxide was detected, but as the nitrate and nitrite concentration declined during January, the monthly mean content of nitrous oxide also declined to below the limit of detection in February. Thus while the soil is aerobic, nitrate derived from organic matter and fertilizer can persist. At the onset of anaerobic conditions this nitrate is rapidly denitrified and nitrous oxide is found in the soil atmosphere until the supply of nitrate is exhausted. No further accumulation of nitrate can occur, however, unless the soil is restored to aerobic conditions. However, why nitrous oxide is detected in one small locality at a different time from another locality while gross environmental conditions seem similar can only at present be explained by the large inhomogeneities inherent in these structurally well-developed clay soils.

LYSIMETER STUDIES

Experimental Methods

Twelve monoliths of a sandy loam soil (Rowland Series) encased in rigid PVC tubes (45 cm diameter, 110 cm deep) were collected in July 1972 and installed at Letcombe during the winter of 1972-73; perennial ryegrass (variety S23) was sown on the lysimeters in September 1972. The method used to collect the monoliths has been described (Cannell et al., 1973). In the first year of the experiment calcium nitrate fertilizer (in total equivalent to 418 kg N/ha) was applied to four lysimeters; the first of six additions (each approximately equivalent to 70 kg N/ha) was made on 4 April 1973 and the others at six-week intervals thereafter. Unfortunately it was impossible to use ^{15}N-labeled fertilizer until the third application on 25 June 1973. Thereafter ^{15}N-labeled fertilizer was used until the end of the first year; subsequently no fertilizer is being added to these

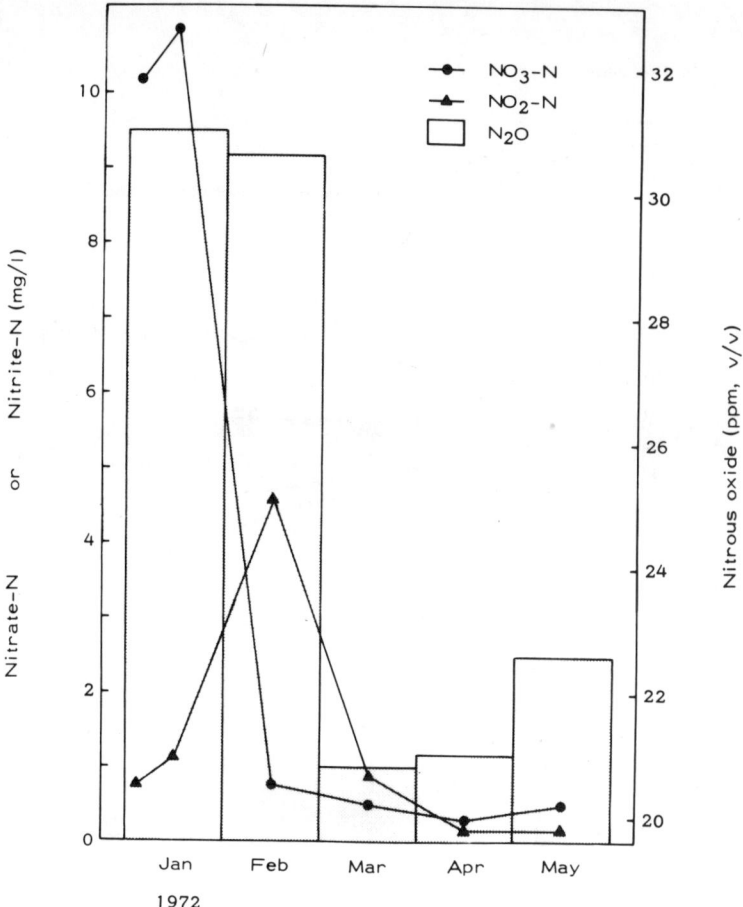

Figure 19.2 Concentration of nitrous oxide in the soil atmosphere of a Lower Lias clay soil compared with the nitrate-N and nitrite-N content of the soil water. (Dowdell and Smith, 1974)

lysimeters so that the fate of the ^{15}N-labeled fertilizer can be observed.

The grass was cut (to a height of 2.5 cm above ground) prior to fertilizer additions and the herbage removed for analysis. Irrigation water was applied to ensure that the minimum amount of water received by each lysimeter was equal to the "50-year" mean annual rainfall.

Two separate drainage systems are installed in each lysimeter (Figure 19.3). The first, free or gravitational drainage, consists of a gravel layer, 5 cm thick, at the base

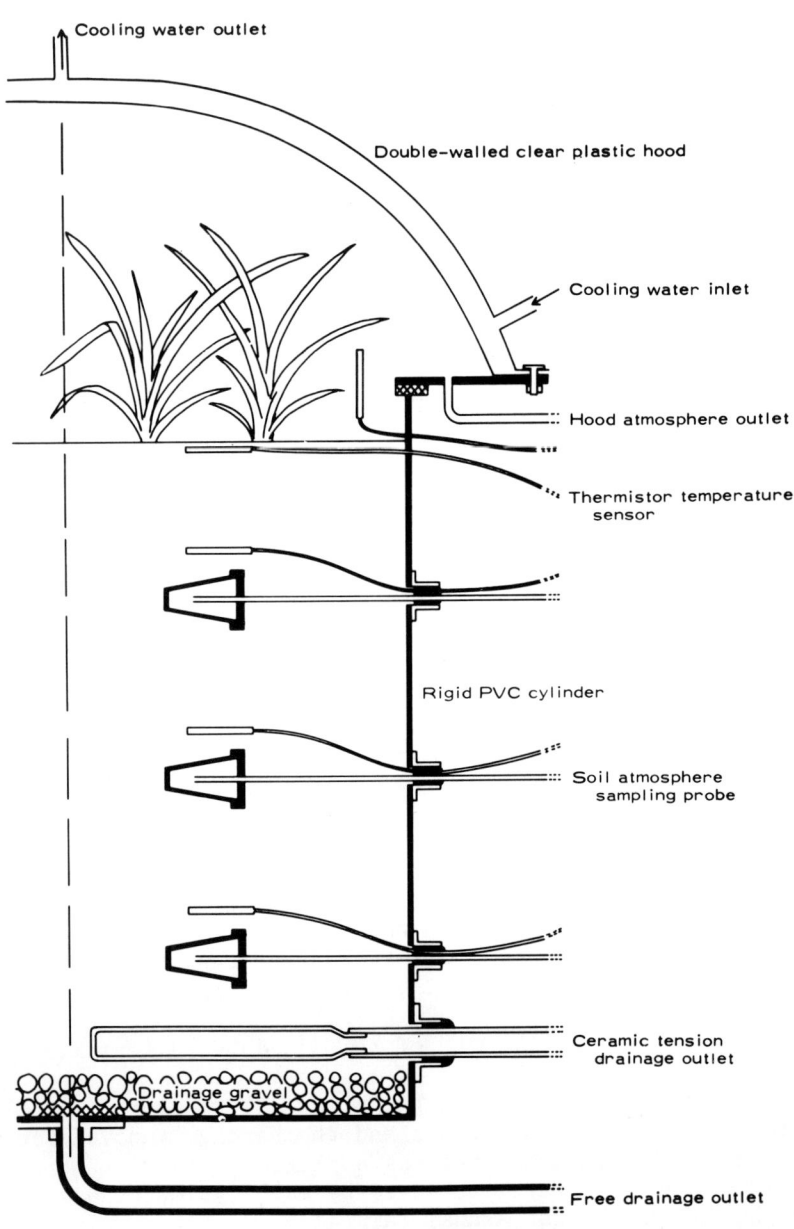

Figure 19.3 Sectional diagram of a monolith lysimeter.

of the soil monolith; the drainage water passes to a collection vessel through an outlet at the base of the lysimeter. The second method, tensioned drainage, uses six ceramic candles (22 cm long, 2.7 cm diameter) installed radially through the lysimeter wall and in the soil immeidately above the gravel and subject to 0.1 bar tension by means of a vacuum pump.

Each lysimeter can be periodically fitted with a clear plastic hemisphere which encloses the entire soil surface and plant canopy. The atmosphere within the hood can be recycled through a system of absorbents which remove water, carbon dioxide and nitrous oxide. Details of the trapping and recovery of nitrous oxide by molecular sieve are given by Dowdell and Crees (1974). Samples of the recovered nitrous oxide and the effluent gas stream from the hood are analyzed for ^{15}N content by mass spectroscopic methods. This technique allows detection on unlabeled N_2O evolved at a rate of 0.001 kg N/ha/day and ^{15}N-enriched N_2O at a rate of 0.011 kg N/ha/day. The loss of labeled nitrogen gas can only be distinguished unequivocally from atmospheric nitrogen if the evolution rate exceeds 7.0 kg N/ha/day.

Results and Discussion

The results presented here are for the first year of the experiment 2 April 1973 to 4 March 1974, but some results up to 4 September 1974 are included.

The grass grown on the nitrogen-treated lysimeters responded well to the fertilizer producing three times more dry matter than the controls. However, the nitrogen content of the fertilized grass (23.8 mg N/gm D.M.) was only slightly higher than that of the controls (21.4 mg N/gm D.M.). These results indicate that although a large amount of fertilizer had been applied, there was no "luxury" uptake of nitrogen.

The fate of nitrogen applied to the lysimeters can be clearly identified since the fertilizer is labeled with ^{15}N. Thus any ^{15}N which is detected in the crop, leachate or gas phase (after allowing for the normal ^{15}N background content) must have come from the fertilizer. A preliminary nitrogen balance is presented (Table 19.1). The results show that during the last eight months of the first year, 43% of the added fertilizer reached the shoots and 8.6% in the early part of the second year. This compares well with other published values for the uptake of the ^{15}N by grasses, e.g., 34-63% (Henzell et al., 1964), 30-80% (Legg and Allison, 1959), 47-65% (Tyler and Broadbent, 1958), 60% (Dilz and Woldendorp, 1960) and 35-60% (Woldendorp, Dilz and Kolenbrander, 1965).

Table 19.1
Lysimeter Experiment--Fate of ^{15}N-labeled Fertilizer
(Dowdell and Webster, 1974)

		mg	%
Total ^{15}N-labeled NO_3-N applied		4809.2	
^{15}N-labeled N contained in			
Herbage	25.6.73 to 4.3.74	2068.0	43.0
Herbage	4.3.74 to 4.9.74	411.2	8.6
Leachate[a]	25.6.73 to 4.3.74	96.2	2.0
Gas phase	25.6.73 to 4.3.74	N.D.	0
Gas phase	4.3.74 to 4.9.74	N.D.	0
Unaccounted for	25.6.73 to 4.9.74	2233.6	46.4

N.D. = None detected.
[a] Results of leachate for period 4.3.74 to 4.9.74 not yet available.

Although the soil is a light sandy loam, only a very small amount of the applied nitrogen (2%) has yet been found in the drainage water. Similar observations have been made by Kolenbrander (1969) who showed that only 2-5% of fertilizer applied to grass was leached through a sandy soil; this was attributed to plant uptake and low summer rainfall.

As yet, significant losses of fertilizer nitrogen by denitrification have not been measured. Observation of the aeration status of the soil has indicated that anaerobic conditions did not occur during the experimental period and that denitrification (as indicated by nitrous oxide) was rare. Thus the major part of the fertilizer has not yet been accounted for, and it must be assumed that it is retained in the soil. The actual amount of nitrogen retained in the soil will only be evaluated accurately when soil analysis is carried out at the end of the experiment. However, analysis of small quantities of soil removed to enable tensiometer insertion (8 March 1973) indicated that 15% of the applied labeled N was present in the top 25 cm of soil. In a lysimeter study of nutrient losses to drainage from soil under grass, Garwood and Tyson (1973) showed that 47-53% of the applied nitrogen could not be accounted for in the first year; however, this residual nitrogen was recovered in the subsequent two years.

The investigation is continuing. The uptake of ^{15}N from the lysimeters treated with fertilizer in the first year will be observed for a further period; an additional group of lysimeters received ^{15}N in the second year.

ACKNOWLEDGMENTS

The author is grateful to Dr. R. Scott Russell and Dr. R. Q. Cannell for many useful discussions, to Miss R. Crees, Mr. K. C. Hall and Mr. C. P. Webster and other members of the Field Studies Department for the conduct of the field and lysimeter studies, and Mr. M. G. Johnson and the staff of the Chemistry and Electronics Department for ^{15}N and other analyses.

REFERENCES

Burford, J. R. and R. J. Millington, 1968: Nitrous oxide in the atmosphere an a red brown earth. *Trans 9th Int. Congr. Soil Sci.* 2, 505-511.

Burford, J. R. and R. C. Stefanson, 1973: Losses of nitrogen from soils. *Soil Biol. Biochem.* 5, 133.

Cady, F. B. and W. V. Bartholomew, 1960: Sequential products of anaerobic denitrification in Norfolk soil material. *Proc. Soil Sci. Soc. Am.* 24, 477-482.

Cannell, R. Q., R. J. Dowdell, R. K. Belford and R. Crees, 1973: Collection of soil monolith lysimeters. Rep. Agric. Res. Council, Letcombe Laboratory for 1972, 41-43.

Cooper, G. S. and R. L. Smith, 1963: Sequence of products formed during denitrification in some diverse western soils.. *Proc. Soil Sci. Soc. Am.* 27, 659-662.

Dilz, K. and J. W. Woldendorp, 1960: Distribution and nitrogen balance of ^{15}N-labeled nitrate applied to grass sods. *Proc. 8th Int. Grassland Congr.*, Reading, 150-152.

Dowdell, R. J. and R. Crees, 1974: Measurement of the nitrous oxide content of the atmosphere. *Lab. Practice* 23, 488-489.

Dowdell, R. J. and K. A. Smith, 1974: Field studies of the soil atmosphere. II. Occurrence of nitrous oxide. *J. Soil Sci.* 25, 231-238.

Dowdell, R. J., K. A. Smith, R. Crees and S. W. F. Restall, 1972: Field studies of ethylene in the soil atmosphere-equipment and preliminary results. *Soil Biol. Biochem.* 4, 325-331.

Dowdell, R. J. and C. P. Webster, 1974: Denitrification and leaching of nitrogen fertilizers. In: *Agriculture and Water Quality*. Ministry of Agriculture, *Fisheries and Food Tech. Bull*. 32, in press.

Garwood, E. A. and K. C. Tyson, 1973: Losses of nitrogen and other plant nutrients to drainage from soil under grass. *J. Agric. Sci., Comb*. 80, 303-312.

Henzell, E. F., A. E. Martin, P. J. Ross, and K. P. Haydock, 1964: Isotopic studies on the uptake of nitrogen by pasture grasses. II. Uptake of fertilizer nitrogen by Rhodes grass in pots. *Aus. J. Agric. Res*. 15, 876-884.

Kolenbrander, G. J., 1969: Nitrate content and nitrogen loss in drainwater. *Neth. J. Agric. Sci*. 17, 246-255.

Legg, J. O. and F. E. Allison, 1959: Recovery of ^{15}N tagged nitrogen from ammonium-fixing soils. *Proc. Soil Sci. Soc. Am*. 23, 131-134.

Smith, K. A. and R. J. Dowdell, 1973: Gas chromatographic analysis of the soil atmosphere: automatic analysis of gas samples for O_2, N_2, Ar, CO_2, N_2O and C_1-C_4 hydrocarbons. *J. Chromatog. Sci*. 11, 655-658.

Smith, K. A. and R. J. Dowdell, 1974: Field studies of the soil atmosphere. I. Relationships between ethylene oxygen, soil moisture content and temperature. *J. Soil Sci*. 25, 217-230.

Tyler, K. B. and F. E. Broadbent, 1958: Nitrogen uptake by rye grass from three tagged ammonium fertilizers. *Proc. Soil Sci Soc. Am*. 22, 231-234.

Wijler, J. and C. C. Delwiche, 1954: Investigations on the denitrifying process in soil. *Pl. Soil* 5, 155-169.

Woldendorp, J. W., K. Dilz and G. J. Kolenbrander, 1966: The fate of fertilizer nitrogen on permanent grassland soils. In: "Nitrogen and Grassland," *Proc. 1st Gen. Meeting, European Grassland Fed*., Wageningen, 53-76.

SECTION III

BIOGEOCHEMICAL CYCLING OF PHOSPHORUS SULFUR AND SELENIUM

CHAPTER 20

A STUDY OF PHOSPHORUS KINETICS
IN A LAKE ECOSYSTEM

D. R. S. LEAN
M. N. CHARLTON

Canada Centre for Inland Waters,
Burlington, Ontario, Canada

Despite the many recent advances in our knowledge of
lake ecosystems, the relationship between energy flux,
algal-bacteria growth rates, secondary productivity, sedimentation and nutrient loading (internal and external) has
not been clearly established. Furthermore, the variables
which influence these processes have not been completely
identified. This paper points out some of the neglected
processes and misleading test procedures currently in use,
then describes an approach toward developing a flow diagram
for phosphorus dynamics in lake waters as a first step toward modeling community-nutrient-energy relationships.
 Previously many processes have been studied successfully
but independently. As a result, even though the geochemist
and the biologist are both interested in measuring mass
transfer processes of phosphorus in lakes, the relative importance of any pathway remains unclear. We must begin to
work on the same system at the same time.
 Any investigation on an entire lake is limited by our
inability to quantify the significant phosphorus compartments, both living and nonliving, and to measure the transfer rates between them as well as net movements or exchange
rates with sediments. To overcome some of these problems,
we conducted our investigation in large enclosures, or
limnocorrals (Kepner Plastics, U.S.A.) shaped as equilateral
triangles (area 25.1 m^2). This simplified, to some degree,

the spatial heterogeneity of a lake and permitted the use of radioisotopes. In this way not only the compartment sizes could be monitored, but the exchange rates between them. The corrals were situated in the Bay of Quinte, a eutrophic elongated bay (lat. 44.1, long. 77.0) which conducts much of the drainage of southern Ontario. Our location is shown in Figure 20.1. This area never stratified or became anoxic. The corrals were installed 15 June 1973 in water where the depth was 4 m. A flotation collar supported a side skirt with a heavy chain on the bottom that extended 0.5 m into the rich muds.

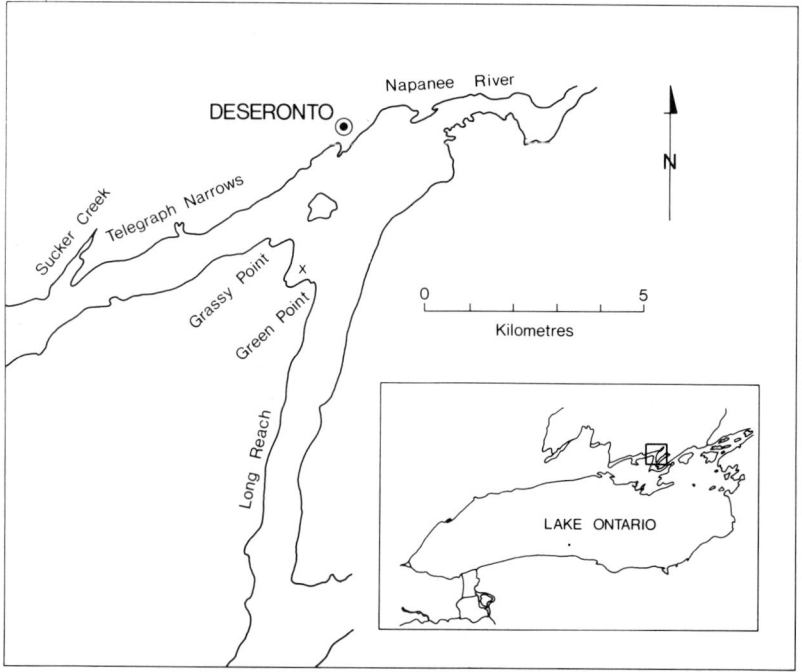

Figure 20.1 Location of study area (x).

We were particularly interested to see if the sediments would continue to release nutrients to the overlying waters when phosphorus loading was reduced. No nutrients were added to corral I. Corral II received only phosphate enrichment while III received both phosphate and nitrate enrichment. Nutrients were added once a week except during the peak biomass period when additions were made daily. Enrichment began 12 July 1973, about three weeks after installation of the corrals, and continued until 17 October.

During 1974 enrichment began on 24 May and continued to 16 September. This amounted to 0.72 and 0.88 g P/m /yr during 1973 and 1974, respectively. Nitrate enrichment in corral III was at a N:P weight ratio of 13.

The measurements of Total Phosphorus and Chlorophyll a are shown in Figure 20.2 and represent a composite of the water column. The methods used have been described previously (Lean et al., 1974). In 1973 the chlorophyll a values were measured by S. Millard (M.Sc. Thesis, in preparation, University of Guelph).

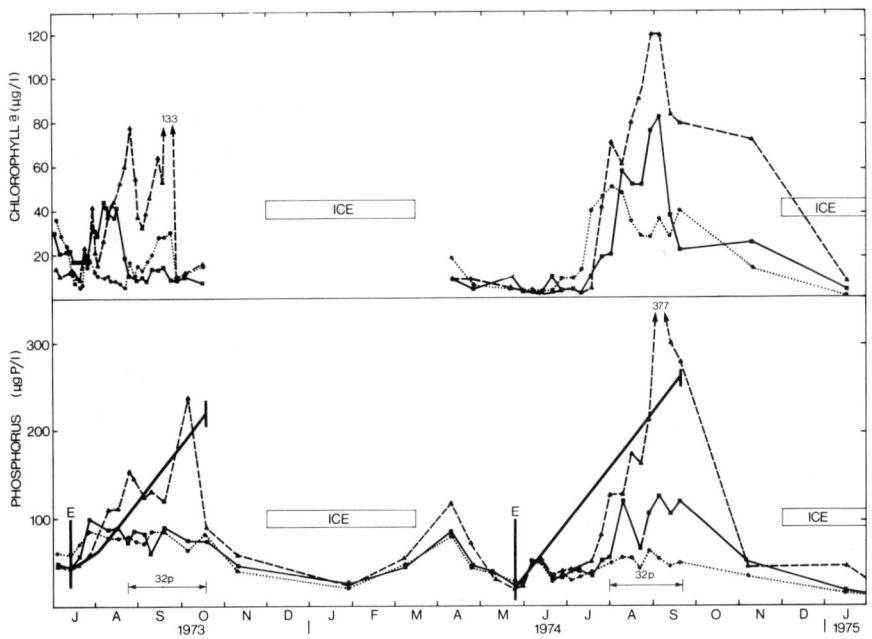

Figure 20.2 Chlorophyll a (above) and phosphorus (below). concentrations in limnocorrals I (····●····), II (— ■ —) and III (--▲--). Solid black line extending from E denotes the theoretical increase in phosphorus in corrals II and III.

Even though corrals II and III were enriched with phosphorus from 24 May 1974 (see theoretical increase curve), it was not until about 55 days later that they were significantly different from the control. "E" denotes the beginning of the enrichment period. During this time the sediments were acting as a sink for phosphorus. Suddenly corral III responded and the total phosphorus even surpassed the theoretical amount. Corral II started to increase, stabilized, then dropped. Soluble Reactive Phosphate (SRP) was not measured in 1973, but from May to September 1974 was usually near 1-3 µg P/l in all three corrals with the exception of the period starting August 1. SRP in corral III increased in a linear fashion until it reached a maximum of 80 µg P/l on 11 September. Corral II showed no such increase. The most striking observation was that even though we added enough phosphorus to corrals II and III to increase the concentration to 220 and 260 µg/l in 1973 and 1974, respectively, the corrals declined to a level equal to the control in the winter. Furthermore, even though corral II received the same phosphorus as III, it was more similar to I.

The net rate of increase in phosphorus in corral III, 16 July to 2 September, (Figure 20.2) was 7.0 µg/l/day (28 mg/m^2/day). The enrichment rate was 2.5 g/l/day; hence, the release rate was 4.5 µg/l/day or 18 mg/m^2/day. Even though the water column neither stratified nor became deoxygenated, one is tempted to speculate that a microzone developed that was sufficiently reduced that phosphate was released. This rate (18 mg/m^2/day) even surpasses Vollenweider's (1968) estimate of 9.6 mg/m^2/day for phorphorus regeneration under anoxic conditions, and also that calculated by Burns and Ross (1972) of 7.4 for Lake Erie under anoxic conditions. Absolute numbers perhaps are misleading. Instead we should consider the fraction released per unit time. Sediments with a higher percent phosphorus in the "available" form would be expected to release more phosphorus under reduced conditions. Burns (personal communication) has examined other Lake Erie data and thinks that the values given above are often exceeded.

Until the increase in phosphorus took place, the chlorophyll did not increase either. This net release of phosphorus during August may be common in ponds and shallow eutrophic lakes (see Anderson, V-J. M., 1974). One hypothesis that may explain this observation is that the release can occur only after the sediments have been sufficiently "conditioned" by the effect of material falling to the sediments and being decomposed. Charlton (1975), using sediment traps, found that the summer sedimentation rates in Corral III were twice those for I and II. This probably caused the observed Eh drop in the surficial sediments: 100 to -10 mv in I, 130 to 0 mv in II, and from 100 to -10 mv in III from June 26 to July 23, 1974 (complete data given by Murphy and Lean, 1975).

We recognize that Eh measurements are difficult to interpret and that sedimentation rates are difficult to measure, but it would seem that the net phosphorus release from the sediments occurred after a sufficient period of time when materials were being decomposed at the sediment interface. If this load can be reduced, the August bloom may not occur. This suggests that both organic and nutrient loading to lakes must be minimized. If this is done, the sediments should act as a sink as illustrated by the winter total phosphorus values, even when disturbed by turbulent mixing.

Corral II was able to sustain a high productivity and biomass (Lean *et al.*, 1974) as long as nitrogen fixation could meet the algae's requirements. This was measured both by the ^{15}N-nitrogen gas technique (C. Liao; C. Liao and D. Lean, papers in preparation) and by acetylene reduction (S. Painter, M.Sc. Thesis, in preparation, University of Guelph). The C/N ratio of the settling material was similar in all three corrals. The biomass decline in II could have been caused by one or both of the cofactors for N-fixation, iron and molybdenum, being exhausted (see Murphy and Lean, 1975). Using techniques commonly used for ^{32}P-PO$_4$ tracer studies (Lean, 1973b), ^{55}Fe was used to demonstrate that iron was in demand in II having turnover times as short as 1-5 minutes during the period of decline, whereas in III there was no shortage of iron and the uptake using ^{55}Fe was slow. (Rapid turnover times occur when concentrations of available iron are low and in demand by the particulate forms.)

This observation has profound implications on the role of nitrogen in eutrophication. Abundant nitrate, ammonia, or other forms of biologically-available fixed nitrogen will influence the species composition of the algae toward the nonnitrogen-fixing forms. When no such forms of nitrogen are available, nitrogen fixation can supply the required nitrogen until the cofactors iron and molybdenum become depleted. Future decisions on the design of sewage treatment plants should consider these observations.

In addition to the net internal loading discussed above, we also measured the exchange rates with the sediments (Lean *et al.*, 1974). During the periods shown as ^{32}P (Figure 20.2), ^{32}P-PO$_4$ was added to the corrals and the transfer rates between the principal compartments measured. These results cannot be discussed fully during this short presentation, but, if we consider only the 1973 experiment in corral I, we can see the total P did not change much during this period. One might expect from this that little was happening, *i.e.*, static biomass and constant level of phosphorus. This proved to be incorrect. The movement to the sediments of radiophosphorus was at a rate of 2.76% per day giving a mean residence time or turnover time of phosphorus in the corral of 25 days. This represents a flux to the sediments of (0.0276 day^{-1} x 70 µg/l) or 1.93 µg/l/day. This is 7.69 mg/m^2/day and this must have been balanced by an equal return from the sediments. Charlton's (1974) August sedimentation rate obtained using sediment traps was 9.8 mg/m^2/day, and when corrected for sediment resuspension was 5.9 mg P/m^2/day. These two independent measurements demonstrate that there was an exchange between the sediments and the overlying water even though the water was not anoxic.

Radiotracer experiments were also conducted periodically to determine the turnover time of phosphate--in other words the time required for an amount of phosphate equivalent to that present to be taken up and replaced. Rapid turnover times reflect a low phosphate concentration under conditions where significant phosphate demand exists. If we were able to reliably measure phosphate (Rigler, 1966; 1968) and the other biologically important forms (Lean, 1973a,b), we could calculate flux rates for phosphorus. In the absence of reliable analytical methods, the compartment size can be estimated only by isotope distribution at an assumed steady state.

One such experiment is shown in Figure 20.3. The analytical measurements are given in Table 20.1. The higher biomass and lower percent SRP for corral III and the Bay of Quinte samples would suggest that the most rapid turnover time for phosphate would occur in these samples. When we examine Figure 20.3, it is obvious that this is not so. The initial slope of the log percent filtrate ^{32}P vs time curve divided by 2.303 approximates the rate constant for transfer to the particulate form. These are 0.42, 0.0089, 0.025 and 0.038 min^{-1} for corrals I, II, III and the Bay, respectively. The corresponding turnover times are simply the reciprocal of the rate constant and are 2.4, 112, 40.4 and 26.1 min. In corral I, for example, an amount of phosphate equivalent to the total amount present is taken up and replaced every 2.4 min.

Table 20.1
Phosphorus Analysis (µg P/l), 18 July 1974

	Corral I	Corral II	Corral III	Bay of Quinte
Total Phosphorus	38	37.5	53.0	55
Total "Soluble" Phosphorus	15	23.3	17.5	12.3
Soluble Reactive Phosphorus	3.8	7.5	3.8	3.28
Percent SRP	10	20	7.2	5.9

It is incorrect to assume that SRP is equivalent to orthophosphate and these values cannot be used to calculate uptake rates of phosphate. If the asymptote level reached for corral I (Figure 20.3) represents a steady-state level, the true phosphate concentration must be less than 5% of the total phosphorous. Sephadex gel filtration of the filtrate (see Lean, 1973a,b) after 4 hours showed that only 30% of the 5% was actually as PO_4. If the total phosphorus value of 38 is correct, the orthophosphate level must be less than 0.6 µg P/l. Longer equilibration times will only provide lower values for orthophosphate. The value obtained as SRP exceeds the orthophosphate level by at least six times.

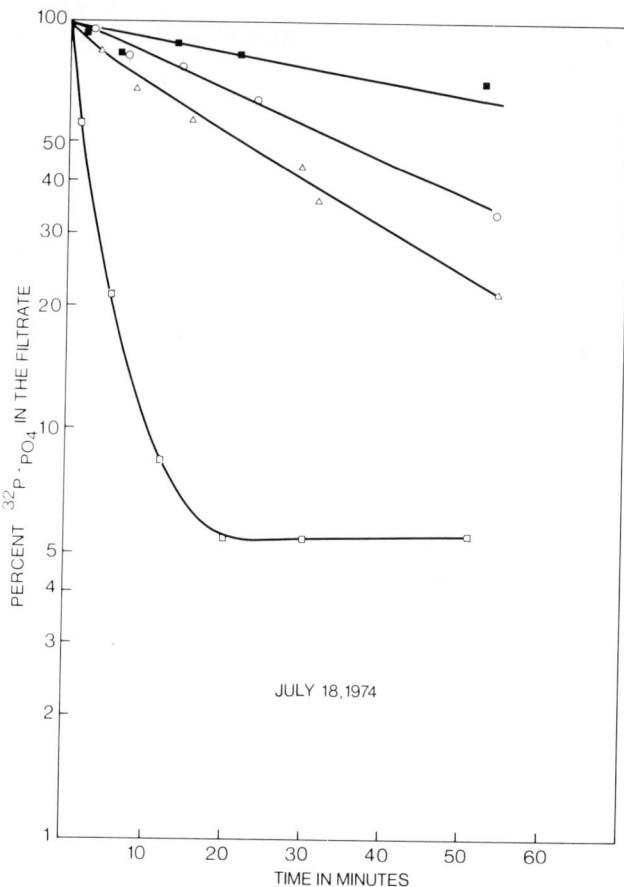

Figure 20.3 Graph showing time course for transfer of ^{32}P-PO_4 between filtrate and seston components for Corral I, control (□); Corral II, P enrichment (■); Corral III, P+N enrichment (○) and the Bay of Quinte outside the corrals (△).

The uptake of phosphate, assuming 0.6 is close to the correct amount, is 0.6 x 0.42 or 0.25 µg P/l/min. This must be balanced by an equal release rate. Assuming the release of organic phosphorus is much slower than the release of phosphate from the particulate form (see Lean, 1973a,b), this pathway will be neglected in this discussion.

Zooplankton grazing was measured using the technique of Haney (1971). The range of values was 18 to 480 with a mean of about 80% per day. This means that a volume of water equivalent to 80% of the entire corral was swept clean of particles less than 30 µ each day. During the period discussed above, the rate was only 21% per day, or expressed as a fraction of the total volume per min, 1.46×10^{-4} min^{-1}, *i.e.*, 14.6 l/min. Peters (1975) found that typically 54% of the phosphorus in the ingested food was excreted, and Peters and Lean (1973) showed this to be almost all orthophosphate. Zooplankton grazing, therefore, can account for a regeneration of 38 µg P/l x 0.85 x 1.46 x 10^{-4} x 0.54 or 0.00255 µg P/l/min. (The fraction of the 38 µg P/l as ingestible particles was 0.85). This amount is insufficient to account for the release rate (0.25), and furthermore when zooplankton were removed from the sample similar kinetics to those with zooplankton present were obtained. Even if we use values 10 times greater for zooplankton grazing, this process seems to be insignificant. The most obvious pathway, therefore, is direct release of phosphate by the particulate material.

This is a difficult concept to grasp. Traditionally it has been thought that phosphate was taken up in proportion to the demands for growth. This is wrong for two reasons. Any surplus phosphorus in the medium is taken up and stored until the polyphosphate pool of the algae is saturated; and secondly, both algae and bacteria are capable of exchanging phosphorus with the media.

Lean and Nalewajko (1975) have studied this phenomenon in four species of algae and Lean (unpublished data) in three species of pelagic bacteria. Organisms were grown in batch culture and either labeled with $^{33}P-PO_4$ or duplicate cultures grown, one of which was labeled with $^{32}P-PO_4$ at the same time the cultures were inoculated. As the biomass increased, the distribution of isotope provided a measure of the compartment size and the net change in the compartments. $^{32}P-PO_4$ was added to portions of the culture or to a duplicate flask at intervals throughout the growth phase. This provided a measure of the influx while the net change was a measure of influx minus efflux. Long before phosphate became depleted in the medium, less than 6% of the phosphate taken up actually stayed in the cell, even at phosphate levels exceeding 30 µg P/l. This observation makes interpretation of Michaelis-Menton kinetics exceedingly difficult.

These experiments also provided additional information. Uptake rates as fast as those observed in lake waters were found in both algal and bacteria cultures. This suggests that organisms are capable of causing the rapid kinetics observed in lake waters.

The ultimate model would relate the uptake or utilization rate of nutrients (C, N and P) to growth rate of algae. Several factors complicate this at the present time:

1. Measurements of primary production are difficult to interpret in light of total daily carbon flux.
2. The biologically important forms of phosphorus (and nitrogen and carbon) cannot be reliably quantified.
3. Algae store surplus phosphorus when it becomes available and use it for growth for several generations.
4. Phosphorus is rapidly exchanged between the particulate and dissolved forms.
5. Phosphorus uptake proceeds as rapidly in the dark as in the light.

The first and most difficult problem is to reliably measure the growth rate of algae. It is common practice to add some ^{14}C-bicarbonate to a bottle and after 2-4 hr calculate the percent isotope incorporated and from this determine the mg C taken up per unit time. Then, using figures for total daily radiation, an integrated value for "net" daily productivity is obtained.

Not nearly as much effort is spent in determining respiration rates. This results in a depletion of the carbon fixed and causes the net change in total particulate C to be close to zero (for 24 hr). On bright days a slight increase may occur, but on rainy days the biomass may even decline.

To illustrate this complication, we have selected a typical experiment from a paper in preparation (Charlton and Lean). We measured production in three ways: ^{14}C bicarbonate, O_2 bottle method (both in 300-ml bottles) and by diurnal O_2. These data are shown in Table 20.2.

The net production using the ^{14}C and the O_2 bottle methods are close to the same value. The total (gross) photosynthesis that occurs is the sum of net production and respiration. The diurnal curve values for this same period are 3.1-9.8 times higher than the bottle values and respiration is 5.2-15.4 times greater. When we look at the 24-hr increase in O_2 we see corrals I and III were identical (+52), while II showed a net decline. Some of the enhancement can be explained by littoral production in the corrals, but

Table 20.2
Production Measurements

Corral	^{14}C Method (m Moles $C/M^2/hr$)	O_2 Bottle Method (m Moles $O_2/M^2/hr$)			Diurnal Curve[a] (m Moles $O_2/M^2/hr$)			Diurnal Curve[b] (m Moles $O_2/M^2/day$)		
		P_G	R	P_N	P_G	R	P_N	P_G	R	P_{N24}
I	10.2	12.8	3.1	9.7	45	22	23	580	528	+52
II	2.6	5.7	2.1	3.6	56	28	28	632	672	−40
III	6.0	10.0	2.6	7.4	78	40	38	1012	960	+52
Bay of Quinte	15.8	19.6	5.8	13.8	61	30	31	850	720	+130

Note: P_N and P_G denote net and gross production and R is the respiration rate in the dark.

[a] The net production was considered to be the O_2 increase during the same time period as the O_2 and ^{14}C bottle experiments and the respiration values calculated from the O_2 drop during the following night.

[b] The gross production is the sum of O_2 increase during the day and respiration rate calculated for 24 hr using the night rate, $P_N = P_G - R$.

since the Bay of Quinte shows the same enhancement as corral I, we are inclined to believe these values. It seems obvious that ^{14}C values are of little use in calculating net productivity. It would seem that algal metabolism is reduced by confinement in bottles for 2-4 hr in these eutrophic waters.

An example of the pitfalls encountered in studies employing bottle methods is provided by Jassby and Goldman (1974). These authors carefully compared ^{14}C productivity and algal biomass in Castle Lake from May to December and found that the biomass did not increase as predicted by the productivity measurements. Although night respiration was mentioned as a possible cause of the discrepancy, the authors preferred to attribute it to cell mortality. Comparisons of diurnal curve bottle production measurements usually show that the former method gives higher (gross) results but production is likely to be matched by respiration--little or no margin may be left for biomass increases (Welch, 1968; Odum and Hoskin, 1958; Melack and Kilham, 1974). Although the ^{14}C method may be the only method sensitive enough to indicate production in some waters, algal metabolism seems to be reduced by confinement in bottles in eutrophic waters, and these data are inappropriate for calculating a useful net productivity.

We have not tried to present any model for phosphorus dynamics within a lake community, but only have illustrated some of the factors that must be understood before we can proceed to a mechanistic relationship between nutrients, energy and productivity in lake communities.

REFERENCES

Andersen, V-J. M., 1974: Nitrogen and phosphorus budgets and the role of sediments in six shallow Danish lakes. *Arch. Hydrobiol.* 74, 528-550.

Burns, N. M. and C. Ross, 1972: Oxygen-nutrient relationships within the central basin of Lake Erie. In: Project Hypo. C.C.I.W. Tech. Report. TS-05-71-208-24. 182 pp.

Charlton, M. N., 1975: Sedimentation: Measurements in experimental enclosures. *Verh. Internat. Verein. Limnol.* 19, 267-272

Haney, J. R., 1971: An *in situ* method for the measurement of zooplankton grazing rates. *Limnol. Oceanogr.* 16, 970-977.

Jassby, A. H. and C. R. Goldman, 1974: Loss rates from a lake phytoplankton community. *Limnol. Oceanogr.* 19, 618-627.

Lean, D. R. S., 1973a: Phosphorus dynamics in lake water. *Science* 179, 678-680.

Lean, D. R. S., 1973b: Phosphorus movement between its biologically important forms in lake water. *J. Fish. Res. Bd. Can.* 30, 1525-1536.

Lean, D. R. S., M. N. Charlton, B. K. Burnison, T. P. Murphy, S. E. Millard and K. R. Young, 1974: Phosphorus: Changes in ecosystem metabolism from reduced loading. *Verh. Internat. Verein. Limnol.* 19, 249-257.

Lean, D. R. S. and C. Nalewajko, 1975: Phosphate exchange by freshwater algae. *J. Fish. Res. Bd. Can.*, In press.

Melack, J. M. and P. Kilham, 1974: Photosynthetic rates of phytoplankton in East African, saline lakes. *Limnol. Oceanogr.* 19, 743-755.

Murphy, T. P. and D. R. S. Lean, 1975: The distribution of iron in a closed ecosystem. *Verh. Internat. Verein. Limnol.* 19, 258-266.

Odu, H. T. and C. M. Hoskin, 1958: Comparative studies on the metabolism of marine waters. *Publ. Inst. Marine Sci. Texas.* 5, 16-64.

Peters, R. H., 1975: Phosphorus regeneration by natural populations of limnetic zooplankton. *Verh. Internat. Verein. Limnol.* 19, 273-279.

Peters, R. H. and D. R. S. Lean, 1973: The characterization of soluble phosphorus released by limnetic zooplankton. *Limnol. Oceanogr.* 18, 270-279.

Rigler, F. H., 1966: Radiobiological analysis of inorganic phosphorus in lake water. *Verh. Int. Verein. Limnol.* 16, 465-470.

Rigler, F. H., 1968: Further observations inconsistent with the hypothesis that the molybdenum blue method measures inorganic phosphorus in lake water. *Limnol. Oceanogr.* 13, 7-13.

Vollenweider, R. A., 1968: Scientific fundamentals of the eutrophication of lakes and flowing waters, with particular reference to nitrogen and phosphorus as factors in eutrophication. OECD-DAS/CSI/68.27.

Welch, H. E., 1968: Use of modified diurnal curves for the measurement of metabolism in standing water. *Limnol. Oceanogr.* 13, 679-687.

CHAPTER 21

DYNAMICS OF PHOSPHORUS, SULFUR AND NITROGEN AT
THE SEDIMENT-WATER INTERFACE

R. O. HALLBERG
L. E. BÅGANDER
ANNA-GRETA ENGVALL

Department of Geology, University of Stockholm
Box 6801, S-113 86 Stockholm, Sweden

INTRODUCTION

In a sea basin with a stagnant water body the oxygen may be completely consumed during the degradation of organic matter, and hydrogen sulfide may then accumulate in the water. This is a common situation in Norwegian fjords, the Black Sea and the Baltic Sea. In the Baltic Sea, water of higher salinity than normal Baltic water occasionally intrudes into the basins from the North Sea. This creates a density gradient which separates the bottom water from the overlying water mass. The bottom water forms a "closed" system with reactions involving an exchange of elements between the bottom water and the sediment. The exchange processes are mainly governed by the bacterial activity in the sediment. Their metabolic activity in turn is dependent on the amount of utilizable organic matter added to the bottom. The bottom water is initially oxygenated, but after some time the oxygen content decreases to a value where reducing processes begin to dominate. During reducing conditions, the sediment is depleted in different nutrient salts such as phosphate and ammonia. They accumulate in the deep water regions of the Baltic. In order to simulate such accumulation processes in the bottom water and to obtain experimental data for computer simulation and prediction of the chemical changes, the present investigation has been performed *in situ*.

METHODS

The conventional method used in investigations of the sediment/water system is to sample the sediment and an appropriate amount of bottom water. The sample can either be split into subsamples which are analyzed or can be set up in the laboratory as an experimental system in order to simulate the natural conditions. In the first case the data generated are representative only for the sampled core and at the actual sampling occasion. However, in many cases this is the only possible method and by sampling several cores from an area, conclusions may be drawn which are representative for larger areas. In the second case, when a laboratory experiment is performed, it is possible to study the mechanism for some dynamic processes, but because of the changes in the chemical, physical and biological characters of the sediment/water system during the sampling prodedure it must to some extent be regarded as artificial. In particular, the data for diffusion processes and kinetics may differ from natural conditions if the original temperature and pressure are not maintained. *In situ* experiments are more readily reproduced and are more representative for the actual conditions since they can be performed with much less disturbance of the original character of the sediment.

In order to simulate different conditions and to study the effect of different factors, the present studies have been carried out in closed systems. The studies include scuba diving as an essential part. To facilitate this procedure, a soft bottom sediment at a water depth of not more than 10 m was chosen. Closed system apparatus consisting of plexiglass boxes were used (Figure 21.1). Sampling of the enclosed water takes place by means of syringes through self-sealing rubber membranes in the walls of the boxes. The boxes are equipped with electrodes for obtaining measurements of pH, Eh and sulfide ion activity (E_S). The measurements are recorded continuously on board a floating laboratory. The water loss at sampling is compensated for by a simultaneous addition of sea water and is considered when the results are calculated.

Natural systems are, in contrast to the box systems, open to the surrounding environment, and it can therefore be argued that the box systems are not representative for the simulated environments. It can, however, also be argued that for nearly all kinds of systems there exist regions or environments in which, at times, the time-invariant conditions closely approach equilibrium even though gradients exist throughout the system as a whole. The concept of

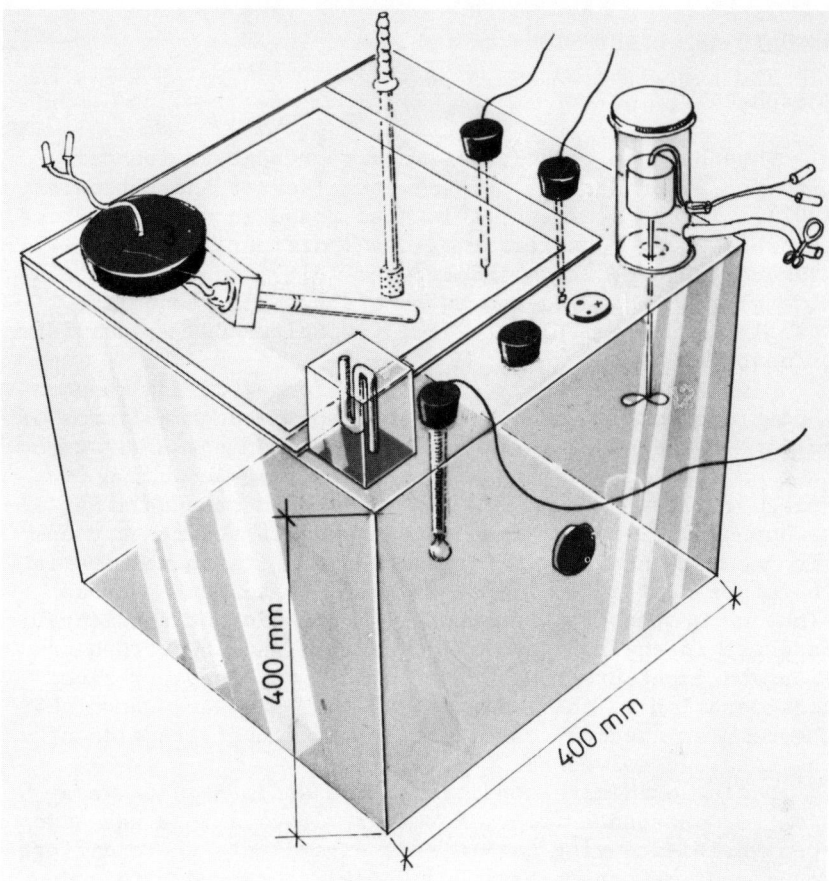

Figure 21.1 The plexiglass box, used in the experiments for in situ measurements, forms a closed system when pushed into the sediment (see Schippel et.al., 1973a).

local equilibrium is a fundamental assumption in thermodynamic theories of irreversible processes. Such local equilibrium conditions may be expected to develop for certain kinetically rapid species and phases at sediment/water interfaces in fresh, estuarine and marine environments (see Stumm and Morgan, 1970, p. 67). In the box system, equilibrium is the time-invariant state while the stationary state is the corresponding state of open systems. Vertical gradients in the water mass and continuous exchange of enclosed water have not yet been investigated but will be in the near future.

RESULTS AND DISCUSSION

Phosphate

Phosphate in solution is mainly present as ion pairs with the major cations of sea water (Kester and Pytkowicz, 1967). In the sediment it is also found in organic matter, adsorbed on hydrous ferric oxides (Mortimer, 1941 and 1971; Mackereth, 1966; Stumm and Leckie, 1970), adsorbed on clay minerals (Chen, 1972) and as minerals, $e.g.$, fondelite, wavellite, strengite, struvite, vivianite and various forms of apatite.

Vivianite [$Fe_3(PO_4)_2 \cdot 8H_2O$] is formed during reduced conditions and is easily demonstrated in batch cultures of sulfate-reducing bacteria where an environment similar to that of reduced sediments is established. According to Nriagu (1972) "Vivianite is probably the most stable Fe(II) orthophosphate solid phase encountered in sedimentary environments. Accurate information about its solubility is therefore central to whether soils and sediments act as sinks or sources for phosphorus, a question of considerable interest in environmental geochemistry and in eutrophication abatement programs." Nriagu has in his paper given consideration to the formation of ion pairs and shown that the complex species constitute a significant fraction of the total dissolved Fe.

During oxidizing conditions, the sediment acts as a sink for phosphate ($e.g.$, Mortimer, 1942). This has also been verified during our *in situ* experiments where different amounts of phosphate have been added to closed oxidized systems (Schippel et al., 1973b). Not all of the phosphate added was taken up by the sediment. About 20% of the phosphate left in solution can be explained by the lack of free cations ($e.g.$, calcium and iron) since phosphate normally is fixed in the sediment by these ions.

During reducing conditions, phosphate is released. The rate of this release has been investigated in the same closed system as described above (see Hallberg et al., 1972; Schippel et al., 1973a). At the beginning of the redox turnover the reaction is very fast and probably of pseudo-zero order but changes subsequently into a first-order reaction (Figure 21.2). The factor influencing the release is assumed to be the activity of the sulfate-reducing bacteria. Their activity is governed by the amount of utilizable organic matter. The correlation between total amount of released phosphate and amount of added organic matter in the box experiments is shown in Figure 21.3. Phosphate

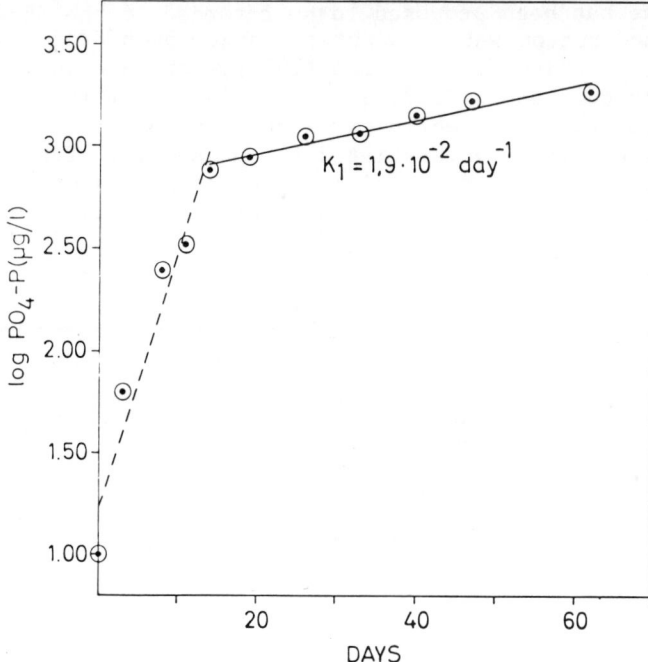

Figure 21.2 Log concentration of phosphate as a function of time. Continuous line represents a first-order reaction (Holm, unpublished results).

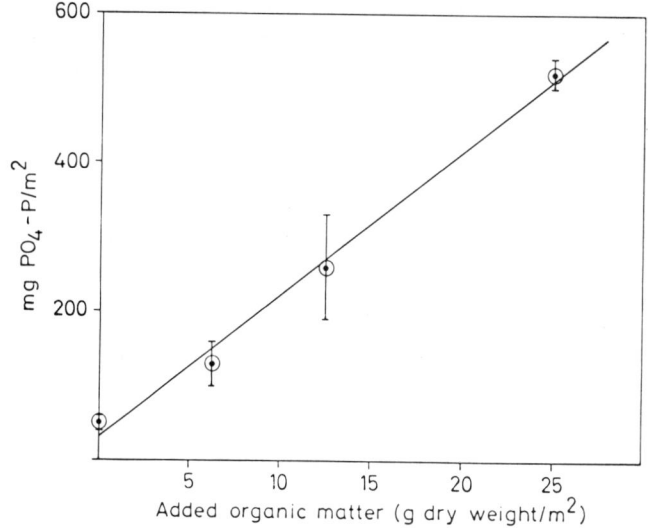

Figure 21.3 Correlation between released phosphate and amount of added organic matter (Schippel, personal communication).

release has been proposed to be retarded if the phosphate-enriched bottom water is not exchanged by hydrodynamic movements. According to Lee (1970), for instance, "it appears that the hydrodynamics of the system is often the rate-controlling step in exchange reactions." This apparently cannot be demonstrated for phosphate. Identical systems with different initial concentrations of phosphate revealed similar types of reactions and release almost identical amounts of phosphate (Figure 21.4).

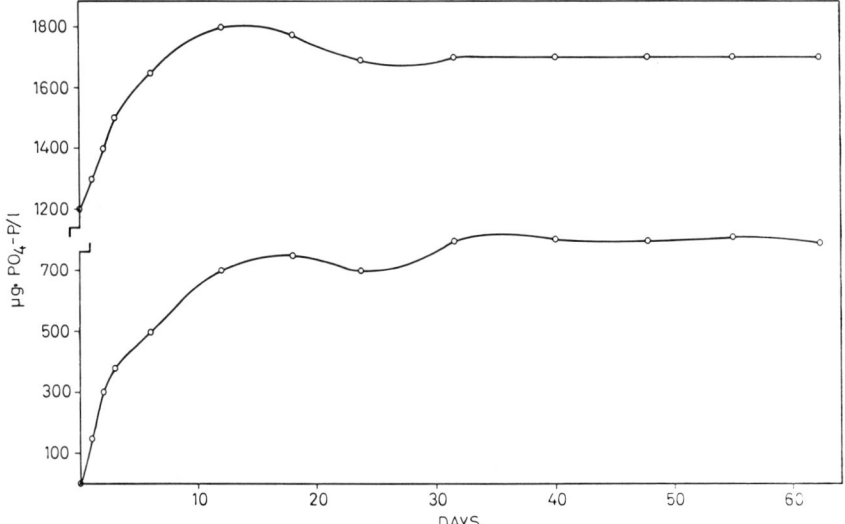

Figure 21.4 Phosphate release in two parallel runs, one with a PO_4 addition at the beginning of the experiment (Schippel et al., 1973b).

Ammonia

The major source of nitrogen in the sediment is the organic matter. During reduced conditions nitrogen is released from the sediment in the form of ammonia. The effect of organic matter on the ammonia release was studied in an experiment using four boxes. One of the boxes was run as a reference and to the other three boxes additions of organic matter were made. The added material consisted of freeze-dried *Cladophora* collected in the neighborhood of the experiment site. Figure 21.5 shows the results of this study. There is no indication of the influence of the addition of organic matter on the ammonia release in the different boxes. During the experiment approximately 10% (about 20 g dry wt/box in the top 5 cm) of the organic matter in the sediment was completely mineralized. The

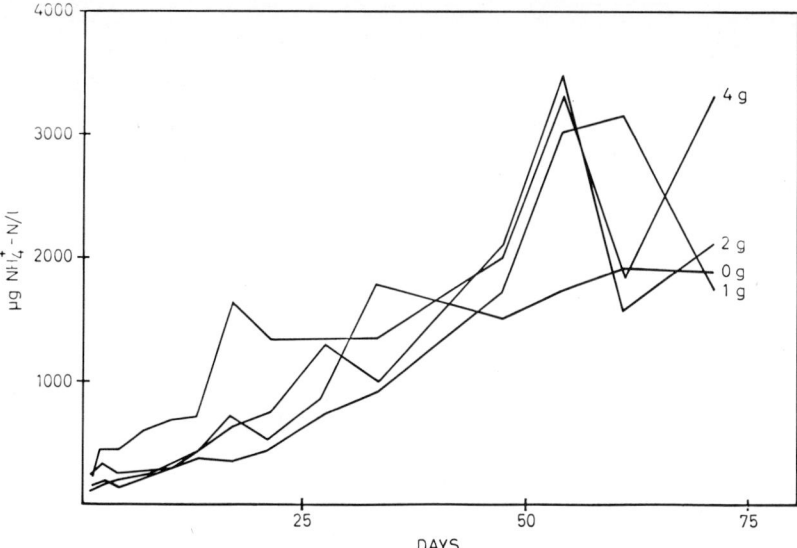

Figure 21.5 The course of ammonia release from the sediment in four experimental boxes with different additions of organic matter   (Anna-Greta Engvall, unpublished results).

lack of effect on the ammonia release in the experiment may be due to the fact that the additions were too small compared to the amount of organic matter mineralized. The original content of organic matter has to be enough for the biological degradation during the whole experiment. The small amounts added compared to the original content could scarcely raise the populations of microorganisms. The degradation products also depend on the composition of the organic matter available, and this was not analyzed in the present investigation.

A study of the effect of illumination on the ammonia release was performed in an experiment with two transparent boxes run parallel to two opaque boxes. The ammonia release in the opaque boxes started immediately, while in the transparent boxes the delay of ammonia release was 35-40 days. The delay of ammonia release was studied in more detail during a short-term experiment in duplicate transparent boxes. This experiment was carried out under good light conditions in the beginning of May. Water sampling from the boxes was performed every morning and evening for one week. During this time the oxygen concentration in the water phase showed a diurnal fluctuation (Figure 21.6).

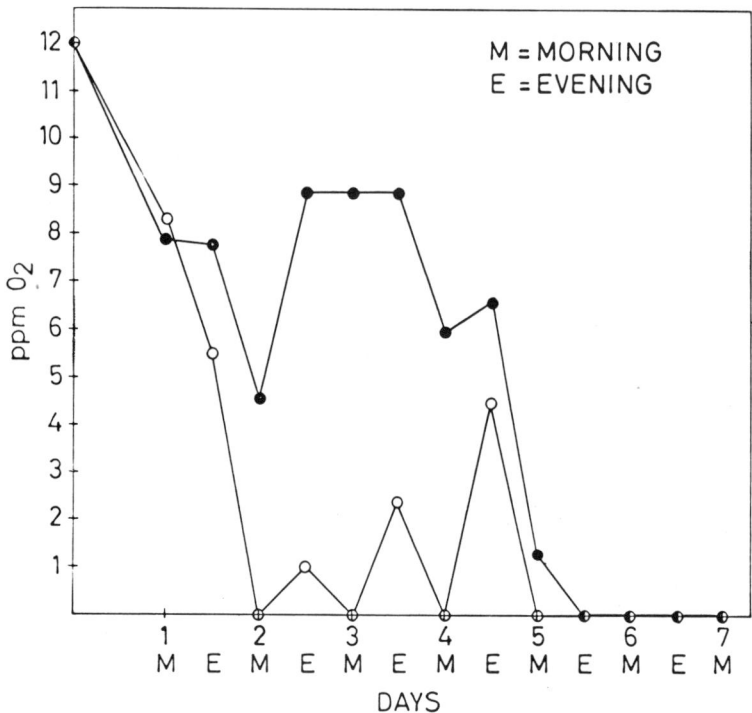

Figure 21.6 The diurnal oxygen variation in two boxes during the first week of an experiment (Lindstroem, personal communication).

The oxygen level was usually higher in the evening than in the morning due to photosynthetic activity. Toward the end of the week, the photosynthetic organisms could not maintain the oxygen concentration at a detectable level and reducing conditions appeared.

Sulfate-Sulfide

Sulfate is the prevailing sulfur source for the sulfate-reducing bacteria, though other compounds like thiosulfate, elemental sulfur and organic sulfur can be utilized. The decrease in sulfate concentration is a better measure than the increase in hydrogen sulfide concentration for determining the rate of sulfate reduction in marine sediments.

The kinetics of bacterial sulfate reduction have been investigated by several workers (*e.g.*, Postgate, 1951; Harrison and Thode, 1958; Kaplan, 1962). These studies showed that the rate of reduction was independent of sulfate concentration above ~ 10 mM. Nakai and Jensen (1964) demonstrated in their experiments first-order reaction kinetics. They calculated the rate constant, K_1, which varied between 1.91 and 2.40 x 10^{-2}/day. Saki (1972) cited in Goldhaber and Kaplan (1974) pointed out that the reaction also could be considered to be zero-order with a rate constant, K_0, which ranges from 0.33 to 0.54 mg S/day. Under this interpretation, K_0 becomes nonlinear at lower concentrations of sulfate, representing a change from zero-order to first-order kinetics, dependent on sulfate concentration. In a theoretical evaluation Rees (1973) states that the process of bacterial sulfate reduction is of first-order only when sulfate concentration is insufficient to saturate enzyme activation sites. At higher concentrations, a zero-order reaction governs the rate of sulfate reduction.

Bågander (unpublished results) has found in *in situ* experiments that the rate of sulfate reduction is a first-order reaction. The reaction can also be interpreted as being a zero-order reaction which changes to first-order at sulfate concentration of about 2 mM (Figure 21.7), which is in accordance with the theoretical calculations of Rees (1973).

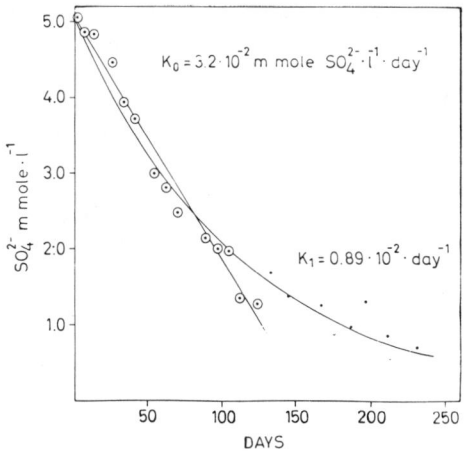

Figure 21.7 Sulfate reduction at the sediment-water interface in a box experiment. The first part of the chemical reaction may be interpreted as a zero-order reaction which changes to first-order at a concentration around 2 mM SO_4^{2-}. The rate constants K_0 (calculated on encircled observations) and K_1 (calculated on all observations) are given (Bågander, unpublished results).

The bacterial sulfur cycle is of special interest as
a factor regulating the pH of sediments. This has been
discussed in a paper by Ben-Yaakov (1973). Presley (1969,
cited in Ben-Yaakov, 1973) has found that the concentration
of dissolved CO_2 in pore waters of recent sediments may
reach the value of 60 m mole kg^{-1} or about 30 times the
total CO_2 concentration in the overlying waters. According
to Ben-Yaakov (1973) "the buffer capacity around the pH
of ocean and pore waters is very low and does not exceed
0.5 mmole kg^{-1} pH^{-1}. That is, an addition of 0.5 mmole
kg^{-1} of CO_2 will reduce the pH of seawater by more than a
pH unit. It is therefore surprising that the concentration
of CO_2 in pore water of marine sediments may reach values
of 60 mmole kg^{-1}, and yet show pH shifts less than one
unit. It is evident that the increase in CO_2 must be
counterbalanced by other processes that tend to increase
the pH of seawater (such as the production of ammonia and
increase in total alkalinity) so that the net balance
results in a fairly constant pH."

Ben-Yaakov (1973) has made a theoretical evaluation
which differs from the model of Thorstensson (1970) in that
it considers the roles of chemical equilibria with solid
phases and does not assume thermodynamic equilibria between
all the species. In one case when total dissolved H_2S
equals total amount of reduced sulfate, a pH of 6.9 is predicted. If on the other hand total dissolved H_2S equals
zero, the pH is predicted to be 8.3. In one of his box
experiments, Bågander (unpublished results) has been able
to verify the sulfur system as one of the important factors
regulating pH shifts in anoxic marine sediments (Figure 21.8).

The decrease in pH from $\sim$ 8 to $\sim$ 6.9 which takes place
at the beginning of the experiment may, however, not be
related to the sulfur system. It occurs before any changes
in the sulfur system of the enclosed water can be detected,
regardless of the oxygen concentration of the water. The change
in pH may be assumed to depend on changes in the carbonate
system but this has not yet been verified. When sulfate
reduction starts, H_2S accumulates in the water phase above
the sediment and the pH is buffered at around 6.9. Since
transparent boxes were used, the intensity of light in combination with the production of hydrogen sulfide created
favorable conditions for the sulfide-oxidizing bacteria,
Chromatium and *Chlorobium*. These bacteria are strict anaer-
obes and were growing on the inside walls of the boxes.
Chlorobium, having optimal growth at higher sulfide concentration than *Chromatium*, was found only in a narrow zone
close to the sediment. In their metabolism these bacteria
use sulfide which is oxidized to elemental sulfur and stored

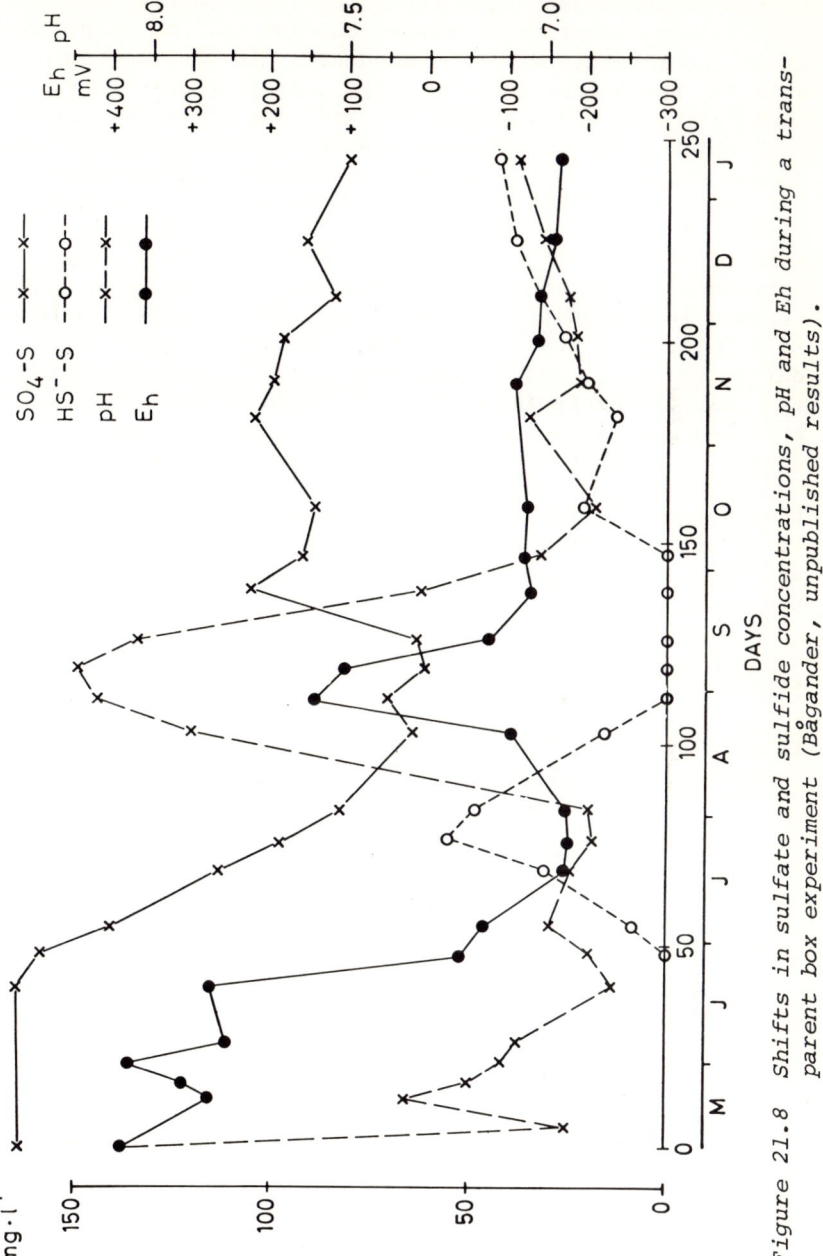

Figure 21.8 Shifts in sulfate and sulfide concentrations, pH and Eh during a transparent box experiment (Bågander, unpublished results).

inside the cell. The sulfide disappears and pH increases to 8.2 in accordance with the data by Ben-Yaakov.

Sulfate reduction continues but the sulfide is oxidized at a higher rate than it is produced. In the absence of sulfide, the bacteria now use the sulfur stored in their cells to yield energy by oxidizing the sulfur to sulfate which gives rise to the observed increase in sulfate concentration. As the metabolism of the bacteria now is directed to using their storage of sulfur, the concentration of sulfide now increases. Again, at a sufficient concentration, the bacteria starts to consume sulfide. This interaction continues throughout the experiment but becomes less significant when the light conditions become less favorable because: a) the insolation time decreases during the fall, and b) by covering of the walls inside the box, the bacteria inhibit the light transmittance. When the sulfide production starts to dominate the system again, pH rapidly decreases to a value of about 6.9.

We have found our methods useful for studying different biogeochemical processes at the sediment/water interface. Other elements that have been studied are silicon, oxygen and iron. Diffusion processes and chelating effects upon the heavy metal cycles are some of the processes which have also been looked into.

REFERENCES

Ben-Yaakov, S., 1973: pH Buffering of pore water of recent anoxic marine sediments. *Limnol. Oceanogr.* 18, 86-93.

Bågander, L. E.: Chemical dynamics of Baltic sediments-- bacterial sulfate reduction (Unpublished results).

Chen, Yi-Shon, 1972: Phosphate interaction with aluminum oxide, kaolinite and sediments. Doctoral dissertation in Engineering (unpublished), Harvard University, Chapter 3.

Engvall, Anna-Greta: Nitrogen exchange at the sediment-water interface (Unpublished results).

Goldhaber, M. B. and I. R. Kaplan, 1974: The sulfur cycle. *The Sea, Marine Chemistry.* 5, 569-655.

Hallberg, R. O., L. E. Bågander, Anna-Greta Engvall, F. A. Schippel, 1972: Method for studying geochemistry of the sediment water interface, Ambio No. 2, 71-72.

Harrison, A. G., H. G. Thode, 1958: Mechanism of the bacterial reduction of sulphate from isotope fractionation studies. *Trans. Faraday Soc.* 54, 84-92.

Holm, N. G.: Chemical dynamics of phosphate and silicon in Baltic sediments (Unpublished results).

Kaplan, I. R., 1962: Sulfur isotope fractionations during microbiological transformations in the laboratory and in marine sediments. Ph. D. thesis, Univ. of California, 213 pp.

Kester, D. R., R. M. Pytkowicz, 1967: Determination of the apparent dissociation constants of phosphoric acid in seawater. *Limnol. Oceanogr.* 12, 243-252.

Lee, F. G., 1970: Factors affecting the transfer of material between water and sediments. Literature Review No. 1. Eutrophication information program, Water Resources Center, Univ. Of Wisconsin, Madison, July 1970.

Mackereth, F. J. H., 1966: Some chemical observations on post-glacial lake sediments. *Phil. Trans. Roy. Soc.*, Sect. B. No. 250, 165-220.

Mortimer, C. H., 1941: The exchange of dissolved substances between mud and water in lakes. *J. Ecol.* 29, 280-329.

Mortimer, C. H., 1942: The exchange of dissolved substances between mud and water in lakes. *J. Ecol.* 30, 147-201.

Mortimer, C. H., 1971: Chemical exchanges between sediments and water in the Great Lakes--speculations on probable regulatory mechanisms. *Limnol. Oceanogr.* 16, 387-404.

Nakai, N., M. L. Jensen, 1964: The kinetic isotope effect in bacterial reduction and oxidation of sulfur. *Geochim. Cosmochim. Acta* 28, 1893-1912.

Nriagu, J. O., 1972: Stability of vivianite and ion pair formation in the system $Fe_3(PO_4)_2 - H_3PO - H_2O$. *Geochim. Cosmochim. Acta* 36, 459-470.

Postgate, J. R., 1951: The reduction of sulphur compounds by *Desulfovibrio desulphuricans*. *J. Gen. Microbiol.* 5, 725-738.

Rees, C. E., 1973: A steady state model for sulphur isotope fractionation in bacterial reduction processes. *Geochim. Cosmochim. Acta* 37, 1141-1162.

Schippel, F. A., L. E. Bågander, R. O. Hallberg, 1973a: An apparatus for subaquatic *in situ* measurements of sediment dynamics. *Sthlm. Contr. Geol.* 24, 6.

Schippel, F. A., R. O. Hallberg, S. Odén, 1973b: Phosphate exchange at the sediment-water interface. *Oikos*, Suppl. 15, 64-67.

Stumm, W., J. O. Leckie, 1970: Phosphate exchange with sediments: its role in the productivity of surface waters. *Advances in Water Pollution Research 2* (New York: Pergamon Press). III 26/1-26/16.

Stumm, W., J.J. Morgan, 1970: *Aquatic Chemistry* (New York: Wiley-Interscience). p. 67.

Thorstensson, D. C., 1970: Equilibria distribution of small organic molecules in natural waters. *Geochim. Cosmochim. Acta* 34, 745-770.

CHAPTER 22

SULFUR AND CARBON ISOTOPIC EVIDENCE
FOR BIOGEOCHEMICAL PROCESSES IN
THE DEAD SEA ECOSYSTEM

A. NISSENBAUM

Isotope Department, Weizmann Institute of Science,
Rehovot, Israel

I. R. KAPLAN

Department of Geology and Institute of Geophysics
and Planetary Physics, University of California,
Los Angeles, California 90024

INTRODUCTION

The area around the Dead Sea has traditionally been associated with sulfur since biblical times (..."Then the Lord rained upon Sodom and upon Gomorrah brimstone and fire from heaven," Genesis XIX, 24). Later historians such as Diodoros Sicilus (*ca*. 50 A.D.) and Strabo (63 B.C. to 24 A.D.) mention the episodal release of stinky gas from Dead Sea water. This gas caused tarnishing of silver and copperware. The appearance of this gas, which is presumably hydrogen sulfide, was correlated with the emergence to the surface of the lake of huge asphalt blocks.

More recently, it was noted by Gay-Lussac (1819) and later confirmed by Bentor (1961) that the Dead Sea is relatively depleted in sulfate and carbonate as compared with both Jordan River (the major water supply to the lake) and ocean water. Sulfate comprises 0.25% of the anion sum in the Dead Sea water as compared with 20% in the Jordan River

and 12% in average ocean water. Bicarbonate is 0.11% of
the anion sum in Dead Sea water, but 27% in the Jordan River
and 0.65% in the ocean. Both Gay-Lussac (1819) and Bentor
(1961) suggested that those anions are removed from the
water column by precipitation as calcium sulfate and calcium carbonate. According to Bentor's (1961) calculation,
gypsum ($CaSO_4 \cdot 2H_2O$) should comprise 39% of the chemical
precipitates in the lake. Neev and Emery (1967) found that
indeed gypsum crystallizes at certain periods from the
upper water column, but that it is nearly absent in sediments from the deep part of the lake. This was attributed
to reduction by sulfate-reducing bacteria. Lerman (1967)
calculated the rate of sulfate reduction in the Southern
Basin of the Dead Sea and found it to be comparable to
rates observed in sediments from the basins off southern
California.

The carbonate cycle of the Dead Sea has not been investigated in any detail until very recently. Friedman (1965)
discussed the origin of aragonite in the Dead Sea. Ben-Yaakov and Sass (private communication) are undertaking
laboratory studies on the behavior of carbonate in brines
under simulated Dead Sea conditions.

The present investigation was undertaken in order to
determine biogeochemical influences on the geochemistry
of the Dead Sea, using the natural abundance of stable isotopes as tracers for such processes.

The Dead Sea Environment

The Dead Sea occupies the deepest part of the Dead
Sea-Arava Rift Valley and, with its surface at 400 m below
sea level, it represents the lowermost point on the surface
of the earth. It has an area of 1000 km^2 and volume of about
143 km^3. The Dead Sea is physiographically divided into a
shallow southern basin with maximum depth of 8 m and the
much deeper northern basin which has a maximum depth of
400 m. The northern basin contains about 95% of the lake
volume. The two basins are separated by an east-west trending sill (at mean depth of 3 m) stretching from the Lisan
peninsula to the western coast.

Water is introduced into the Dead Sea through the Jordan River. Some small near-shore seepages and possibly
some underwater springs also contribute salts. The Dead
Sea has no outlet for its water, and the loss of water occurs
only by evaporation.

The Dead Sea brine is characterized by a unique chemistry. Its total dissolved salt content is around 310 g/l

Phosphorus, Sulfur and Selenium Cycling 311

(23% w/v), with specific density of around 1.227 ± 0.07. The major anion is Cl (*ca.* 220 g/l) followed by Br (*ca.* 5 g/l). The major cation is Mg (44 g/l) followed by Na (38 g/l), Ca (17 g/l) and K (8 g/l). The Dead Sea is also highly enriched in Mn, Cu, Zn and Pb (Nissenbaum, 1975a).

The water column in the northern basin is divided into two parts, the upper water mass (UWM) between 0-80 m, and the lower water mass (LWM) below 80 m. The transition between the two water bodies is gradual and extends over a 20-40 m zone. The UWM is somewhat less saline that the LWM (329 g/l *vs* 332 g/l). The UWM is microaerobic and contains some oxygen, at least to 40 m depth. The LWM is completely anoxic and contains 0.5-1.0 mg H_2S/l.

The sediments in the northern basin are composed of about equal parts of chemical precipitates (mostly aragonite) and of detrital clay minerals. Gypsum is found in the shallow sediments only. The organic carbon content is around 0.5% (Nissenbaum *et al.*, 1972a).

The biota of the lake water column is composed of several species of obligate halophylic bacteria and a single species of the green alga, *Dunaliella* (Nissenbaum 1975b).

MATERIALS AND METHODS

Sampling locations for sediments and water are given in Figure 22.1. A concise description of the sediments used for this study is given in Nissenbaum *et al.*(1972a). Water samples for analysis were collected with an all-plastic 4-liter Van Dorn-type water sampler. The waters were transferred on deck into polyethylene bottles and transported into the laboratory for analysis. pH was measured on deck immediately after collection. Dissolved H_2S was also measured on deck by flushing the H_2S out of the acidified water sample into a silver nitrate-containing trap. Bicarbonate was measured by titration and sulfate by precipitation as $BaSO_4$.

The sediments were collected by a grab sampler (provided by Dr. D. Neev, Geological Survey of Israel). After collection and transportation into the laboratory, the samples were kept refrigerated at all times prior to analysis.

Isotopic analyses for dissolved bicarbonate were made on $BaCO_3$, which was obtained by acidifying the water, flushing out the CO_2 with nitrogen and absorbing the CO_2 in a series of traps containing $Ba(OH)_2$. The isotopic analysis of the sulfate was done on $BaSO_4$, which was precipitated from the water. The detailed procedures are the same as described by Kaplan *et al.* (1963).

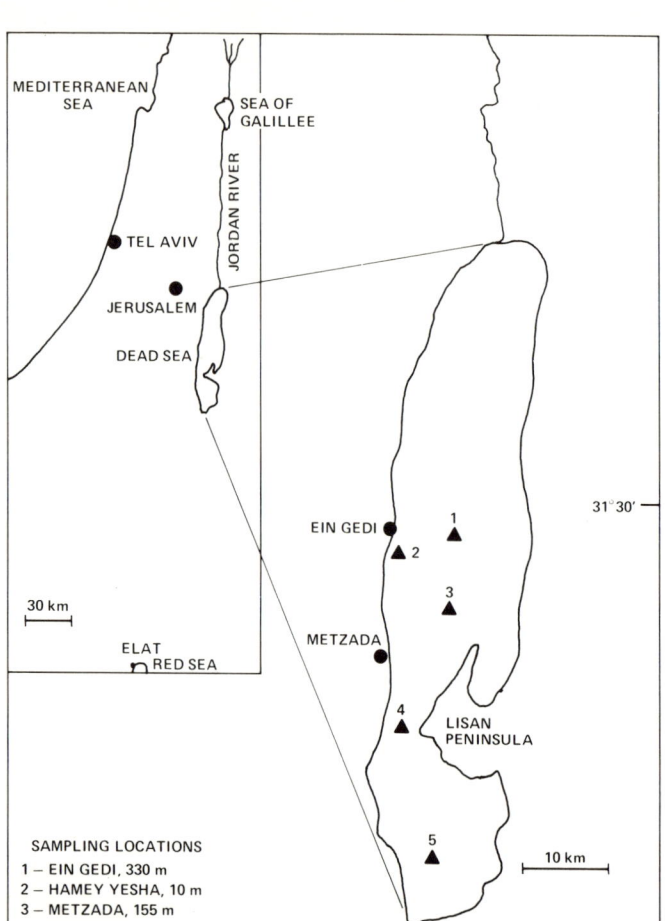

Figure 22.1 Location map of sampling stations in the Dead Sea.

Analysis of δC^{13} in sediment carbonate was made by acidification with H_3PO_4 following the method of McCrea (1950). The quantity of evolved CO_2 was measured and used in the calculation of the amount of carbonate present in the sediment. The sediments were fractionated according to grain size, using standard techniques of soil analysis. Four fractions were obtained (> 74μ, 20-74μ, 2-20μ and <2μ) and each was separately analyzed. Knowledge of the relative distribution of each grain size group and its isotopic composition allowed the calculation of the isotopic composition of the whole sample. Analysis for sulfur species in the sediment followed the procedure outlined by Kaplan et al. (1963).

Isotopic measurements were performed on a dual collector, dual inlet, Nuclide mass spectrometer. The data is expressed in the δ notation:

$$\delta S^{34}, C^{13}, O^{18} \; (\permil) = \left[\frac{R_{sample} - R_{standard}}{R_{standard}} - 1 \right] \times 1000$$

where R is the S^{34}/S^{32}, C^{13}/C^{12} or O^{18}/O^{16} ratio. The standards are the Chicago PDB for carbon and oxygen and the Canyon Diablo meteorite for sulfur.

RESULTS AND DISCUSSION

The analytical and isotopic data on the water samples are given in Table 22.1. The distribution of sulfur isotopes in the sediments is given in Table 22.2, and the data on carbon and oxygen isotopic composition of the sediments are given in Table 22.3.

Sulfur Cycle

The concentration of sulfur in the water column shows a decrease from around 540 mg/l at the surface to 405 mg/l at 300 m (Figure 22.2). Occasional low values for sulfate

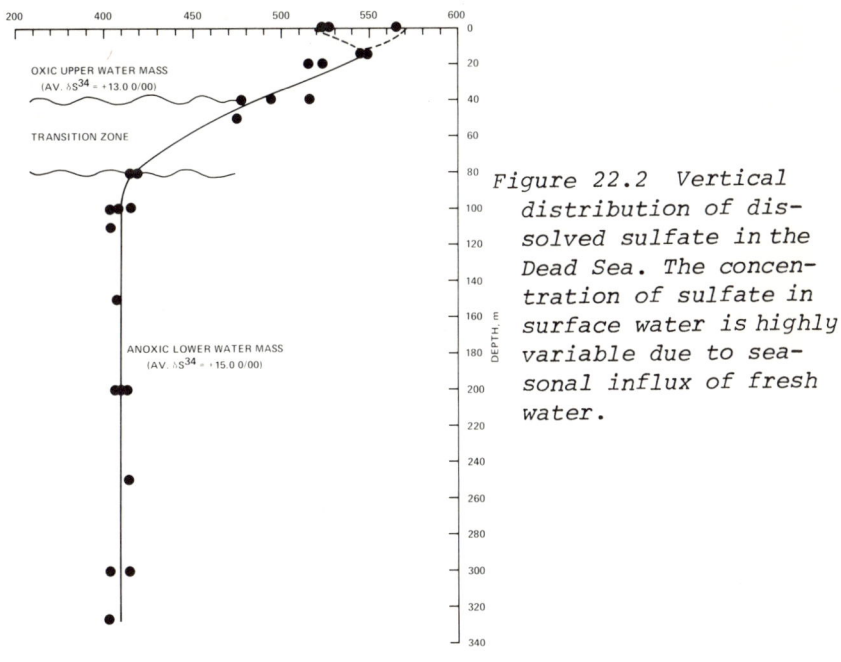

Figure 22.2 Vertical distribution of dissolved sulfate in the Dead Sea. The concentration of sulfate in surface water is highly variable due to seasonal influx of fresh water.

Table 22.1
Concentration and Isotopic Composition of Dissolved Sulfate
and Bicarbonate in Dead Sea Water
[Concentrations given in mg/l; δC^{13} and δS^{34} are reported relative to the PDB and Meteoritic sulfur standards]

Profile	Water Depth (meters)	Cl	SO_4	δS^{34}‰	HCO_3	δC^{13}‰	pH
Ein Gedi	0	194300	529	+13.6	192	−9.7	6.5
Ein Gedi	15	207400	543	+13.4	187	−6.4	6.4
Ein Gedi	40	214400	493	+15.9	178	−15.4	6.3
Ein Gedi	80	220500	416	+15.8	168	−11.3	—
Ein Gedi	100	220500	409	+15.2	165	−6.9	6.1
Ein Gedi	200	222300	409	+15.7	174	−7.8	6.0
Ein Gedi	300	222300	402	(−)	182	−9.9	5.9
Metzada	0	195100	536	+13.6	183	−9.9	6.5
Metzada	15	194200	543	+12.6	187	−9.5	—
Metzada	40	214400	479	(−)	175	−12.7	6.4
Metzada	80	220500	416	+14.3	173	−9.1	6.1
Metzada	110	221300	402	+15.6	166	−9.2	6.0
Metzada	155	222200	409	+15.4	160	−10.4	—
Lisan Straits	3	207300	528	+13.2	71	−10.5	
Hamey Yesha	0.5	205600	543	+12.4	85	−8.3	6.7
	10	208200	528	+10.3	73	−8.7	6.7
Southern Basin	8	209100	480	+12.8	123		6.6

Table 22.2
Distribution of Sulfur Isotopes in Dead Sea Sediments
[Results reported in δS^{34} ‰]

Location and Depth	Pore Water Sulfate	Pore Water H_2S	Acid Volatile Sulfide	Sulfate	Organic Sulfur
Ein Gedi, 330 m	+20.6	-16.3	-16.3	+12.7	-19.6
Metzada, 165 m	+10.0	-12.7	-15.9	+13.1	-19.6
Hamey Yesha, 10 m	+10.1	N.D.	N.D.	+15.6	N.D.
Southern Basin, 8 m	+11.1	N.D.	N.D.	-4.0	N.D.
Lisan Strait, 3 m	N.D.	N.D.	N.D.	+15.8	N.D.
Sediment Trap, opp. Metzada, 80 m	N.D.	N.D.	N.D.	+14.8	N.D.

N.D.=Not Detected.

Table 22.3

Distribution of Carbon and Oxygen Isotopes in Different Size Fractions of Carbonate from Dead Sea Sediments (Results are given relative to the PDB standard)

Location	Fraction Size	δC^{13} (‰)	δO^{18} (‰)	Calculated Average for Total Sample	
				δC^{13} (‰)	δO^{18} (‰)
Ein Gedi, (330 m)	>74μ	-5.5	-5.2	-3.4	-4.2
	20-74μ	-3.5	-5.4		
	2-20μ	-3.4	-3.1		
	<2μ	-3.4	-3.6		
Metzada, (120 m)	>74μ	-2.6	-3.2	-0.3	+0.2
	20-74μ	-3.5	-4.4		
	2-20μ	-0.5	-0.3		
	<2μ	+0.8	+1.7		
Southern Basin, 8 m	>74μ	-3.4	-3.8	Could not be calculated due to incomplete isotopic information	
	20-74μ	-6.0	-8.0		
	2-20μ	N.D.	-2.5		
	<2μ	-2.1	-0.1		
Hamey Yesha (10 m)	>74μ	-4.3	-4.5	-1.9	-2.0
	20-74μ	-2.9	-3.4		
	2-20μ	-2.0	+2.4		
	<2μ	-0.1	+0.5		

N.D.=Not Determined.

in the surface samples are due to dilution of Dead Sea water by a superficial layer of fresh water introduced by floods, which can also be observed from the lowering of chloride ion concentration (Table 22.1). The sulfate decreases gradually through the UWM and the intermediate zone, and seems to have fairly regular depth distribution in the LWM. Concomitant with the change in concentration of dissolved sulfate, we observe an increase of the δS^{34} values from $\simeq$ +12.0‰ in the UWM to $\simeq$ +15.0‰ in the LWM. The isotopic composition of sulfate in samples from the Lisan Straits and the Southern Basin are similar to the values for the UWM from the Northern Basin (Tables 22.1 and 22.2).

The source of sulfate in the UWM could be water brought in through the Jordan River and/or sulfate brought in by runoff. The observed isotopic values would than be dependent on the relative contribution from each source which have the values given in Table 22.4. Another possible contribution could be the diffusion and oxidation of H_2S from the LWM (see below).

The source of sulfate in the LWM is more difficult to explain. The observed average value of δS^{34} of +15‰ could conceivably represent depletion of S^{32} by microbial sulfate reducers and removal of the released H_2S either by the sediment or by diffusion into the UWM, and subsequent enrichment of the remaining sulfate in S^{34}. Because of lack of information on the rate of sulfate reduction and the age of the LWM, it is not possible to use a kinetic model to calculate the enrichment factor.

It has usually been assumed that the salts in the Dead Sea are derived from several sources, namely 1) the Jordan River 2) submarine springs flowing into the lake, and 3) the residual brines of an ancient lake (Bentor 1961; Neev and Emery 1967; Starinsky 1974). Most of the more recent reports tend to support sources 2 and 3 as the major sulfate contributors. Although the isotopic composition of these sources is not known, some information can be gleaned by studying the δS^{34} in springs and sediments in the Jordan Valley (Table 22.3). All the fresh water springs are depleted in S^{34} compared with the Dead Sea. (Ein Bokek is especially depleted, probably due to oxidation of pyrite in the Ein Bokek oil shales near its source.) Even the saline springs, except the Moomilla spring (with δS^{34} = +17‰), apparently are not suitable as sources based on isotopic evidence. The Moomilla Borehole represents a very small seepage on the flank of the Sodom Salt Mountain, and its sulfate content is probably derived from the dissolution of the anhydrite.

Table 22.4
Concentration and Isotopic Composition of Sulfate in Sources
which may Contribute Water to the Dead Sea

Water Source	SO_4 (mg/l)	δS^{34} (‰) SO_4	Comments
Jordan River	120 to 150	+ 6.5 to + 9.0	The major source of sulfate
Rainwater	9 to 12	+ 3.5 to + 8.5	Concentration is seasonally and annually variable.
Mayan David	31 to 35	+8.4	Small fresh water spring
Nahal Arugot	53 to 96	+ 7.5 to +10.9	Small fresh water spring
Ein Bokek	351 to 353	- 0.8 to + 1.3	Small fresh water spring
Ein Feshka	71	+11	Group of moderately saline springs
Zohar Springs	670 to 714	+ 9.7 to +11.2	Group of highly saline springs
Mommilla Borehole	1390	15.2	Seepage near the shore. Tentatively assumed to be similar to submarine springs.
Dead Sea Upper Water Mass	560 to 470	+12.4 to +14.1	Average δS^{34} at about +13.0‰
Dead Sea Lower Water Mass	409 to 416	+14.1 to +15.6	Average δS^{34} at about +15.0‰

At present, the LWM is anoxic and contains small amounts of hydrogen sulfide (Table 22.5). Contrary to the report by Neev and Emery (1967), we have not been able to detect sulfide in the UWM The concentration of H_2S found by us in the LWM is about one order of magnitude smaller than those reported by Richards (1965) for the Black Sea. It is possible that the low concentration of dissolved H_2S is due to lack of suitable hydrogen acceptors in the Dead Sea. There is no doubt, however, that the H_2S is produced by biological processes. The isotopic separation between the sulfate and dissolved sulfide is in the range of that observed during the biological sulfate reduction in nature (Kaplan *et al.*, 1963). We also believe that the difference in δS^{34} between dissolved H_2S in the water and the H_2S dissolved in the pore waters of the sediment (Tables 22.2 and 22.5) indicates a slower rate of reduction of sulfate in the water column. Lerman (1967) has suggested that the H_2S may be partly fixed by iron and may partly escape into the atmosphere. However, as we have detected no sulfide either in the UWM or in the air above the Dead Sea, it seems that the sulfide which diffuses to the intermediate zone is oxidized to sulfate. We have no evidence to indicate whether this is a biological process or whether it occurs by inorganic oxidation with dissolved oxygen.

A process which removes sulfate from the UWM is the precipitation of gypsum, which occurs mostly in the summer (Neev and Emery, 1967). From mass balance calculations by Bentor (1961) and calculations of gypsum solubility in Dead Sea brine by Starinsky (1974), it is apparent that the Dead Sea is supersaturated with regard to calcium sulfate minerals. Isotopic analyses of gypsum from the sediments of the shallow part of the lake (Table 22.2), shows that the sulfate is slightly enriched in S^{34} compared with the lake water. An exception may be the Southern sample, but we believe that the low δS^{34} value for this sample is due to oxidation of sulfide, either *in situ* or during sample handling. Analysis of very recently precipitated gypsum from a sediment trap at 80 m depth gave δS^{34} = +14.8‰ . If we assume the gypsum to form from surface waters, then the separation of SO_4-S isotopes between the water and gypsum is about 2.5‰, which is the same as the value reported by Holser and Kaplan (1966) for the Laguna Ojo de Liebre in Baja California.

Although gypsum is precipitated from the water column, it is almost entirely absent from the deeper anaerobic sediment. The small quantities observed in these deep sediments, are slightly depleted in S^{34} in comparison with the

Table 22.5
Coexisting Sulfate and Sulfide in Dead Sea Water

Location	Depth (m)	SO_4 (mg/l)	H_2S mg/l	$\delta S^{34}(SO_4)$	$\delta S^{34}(H_2S)$	$\alpha(SO_4-H_2S)$
Ein Gedi	100	402	0.35	+13.7‰	−21.6‰	1.035
Ein Gedi	200	409	0.23	+14.2‰	−21.7‰	1.036
Ein Gedi	300	416	0.56	+14.1‰	−21.6‰	1.036
Metzada	120	409	0.31	+15.0‰	−19.6‰	1.035

lake water. The isotopic evidence suggests that sulfide production is entirely biological. The concentration of sulfate in the pore water is very small in comparison with the overlying water. The percentage of dissolved sulfate in the pore water relative to the overlying water are EG: 8.0%; M: 7.8%; S: 11.5% and HY: 8%. The δS^{34} of the pore water sulfate ranges from +10 to +20‰. The isotopically lighter values are rather unusual, as we would expect sulfate reduction to enrich the residual sulfate in δS^{34} (Nissenbaum et al., 1972b), as is observed for the EG sample. Kaplan et al. (1963) described cases from the Santa Monica and Santa Barbara basins where the lower δS^{34} content of pore water sulfate was ascribed to oxidation of sulfide after collection. Thode et al. (1960) described similar phenomena from the Orinoco River Delta and explained it as the results of *in situ* oxidation.

The sulfide which forms by the reduction of sulfate may become fixed by the sediment as iron sulfide, or remain dissolved in the pore water. We have found sulfide only in pore waters of the deep anaerobic sediment, even though interstitial sulfate is exhausted in the shallow sediments. Acid volatile sulfides, which represent poorly crystalline iron sulfides, were also found only in the deep samples, and in small quantities (EG: 0.022% by weight, M: 0.042% by weight). Isotopically, both pore water H_2S and the iron sulfide have the same composition in EG, whereas the M samples show acid-volatile sulfide to be slightly enriched in S^{34}. The isotopic fractionation factors between sulfate and sulfide are in the range of 1.0025 to 1.0029 for the sediment (Table 22.2) and 1.0035 for the water column (Table 22.5); these values are similar to those found in the basins off southern California (Kaplan et al., 1963).

We have not been able to detect any pyrite in the Dead Sea sediments, except as infrequent coating of twigs in very near-shore deposits. As we have not been able to detect elemental sulfur in Dead Sea sediments, we believe that the absence of pyrite is due to the deficiency of this sulfur species as suggested for Lake Mendota (Nriagu, 1968) and some sediment layers in the Black Sea (Berner, 1974).

The isotope data indicate that the organisms do control the critical steps in the sulfur cycle in the Dead Sea in the same way as has been demonstrated for other marine basins or meromictic lakes. The exact nature of the sulfate-reducing organisms is not known. Nissenbaum (1975b) found anaerobic Clostridium-like sulfate reducers in sediment from 330 m depth.

Due to lack of quantitative data, it is presently impossible to obtain information on the role of microorganisms in the sulfur cycle or the flux rates involved. In Figure 22.3, we have attempted to describe the major processes pertaining to the sulfur transformation in the lake and its sediments.

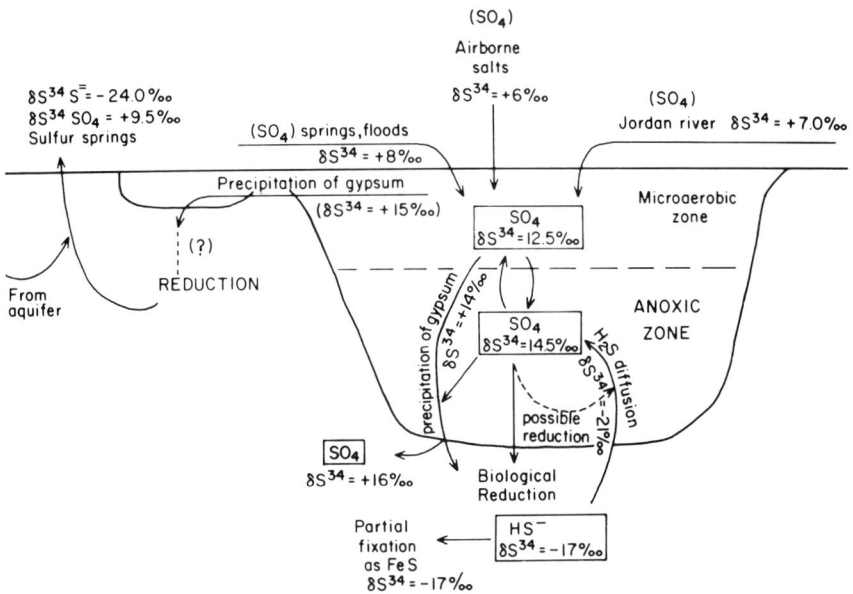

Figure 22.3 Tentative sulfur cycle for the Dead Sea showing δS^{34} data of various components.

Carbon Cycle

The concentration of dissolved bicarbonate in Dead Sea water ranges from 70 to 190 mg/l. These results were obtained using the classical oceanographic titration techniques. It is known, however, that concentrated brines respond differently than seawater during the titration. The reasons for this are partly technical (high liquid junction potential on the electrode) and partly chemical, as for example, in the case of pH change caused by hydrolysis of alkaline earth complexes. Neev and Emery (1967) reported that the extraction of total dissolved CO_2 by addition of acid yields CO_2 values which are only one-third to one-fourth of the amount obtained by titrimetry. We have also observed the same phenomenon, and thus we believe that the data reported in the literature and in Table 22.1 may be too high. The Dead Sea is therefore highly impoverished in dissolved carbonate species, but the low accuracy of the analytical method used prevents discussion of the temporal and spatial distribution of the carbonates. It is therefore conceivable that a low activity of HCO_3^- may in part limit photosynthetic production. The concentration of dissolved organic carbon in the water has been reported by Neev and Emery (1967) to range from 8.3 mg/l in the surface to 4.2 mg/l at 75 m and 7.9 mg/l at 130 m depth.

The δC^{13} of the dissolved carbonate is given in Table 22.1. The results range from -6.5 to -15.4‰. The highest depletion of δC^{13} is at the 40 m depth. The dissolved carbonates are much lighter than would be expected if equilibrium with atmospheric CO_2 were in effect. We suggest that the dissolved carbonates represent partial re-equilibration of bicarbonate introduced by the Jordan River ($\delta C^{13} = -13‰$) with atmospheric CO_2. Bicarbonate derived from oxidation of dissolved or suspended organic matter, which has a δC^{13} value of -24‰, seems to be of limited importance in controlling δC^{13} in the water column, except in the interface between the anoxic and oxic zones at 40 m depth.

The Dead Sea is probably saturated with regard to both calcite and aragonite (Sass and Ben-Yaakov, private communication). The chemical precipitation of aragonite from Dead Sea water accompanied by "whitening" of the water has been described (Neev and Emery, 1967; Friedman, 1965). In addition to aragonite, Dead Sea sediments contain some calcite. Friedman (1965) found that in near-shore sediments, the aragonite is enriched in C^{13} and O^{18} compared to the calcite. Friedman (1965) suggested that while the aragonite is a chemical precipitate from sea water, the calcite is formed by bacterial oxidation of organic matter with gypsum, a process which would preferentially concentrate the light isotope in the calcite. However, the samples which Friedman used for calcite and aragonite determinations were collected at the lake edge, and probably do not represent processes presently occurring within the lake. Results similar to those of Friedman (1965) were obtained by Katz, Kolodny, Nissenbaum and Zak (in preparation) for calcite-aragonite pairs from the Pleistocene Lisan Lake, which is the precursor of the present Dead Sea.

The data in Table 22.3 illustrate the isotopic inhomogeneity of the Dead Sea sediments. With increasing grain size, the carbonate becomes depleted in C^{13} and O^{18}. Microscopic observation indicates that the coarser fractions contain a lot of detritus. Some of the δC^{13} values obtained

for suspended matter in the Jordan River are rather close
to the values for the larger than 20µ size fractions. The
smaller size fractions presumably represent more significant contributions of authigenic carbonate, and the δO^{18}
values are closer to those we would expect from direct
carbonate precipitation from Dead Sea water. We have found
no evidence to support Friedman's (1965) suggestion that
some of the calcite may be due to respired organogenic CO_2.

We therefore suggest that biogeochemical processes play
a minor role in the geochemistry of carbon in the Dead Sea.
It is possible that the removal of CO_2 by photosynthesis
may play some, but probably unimportant, role in triggering
massive carbonate deposition. The disappearance of gypsum
from the bottom sediments may be partly responsible for
precipitation of authigenic carbonate in the sediment.

ACKNOWLEDGMENTS

The authors wish to thank Dr. D. Neev and Mr. D. Argas
(Geological Survey of Israel) for their invaluable help in
sample collection. Dr. I. Shnerb (Dead Sea Potash Works)
provided the analysis of sulfate and carbonate. One of us
(A. Nissenbaum) wishes to thank the U.S. National Research
Council and NASA, Ames Research Center, for providing support
during preparation of this manuscript.

Publication No: #1460   Institute of Geophysics & Planetary
                          Physics, University of California
                          at Los Angeles, California 90024

REFERENCES

Bentor, Y. K., 1961: Some geochemical aspects of the Dead
  Sea. *Geochim. Cosmochim. Acta* 25, 239-260.

Berner, R. A., 1974: Iron sulfides in Pleistocene Black
  Sea sediments and their Paleo-oceanographic significance.
  In: *The Black Sea: Geology, Chemistry and Biology*,
  E. T. Degens and D. S. Ross, eds. (Tulsa, Okla.) *Am.
  Assoc. Pet. Geol. Memoir* 10, 524-531.

Friedman, G. N., 1965: On the origin of aragonite in the
  Dead Sea. *Israel J. Earth Sci.* 14, 79-85.

Gay-Lussac, J. L., 1819: Analyse de l'Eeau de la Mer Morte.
  Ann. de Chimie et de Physique 11, 195-197.

Holser, W. T. and I. R. Kaplan, 1966: Isotope geochemistry
  of sedimentary sulfates. *Chem. Geol.* 1, 93-135.

Kaplan, R. R., K. O. Emery and S. C. Rittenberg, 1963:
The distribution and isotopic abundance of sulphur in
recent marine sediments off Southern California. *Geochim.
Cosmochim. Acta* 27, 297-331.

Lerman, A., 1967: Model of chemical evolution of a chloride
lake--The Dead Sea. *Geochim. Cosmochim. Acta* 31, 2309-
2330.

McCrea, J. M., 1950: On the isotopic chemistry of carbonates and a Paleotemperature scale. *J. Chem. Phys.* 18.
849-857.

Neev, D. and K. O. Emery, 1967: The Dead Sea. *Geol.
Survey of Israel Bull.* 41, 147 pp.

Nissenbaum, A., M. J. Baedecker and I. R. Kaplan, 1972a:
Organic geochemistry of Dead Sea sediments. *Geochim.
Cosmochim. Acta* 36, 709-727.

Nissenbaum A., B. J. Presley and I. R. Kaplan, 1972b:
Early diagenesis in a reducing fjord, Saanich Inlet,
B.C. - I. Chemical and isotopic changes in major components of interstitial water. *Geochim. Cosmochim. Acta*
36, 1007-1027.

Nissenbaum A., 1975a: Minor and trace metals in Dead Sea
water: A preliminary report. (Unpublished report -
Geoisotope Group, Weizmann Institute of Science, Rehovot,
Israel).

Nissenbaum A., 1975b: The microbiology and biogeochemistry
of the Dead Sea. *Microbial Ecology* 2, 139-161.

Nriagu, J. O., 1968: Sulfur metabolism and sedimentary
environment: Lake Mendota, Wisconsin. *Limnol. Oceanogr.*
13, 430-439.

Richards, F. A., 1965: Anoxic basins and fjords. In:
Chemical Oceanography, vol.1, J. P. Riley and G. Skirrow,
eds., (New York: Academic Press), pp. 11-645.

Starinsky, A., 1974: Relationship between Ca-chloride
brines and sedimentary rocks in Israel. Unpublished
Ph.D. Thesis (in Hebrew), Hebrew University of Jerusalem.

Thode, H. G., A. G. Harrison and J. Monster, 1960: Sulphur
isotope fractionation of recent sediments of NE Venezuela.
Bull. A.A.P.G. 44, 809-1817.

CHAPTER 23

STABLE ISOTOPE FRACTIONATION BY
CLOSTRIDIUM PASTEURIANUM

E. J. LAISHLEY, R. G. L. MCCREADY AND R. BRYANT

Department of Biology, The University of Calgary

H. R. KROUSE

Department of Physics, The University of Calgary,
Calgary, Alberta T2N 1N4, Canada

INTRODUCTION

Over the past quarter century, numerous investigations and many thousands of stable isotope abundance determinations have verified the ability of microorganisms to exert isotopic selectivity during metabolism. It is interesting that the majority of these studies have been devoted to sulfur isotope fractionation, and *Desulfovibrio desulfuricans* has been the most frequently used organism.
In our laboratory over the years, we have tended to study organisms which were not classified as sulfate reducers, e.g., *Salmonella sp.* (Krouse et al., 1967-68). More recently we have directed our attention to *Clostridium pasteurianum* strain W5 for a number of reasons.
Academically, the organism has been extensively studied to elucidate the biochemical mechanism of nitrogen fixation (Dalton and Mortenson, 1972; Streicher and Valentine, 1973). It grows readily in a chemically-defined medium so that parameters affecting the regulation of its metabolism can easily be investigated. This strict anaerobe ferments various carbohydrates to butyric acid, acetate, CO_2, and hydrogen (Jungermann et al., 1973). It has been

shown that this organism also participates in the sulfur cycle (McCready et al., 1975).

From the biogeochemical viewpoint, interactions among biological cycles introduce complexities that are difficult to differentiate. Although such interactions have been considered by agriculturalists, detailed investigations of their nature have not been undertaken from the viewpoint of environmental assessment. In the Province of Alberta such considerations are important. Because of the sour gas industry, in excess of 1000 long tons of SO_2 are discharged into the atmosphere each day. Future development of the Athabasca Tar Sands could double the atmospheric emissions. The SO_2 may react directly or undergo oxidation before incorporation in the biosphere. Under such conditions, it is insufficient to confine investigations solely to the sulfur cycle when assessing the impact on the environment. Clearly other biological cycles may be perturbed by the influx of anthropogenic sulfur. In this regard, *C. pasteurianum* is appropriate for study since it represents one biological system which is involved in parts of the carbon, nitrogen and sulfur cycles.

Although clostridia are usually found in soils, they have also been isolated from springs in Western Canada. In these springs, synergetic relationships exist whereby a bacillus reduces $SO_4^=$ to $SO_3^=$ while a clostridium reduces the product $SO_3^=$ to sulfide (Šmejkal et al., 1971).

From the above discussion it is obvious why we chose to study simultaneously carbon, sulfur, and nitrogen isotope fractionation by this organism. It was hoped that the isotopic data might provide an evaluation of how the environmental stress of high concentrations of oxidized sulfur compounds affect the carbon and nitrogen cycles and consequently soil fertility. Our experimental variables are listed in Figure 23.1. To date, we have carried out some 90 experiments and over 2000 isotopic determinations. The majority of runs have been devoted to $SO_3^=$ reduction because of the synergism found in nature and the evolution of H_2S which is toxic to the environment.

MATERIALS AND METHODS

For ammonia growth conditions of *C. pasteurianum*, 1% sucrose-synthetic salts media with the appropriate additions of Na_2SO_4 or Na_2SO_3 were prepared as described by McCready et al. (1975). For some experiments the complex medium of Laishley et al. (1971) was altered by replacing the glucose with 2% sucrose and adding appropriate concentra-

S, C and N isotope fractionation by *C. pasteurianum*

EXPERIMENTAL VARIABLES

- $[SO_4^=]$, $[SO_3^=]$, MIXTURES
- OTHER S- COMPOUNDS
- N_2 - FIXATION vs. NH_4^+ GROWTH
- TEMPERATURE
- CARBON SOURCES
- CELL FREE EXTRACTS vs. WHOLE CELLS

Figure 23.1 The experimental variables in isotopic studies with *C. pasteurianum*.

tions of $SO_3^=$. Nitrogen-fixing conditions were obtained by bubbling purified nitrogen gas during growth of the culture in the above synthetic salts medium with the sucrose increased to 2% and the ammonium chloride omitted.

The evolved H_2S was precipitated in a trap containing silver nitrate solution and converted to SO_2 for sulfur isotope analyses as described by McCready et al. (1975).

After removal of sulfide, CO_2 was collected from the fermentation by a second trap containing a saturated solution of $Ba(OH)_2$. After 3 min sampling time, $BaCO_3$ was collected by millipore filtration, washed several times and then dried at 105°C for 12 hr. The dried sample of $BaCO_3$ was treated with hypophosphoric acid to generate CO_2 for carbon isotope analyses by mass spectrometry. The carbon isotope compositons of the total sugar molecule and cells were determined by completely combusting the sugar to CO_2 with a pressure of 10 cm O_2 and circulation through CuO at 800°C.

To study the effect of sulfate and sulfite on the fractionation of nitrogen isotopes by *C. pasteurianum* during nitrogen-fixing conditions, large aliquots of culture were collected at appropriate stages of growth by centrifugation, washed and then lyophilized. Forty milligrams of lyophilized cells were digested and distilled by the micro-Kjeldahl procedure of Monro and Fleck (1969) to obtain the intracellular nitrogen for isotopic analysis. The distilled ammonia was collected in 5 ml of 0.1 N H_2SO_4 as $(NH_4)_2SO_4$ and oxidized to N_2 gas by the alkaline hypobromite conversion

method of Rittenberg (1946). The nitrogen gas was passed
through hot CuO and hot Cu to oxidize organic impurities
and react nitrogen oxides and oxygen before analysis by
mass spectrometry.

Growth was measured by a Klett-Summerson colorimeter
using a 540-nm filter.

The isotopic data are expressed in the usual δ-notation
where

$$\delta\ X_2\ \text{in}\ \permil = \left[\frac{(X_2/X_1)\ \text{product}}{(X_2/X_1)\ \text{initial substrate}} -1 \right] \times 1000$$

X_2 = heavy isotope C^{13}, N^{15} or S^{34}
X_1 = light isotope C^{12}, N^{14} or S^{32}

SULFUR ISOTOPE FRACTIONATION

Sulfur Isotope Fractionation
during Growth on Sulfate

We reported earlier (McCready et al., 1975) that during
growth on minimal-salts-sucrose-NH_4^+ media plus various concentrations of Na_2SO_4, (10^{-4}-$10^{-2}M$), C. pasteurianum grew
normally and H_2S was evolved only during late stages of
growth on $10^{-2}M$ $SO_4^=$. The δS^{34} values for the H_2S varied from
0 to +2 ‰. We concluded that the H_2S was generated by
intracellular turnover and therefore the isotopic selectivity of incorporation of sulfur into the cell was very small.
Since that report, we have carried out further experiments
in which the isotopic composition of the cellular material
was also determined as summarized in Table 23.1. The
range of values tends to confirm our earlier conclusions.

In recent experiments under N_2-fixing conditions, we
were able to obtain sufficient H_2S for isotopic analyses
over the total range of 10^{-4} to $10^{-2}M$ $SO_4^=$. Again, it was
found that the isotope fractionation during $SO_4^=$ reduction
was small.

Sulfur Isotope Fractionation
during Growth on Sulfite

These investigations have proved to be most exciting
primarily because of the inverse isotope effects encountered
(McCready et al., 1975). Figure 23.2 presents the typical
isotopic pattern for H_2S released during $SO_3^=$ reduction with
growth on ammonia. During the lag phase, there is a "normal"

Table 23.1
δS^{34} Values for the Intracellular Sulfur of Cells Grown on Sulfate and Ammonia

$SO_4^=$ Concentration	Run No.	δS^{34} (‰)
$10^{-4} M$	8	+1.8
	9	+1.7
	11	+0.3
$10^{-3} M$	13	+1.1
	14	+1.3
	54	+1.5
$10^{-2} M$	5	+1.5
	6	+0.7
	15	+2.2
$6 \times 10^{-2} M$	57	+0.8
	58	+0.3
Average		+1.2 ± 0.6

kinetic isotope effect whereby the product H_2S is enriched in the lighter S^{32}. However in the logarithmic growth phase, the H_2S produced at any instant becomes progressively more enriched in the heavier S^{34}. By the end of logarithmic growth, an "inverse" kinetic isotope effect had developed whereby the H_2S evolved is isotopically heavier than the unreacted $SO_3^=$ and intermediates in the culture vessel. As the growth enters the stationary phase, the H_2S again becomes isotopically lighter with time and a normal kinetic effect is restored.

It is clear that had we terminated this experiment after 8 hours or 20% conversion to H_2S, we would have reported a normal kinetic isotope effect. It is noteworthy that many reports in the literature represent very small percentage conversions and usually one sample of the accumulated H_2S has been taken for each run. Figure 23.2 illustrates that the sampling of successive fractions of the evolved H_2S coupled with extending the conversion as far as possible provides a more informative description of the system. It is interesting that inverse isotope effects were also observed with a clostridium isolated from thermal springs although the isotopic variations were smaller than those of the current study (Šmejkal et al., 1971). This suggests that inverse isotope effects in nature may not be so rare.

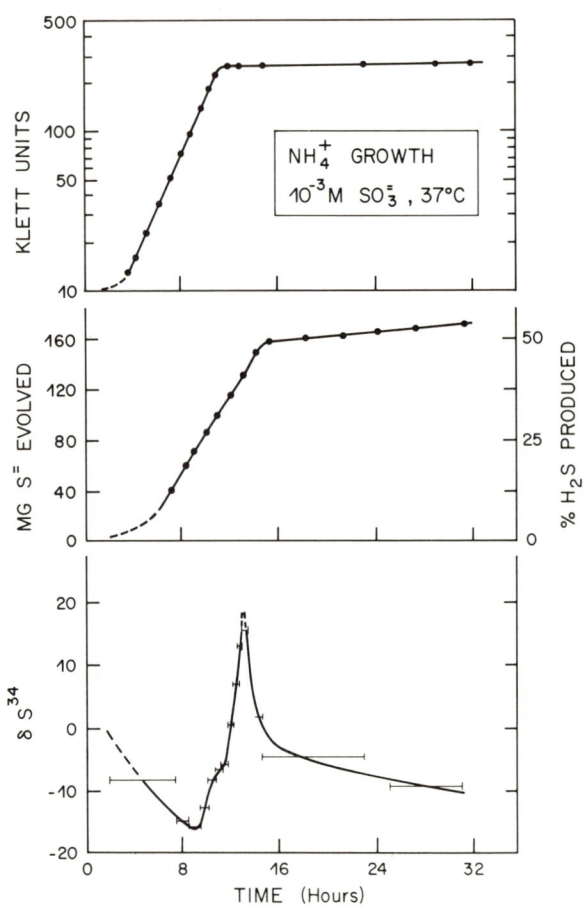

Figure 23.2 Typical experimental results obtained during the reduction of sulfite by *Clostridium pasteurianum* grown on sucrose medium supplemented with NH_4Cl. Inoculum, 12-hr culture.

Now let us attempt to interpret the isotopic behavior. Since lighter isotopic bonds vibrate faster and are therefore more energetic, they are expected to rupture preferentially, giving the "normal" kinetic isotope effect. Since bond formation is also involved in the process, inverse effects are not unexpected but are difficult to explain. We considered models which could give rise to the isotope plot of Figure 23.2. Figure 23.3 shows that if there is a branching of steps in the process giving rise to parallel reaction paths, two normal kinetic isotope effects can

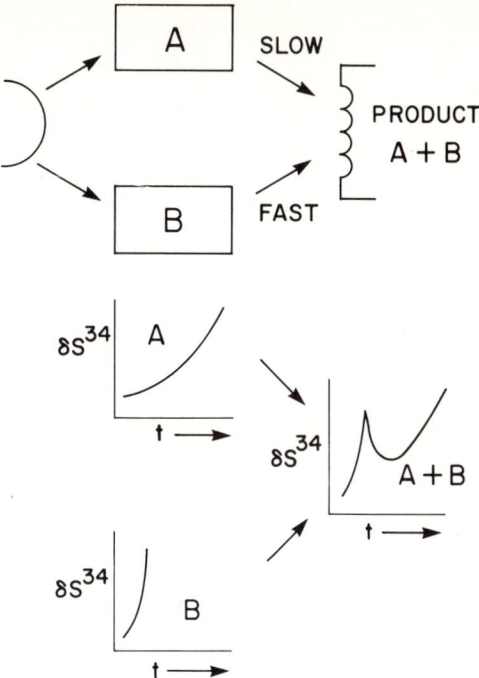

Figure 23.3 A simple model to explain the inverse isotope effects observed during the reduciton of sulfite by *Clostridium pasteurianum*.

create an overall inverse isotope effect. As a reaction proceeds with a normal kinetic isotope effect, the remaining reactant and the product become progressively heavier isotopically with time. Path A is slower than B, and the combination of A + B gives a positive δS^{34}-peak. This is an oversimplified argument but it suggested that two pathways existed for the $SO_3^=$ reduction. Consequently we postulated (McCready *et al.*, 1975) the existence of two mechanisms of $SO_3^=$ reduction as shown in Figure 23.4. The assimilatory pathway on the right was the accepted pathway for this organism. The dissimilatory pathway on the left was associated with the sulfate reducer *Desulfovibrio sp.* (Kobayashi, *et al.*, 1969). Kobayashi *et al.* (1969) recently identified the intermediates of the dissimilatory pathway during $SO_3^=$ reduction by *Desulfovibrio sp.* Further, Lee *et al.* (1973) found the assimilatory pathway to exist in dissimilatory $SO_4^=$-reducing organisms. It therefore seemed reasonable to postulate that both pathways also existed for *C. pasteurianum*. Further evidence has been obtained

334 Environmental Biogeochemistry

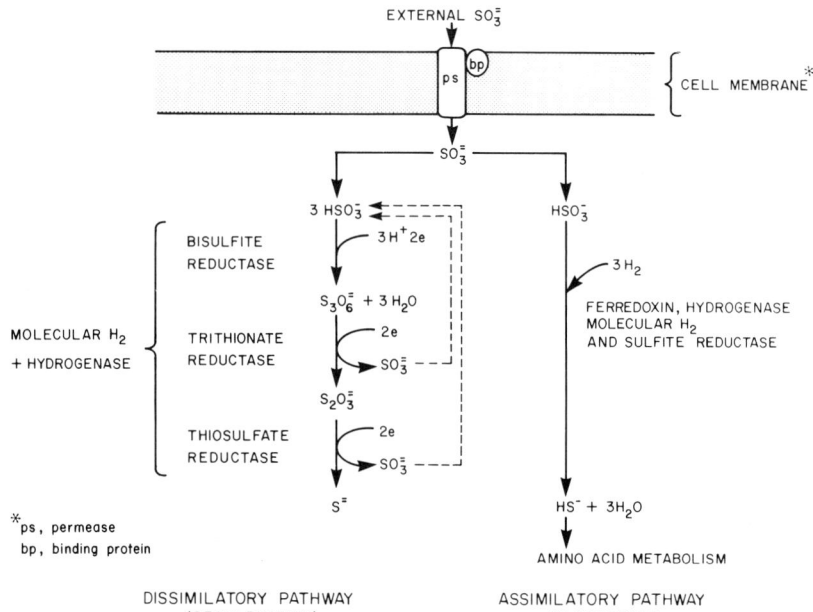

Figure 23.4 Possible assimilatory and dissimilatory pathways for sulfite reduction by Clostridium pasteurianum.

to verify the existence of two pathways but their sheer complexity makes it difficult to correlate their features with the isotopic data.

We have not been able to extend our data to larger percentages of conversion because this organism initiates sporulation and curtails its activity.

During growth on $SO_3^=$ in the range of 10^{-4} to $10^{-2}M$, the generation times of the cultures increased with increased $SO_3^=$ concentration and reached a maximum of 180 min on 10^{-2} M $SO_3^=$ during ammonia growth (McCready et al., 1975).

In addition, the morphology of cells grown on $10^{-2}M$ sulfite was found to be radically different from that of sulfate-grown cells in that cell shape was altered (Figure 23.5), and the cytoplasm did not contain the electron translucent carbohydrate granules (Laishley et al., 1974). At lower sulfite concentrations (10^{-3} and $10^{-4}M$) the cells showed the cytoplasmic changes noted above but their cell shape was not modified as drastically.

It was found that patterns similar to Figure 23.2 occurred for all concentrations of $SO_3^=$ studied in the range

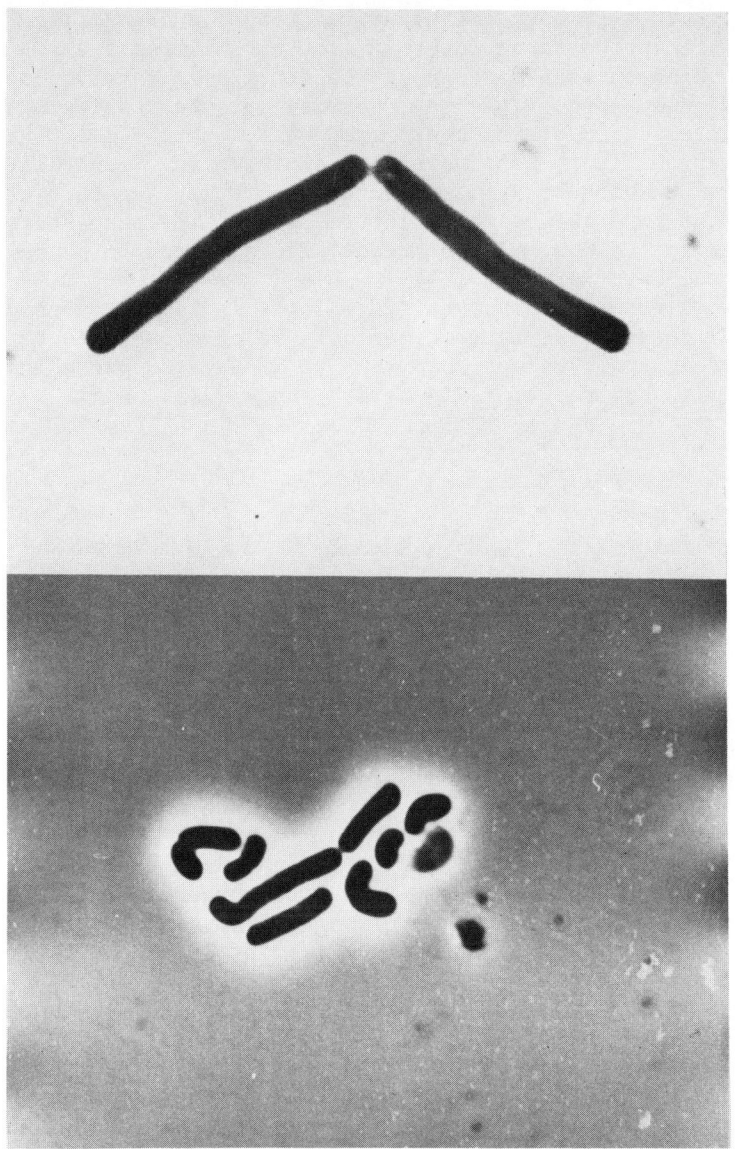

Figure 23.5 Alteration of the cellular morphology of
Clostridium pasteurianum during growth on $10^{-2}M$ $SO_3^=$.
Upper figure, cells from the lag phase showing elongation
prior to division. Lower figure, cells from the mid-
logarithmic growth phase on $10^{-2}M$ $SO_3^=$. Inoculum, 8-hr
culture (X 2500, phase contrast).

10^{-4} to $10^{-2} M$. Both the negative troughs and positive peaks tended to occur later in time with increasing $\overline{SO_3^=}$ concentration as a consequence of the growth rate.

When $\overline{SO_3^=}$ reductions were carried out under NH_4^+ growth conditions with other sugars (glucose, fructose), patterns similar to Figure 23.2 were also encountered. Generally, runs with monosaccharides produced a broader negative trough when the δS^{34} values were plotted against the percentage conversion to H_2S. These trends also appeared to be related to the growth rate.

At a lower temperature (25°C), the isotopic maxima generally occurred at later times because of slower growth and had greater δS^{34} values than runs at 37°C. (Compare Figure 23.2 with Figure 23.6; Figure 23.7 and 23.11.)

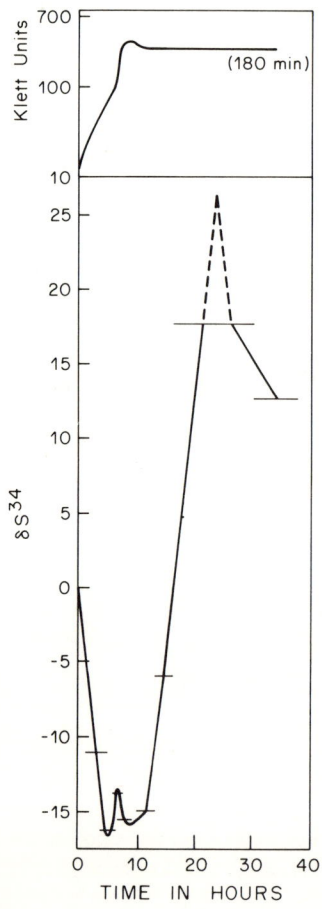

Figure 23.6 The growth and sulfur isotope pattern obtained during growth of Clostridium pasteurianum on 10^{-3} M $\overline{SO_3^=}$ at 25°C.

When N_2-fixing conditions are used instead of NH_4^+ during $SO_3^=$ reduction, growth is generally slower. The inverse sulfur isotope effects are still present but the δS^{34} maxima occur at later times because of the slower growth (Figures 23.7 and 23.11).

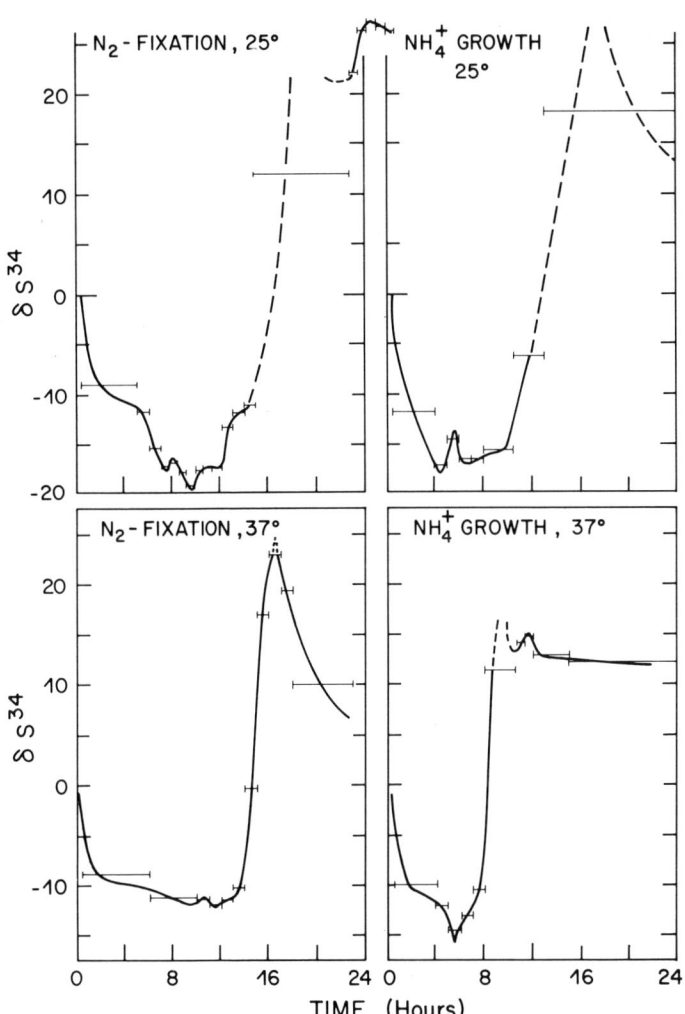

Figure 23.7 A comparison of the isotopic patterns for evolved sulfide during growth of *Clostridium pasteurianum* on $10^{-3}M$ $SO_3^=$ under N_2-fixing and ammonia growth conditions at 25°C and 37°C. Inocula were from mid-logarithmic growth phase: ammonium, 8-hr culture; N_2-fixing, ~12-14-hr culture.

The choice of "time" or "percent reaction" for the x-axis in presenting the isotopic data depends upon what information is desired. By choosing time as a variable, it is easier to compare isotopic data for a number of elements simultaneously and relate them to growth. On the other hand, using the percentage reaction as a variable permits the application of suitable equations to determine isotopic balance and elucidate the kinetic isotope effects in individual steps of the conversion. If a number of intermediates accumulate during the reaction, the isotopic behavior of the product formation will not° be consistent with that of substrate disappearance. Generally, for a given substrate, the isotopic patterns obtained for evolved H_2S tend to duplicate each other more closely if "percent H_2S production" is chosen as the x-axis rather than "time."

We decided to carry out $SO_3^=$ reduction experiments under conditions radically different from the synthetic salts media by growing the cells on a peptone-yeast extract with 2% sucrose. The data from one of these experiments is presented in Figure 23.8. Growth on the complex medium produced the highest cell yield and the greatest percentage

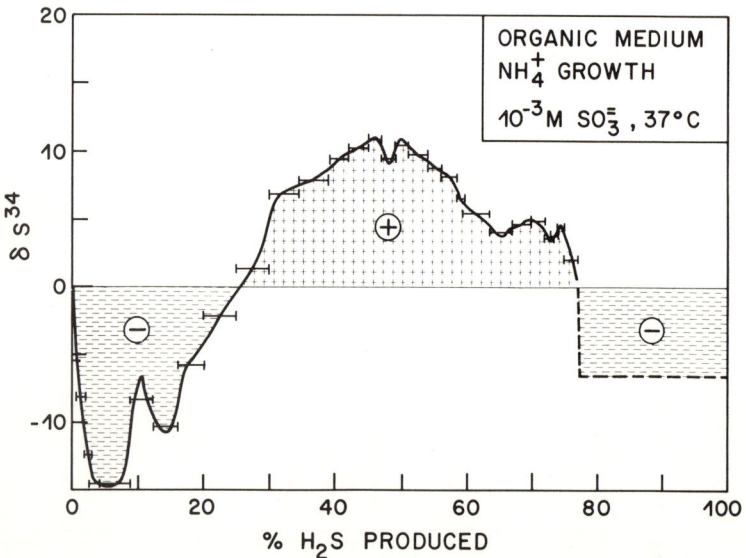

Figure 23.8 The isotopic pattern obtained from the H_2S produced by Clostridium pasteurianum during growth on $10^{-3}M$ $SO_3^=$ in the complex organic medium. Inoculum, 8-hr culture.

production of H_2S (77%) to date. The release of S^{34}-enriched sulfide extended over a longer time (and greater percentage reaction) but it did not achieve as high a δ-value as with experiments on synthetic media.

It is significant that the inverse isotope effect still persists even with a complex organic medium containing sulfur-amino acids. In Figure 23.8, isotopic balance dictates that had the conversion continued, a net release of S^{34}-depleted H_2S would result. In about one-third of our runs, the H_2S possessed negative δ-values near the end of the experiment (c.f. Figure 23.2).

It is also interesting that "fine structures" have been found in the isotopic patterns for a number of experiments. A less negative δS^{34} peak is often encountered in the minimum trough (Figures 23.6, 23.7 and 23.8) and a less positive δS^{34} dip near the δS^{34} maximum (Figures 23.7, 23.8 and 23.11) (see also McCready et al., 1975).

It was also found that the age and the physiological state of the inoculum noticeably altered the growth rate and the percentage conversion realized. In our earlier experiments (Figures 23.2 and 23.6), 12-hr cultures grown on limited ammonia were used as the inoculum. In later experiments (Figures 23.7 and 23.11), and 8-hr inoculum which was not ammonia-limited was used. The latter realized larger conversions, shorter generation times, and shifted the maximum δS^{34} peak to higher percentage H_2S production.

Sulfur Isotope Fractionation
with Mixtures of $SO_3^=$ and $SO_4^=$

It was of interest to study the effect of various $SO_4^=$ concentrations (10^{-4} to $6 \times 10^{-2} M$) on the growth and isotopic fractionation during $SO_3^=$ reduction since we were concerned about possible oxidation of sulfite to sulfate during our experiments, and both valence states of sulfur are present in the environment. All inocula for these studies were 8-hr nonammonia-limited. $SO_4^=$ concentrations up to $10^{-2} M$ decreased the generation times of the cultures to 50-55 minutes in comparison to 85 minutes on $10^{-3} M$ $SO_3^=$ alone. The highest $SO_4^=$ concentration ($6 \times 10^{-2} M$) however increased the generation time to 75 min. At all $SO_4^=$ concentrations, the inverse isotopic pattern was observed, but the location of the δS^{34} maximum and its value depended upon $SO_4^=$ concentration. Figure 23.9 summarizes this behavior. It is seen that a larger δS^{34} maximum is generally located at a higher percentage H_2S production. It is also noted that the maximum δS^{34} was obtained at $SO_4^=$ concentrations

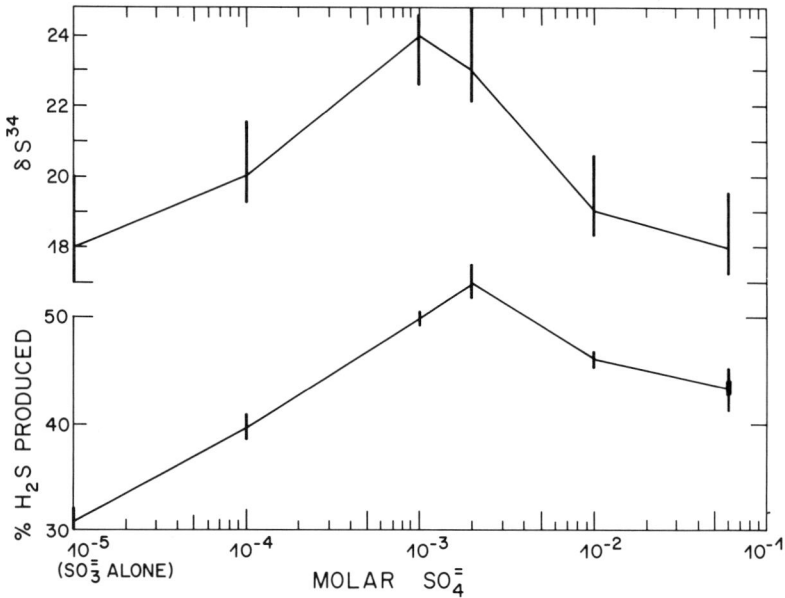

Figure 23.9 The correlation observed between the location of the δS^{34} maximum in terms of percent H_2S production and the maximum δS^{34} value during $SO_3^=$ reduction by Clostridium pasteurianum on mixtures of sulfate and sulfite. All inocula were 8-hr cultures.

of 10^{-3} to 2×10^{-3} M and it is in this range that the maximum percent conversion of the $SO_3^=$ was realized on defined media. The presence of $SO_4^=$ did not significantly alter the isotopic pattern of the evolved H_2S, but it altered the rate of growth. We believe that this growth alteration gave rise to the shifts summarized in Figure 23.9.

We further checked the influence of $SO_4^=$ on $SO_3^=$ reduction with a preliminary stable isotope labeling experiment in which $SO_4^=$ enriched in S^{34} was added to $SO_3^=$. The isotopic pattern of the evolved H_2S did not differ significantly from a control experiment with $SO_3^=$ and $SO_4^=$ of similar S^{34}/S^{32} compositions. We also examined isotopically the incorporated sulfur in these experiments. We were surprised to find that although $SO_4^=$ stimulated growth, only 15% of the intracellular sulfur came from $SO_4^=$ while the remaining 85% was incorporated from $SO_3^=$.

CARBON ISOTOPE FRACTIONATION

Our carbon isotope data are not as extensive and have not been taken with the detail associated with our sulfur investigations. Aside from the technical difficulties associated with the collection of the copious quantities of CO_2 produced during our $SO_4^=$ and $SO_3^=$ investigations, there are also problems in ascertaining mass balance since carbon is involved in many metabolic reactions. We are in the process of scaling down the fermentation volume to specifically examine carbon isotope fractionation quantitatively. Table 23.2 and Figure 23.10 summarize δC^{13} values encountered for CO_2 released under a variety of growth conditions. In each case the δC^{13} is expressed in terms of the mean isotopic composition of the whole sugar molecule. This may differ slightly from the C^{13}/C^{12} composition of the C_3 and C_4 carbons of the hexoses which would be involved in phosphoclastic production of CO_2.

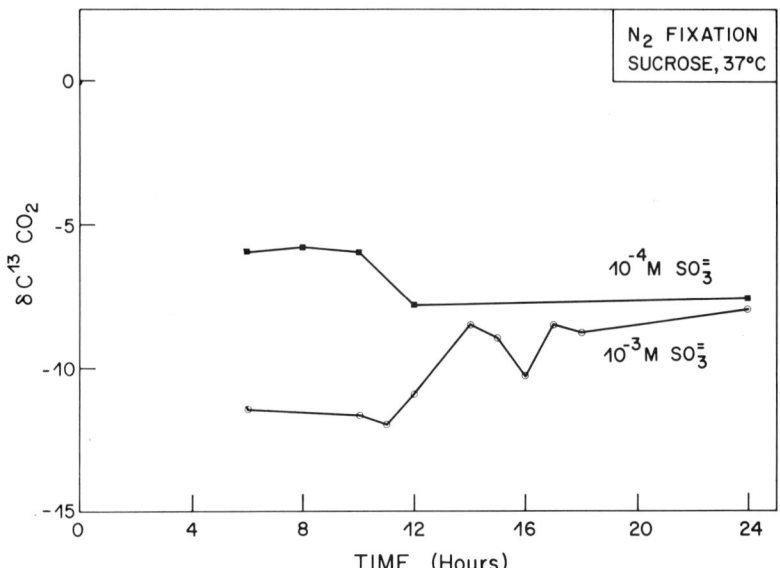

Figure 23.10 *The effect of sulfite concentration on the carbon isotope fractionation obtained with* Clostridium pasteurianum *during growth on sucrose under nitrogen-fixing conditions at 37°C. Inocula were from the mid-logarithmic growth phase (12-14 hr).*

In runs 44 and 45 (Table 23.2), the total produced CO_2 was collected as successive fractions. In all other runs including those in Figure 23.10, the CO_2 was collected

Table 23.2
δC^{13} Values for CO_2 Evolved by *Clostridium pasteurianum* under Various Growth Conditions

Experimental Conditions	Time in Hours	δC^{13}	
$10^{-3}M$ $SO_3^=$, 25°, glucose (Runs 44 and 45)	0-3	-7.5	-10.0
	3-5	-10.4	-9.2
	5-8	-12.3	-10.9
	8-12	-15.4	-13.4
	12-16	-12.6	-15.6
	16-24	-11.0	-13.6
	24-28	-13.6	-12.9
	28-29	-13.6	-11.3
	29-31	-11.4	-14.7
$10^{-3}M$ $SO_3^=$ & $SO_4^=$, 37°, sucrose (Run 46)	5	-5.7	
	6	-9.3	
	7	-10.5	
	16	-7.1	
$10^{-3}M$ $SO_3^=$, $10^{-4}M$ $SO_4^=$, 37°, sucrose (Run 47)	5.5	-6.7	
	7	-10.6	
	8	-10.4	
	12	-2.2	
	24	-9.3	
$10^{-3}M$ $SO_3^=$ $10^{-2}M$ $SO_4^=$, 37°, sucrose (Run 48)	3	-12.5	
	4	-9.5	
	5	-9.7	
	6	-7.8	
	7	-9.6	
	8	-12.3	
	10	-11.2	
	12	+6.7	
	24	-6.0	
$10^{-3}M$ $SO_3^=$ $2 \times 10^{-3}M$ $SO_4^=$, 37°, sucrose (Run 49)	3	-13.1	
	4	-12.3	
	5.5	-10.9	
	7.5	-10.4	
	10	-6.6	
	12	-7.9	
	24	-9.9	
$10^{-3}M$ $SO_3^=$ $6 \times 10^{-2}M$ $SO_4^=$, 37°, sucrose	3.5	-2.8	
	4.5	-3.6	
	5.5	-5.8	

Table 23.2 (cont.)

Experimental Conditions	Time in Hours	δC^{13}
	6.5	−10.0
	7.5	−7.6
	8	−7.5
	13.5	−8.0
	24	−1.8
	29.5	+0.1
Complex Medium, $10^{-3}M$ SO_3, 37°, glucose	6	−15.5
	7	−14.4
	8	−17.1
	10.5	−14.8
	12.5	−15.9
	13.5	−16.3
	25	−11.3

sporadically over short time intervals. It is seen that the evolved CO_2 can be depleted in the heavier C^{13} by as much as −17‰ with a mean δC^{13} value for all runs of −11‰. Although the mechanisms of CO_2 production are quite different, these data compare to the carbon isotope fractionation encountered during decarboxylation of lactate by *Desulfovibrio desulfuricans* (Kaplan and Rittenberg, 1964) and other *Desulfovibrio sp.* and a *Clostridium* (Smejkal et al., 1971). Kaplan and Rittenberg (1964) also found that with $SO_3^=$, the carbon isotope fractionation appeared proportional to the rate of metabolism of lactate. In qualitative terms, a similar statement can be made about our data. For example, in Figure 23.10, growth on the $10^{-4}M$ $SO_3^=$ was slower than on $10^{-3}M$ $SO_3^=$ under N_2-fixing conditions. The slower growth should correspond to slower metabolism of the sugar and it is seen that the isotope fractionation is about 3‰ smaller for the lower rate.

We have also examined cellular carbon at the termination of two $SO_3^=$ reduction experiments and found δC^{13} values of −0.9 and +0.4‰. These contrast to the findings of Kaplan and Rittenberg (1964) where *D. desulfuricans* cells had δC^{13} values similar to the evolved CO_2. One would not expect the evolved CO_2 and the incorporated carbon to behave the same isotopically since different parts of the parent molecule are involved in the two processes. Clearly for unambiguous interpretation, it is necessary to know the isotopic composition of the different parts of the substrate molecule. If for our sugars, the isotopic composition of the molecule is uniform, we could conclude that cellular incorporation of carbon exhibits very small isotope effects.

NITROGEN ISOTOPE FRACTIONATION

To date, we have only examined the nitrogen isotope fractionation during nitrogen fixation at 25°C at various growth stages on 10^{-3} M $SO_4^=$ and $10^{-3} M$ $SO_3^=$ as summarized in Table 23.3. The high δN^{15} values during the lag phases of both runs and the early log phase of the $SO_3^=$ experiment must be identified with the NH_4^+ contamination carried over with the inoculum. This is seen in the portion of the growth curve indicated by an asterisk in Figure 23.11. This type of inflection has been examined by Seto and Mortenson (1974) and identified with the induction of the nitrogenase complex as a result of the shift from low NH_4^+ growth to N_2-fixing. The isotope data fell well within the range of -9‰ to +4‰ found by Hoering and Ford (1959) for fixation of nitrogen by four different *Azotobacter sp*. On the basis of the average fractionation for the individual species, Hoering and Ford (1959) concluded that the ratio of the rates of $N^{14}N^{14}$ and $N^{14}N^{15}$ fixation was 1.000 ± 0.001 and consequently the rate-determining step in the reaction did not involve a change in bonding to nitrogen. It would seem that the same conclusions apply to our data even though the *Azotobacter* is a strict aerobe and the *Clostridium* is a strict anaerobe. The similarity of the nitrogen isotope data is not surprising since the nitrogenase complexes in both organisms are comparable.

A very interesting finding during the course of these investigations was that under N_2-fixing conditions, increasing concentrations of $SO_4^=$ or $SO_3^=$ enhanced growth. In going from $10^{-4} M$ to $10^{-3} M$ of either $SO_4^=$ or $SO_3^=$ at 37°C, the generation time was decreased by one third. At 25°C, the effects were not as large, the decrease being only 10%. Even growth in the traditional Winogradsky medium realized similar trends in the growth rate. This contrasts to the findings during growth on ammonia as summarized in sections 1-1 and 1-2.

WHOLE CELLS *VS* CRUDE CELL EXTRACTS

As complete conversions of the oxidized sulfur compounds were not obtained with whole cells, and we were interested in unraveling the complex pathways proposed in Figure 23.4, studies were started with crude cell extracts. When *C. pasteurianum* was grown in the presence of $10^{-3} M$ $K_2S_2O_3$ or $K_2S_3O_6$, growth was equivalent to that on $SO_4^=$, but no H_2S was produced from the $S_2O_3^=$ and only 1.6% of the $S_3O_6^=$ was **evolved** as sulfide. In contrast, in preliminary experiments

Table 23.3

Nitrogen Isotope Fractionation During Nitrogen-Fixing Growth of *Clostridium pasteurianum* at 25°C

Growth Phase	Klett Unit Range	δN^{15} of Cellular Nitrogen	
		$10^{-3} M\ SO_3^=$	$10^{-3} M\ SO_4^=$
Lag	10	+9.0	+5.7
Early logarithmic	33–45	+4.9	0.0
Mid-logarithmic	78–90	0.0	−1.1
Late-logarithmic	159–165	+0.2	+2.9
Stationary	300–378	−0.1	−1.6

δN^{15} of N_2 source = 0.0‰.
δN^{15} of $(NH_4)_2 SO_4$ (initial inoculum) = +10.9‰.

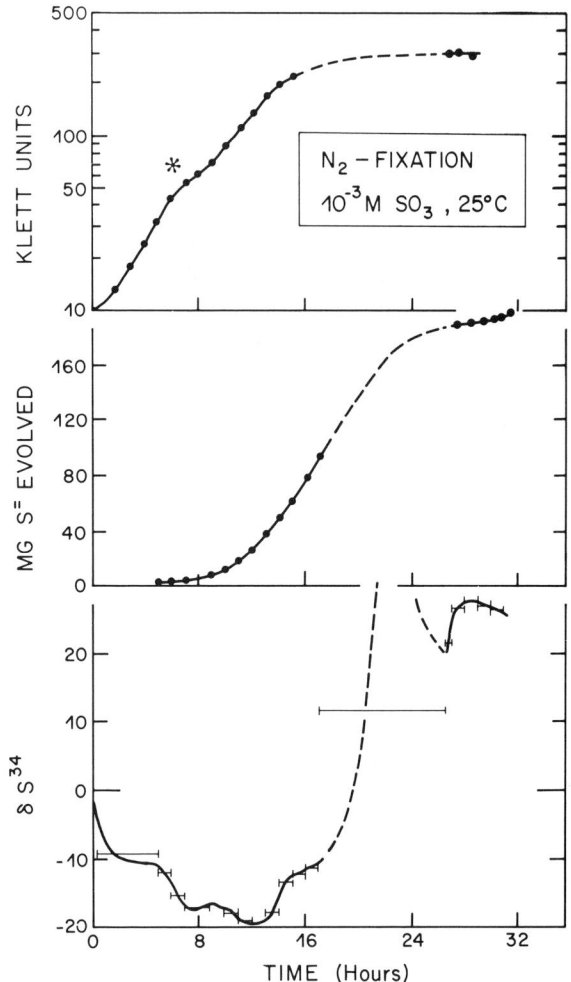

Figure 23.11 Typical experimental results obtained during the reduction of sulfite by N_2-fixing Clostridium pasteurianum at 25°C. Inoculum, mid-logarithmic growth (12-14 hr). Note the inflection in the growth at 7 hr which indicates the induction of the nitrogenase systems.

using crude cell extracts, 80% or better conversions have been obtained with $SO_3^=$, $S_2O_3^=$ and $S_3O_6^=$. With $SO_3^=$ and $S_2O_3^=$, the conversions approximate first-order kinetics for the bulk of the reduction and are consistent with a kinetic isotope effect of k_{32}/k_{34} of 1.005. These results are surprising. For example, they suggest that when the ligand sulfur of $S_2O_3^=$ goes to H_2S, the central atom immediately follows. These small kinetic isotope effects contrast with the large kinetic isotope effects found by Agarwala et al. (1965) for the acid decomposition of thiosulfate. The preliminary isotopic data for $S_3O_6^=$ are interesting in that the behavior is more complex and an inverse isotope effect is evident although the isotope variations are significantly smaller than those encountered with $SO_3^=$ reduction using whole cells. A complicating factor is the presence of both the assimilatory and dissimilatory sulfite reductases in these extracts. Attempts are being made to selectively inhibit or remove the assimilatory enzymes from these extracts in order to specifically study the dissimilatory system.

In a complementary project we have examined the decarboxylation of pyruvate (phosphoclastic reaction) with crude cell extracts in order to elucidate the transformations of carbon in the system. The kinetics of the conversion were first order with respect to the carboxyl group of pyruvic acid and a normal kinetic isotope effect was found with k_{12}/k_{13} ranging from 1.010 to 1.013. This isotope fractionation is of the same order as that found for the CO_2 released by whole cells utilizing various sugars suggesting that the carbon isotope effects may be identified with the decarboxylation reactions.

SUMMARY

C. pasteurianum is a common soil organism that was predominantly recognized for its role in N_2 fixation, and consequently its possible participation in other biological cycles was not thoroughly explored. Clearly, it can participate actively in the sulfur cycle with interesting attending isotope effects. These findings suggest that sulfur cycle investigations should not be confined to the classical sulfate reducers. Indeed broader surveys of the biogeochemical capabilities of many individual organisms seem warranted in these days of increased emphasis on environmental research.

ACKNOWLEDGMENTS

This investigation was supported by the National Research Council Grants A 5058 (EJL) and A 8176 (HRK) and the University of Calgary Interdisciplinary Sulphur Research Group (UNISUL). The authors also acknowledge the technical assistance of Miss N. Enojo, Miss J. Pontoy and Miss E. Bigornia.

REFERENCES

Agarwala, U., C. E. Rees and H. G. Thode, 1965: Sulfur isotope effects in the hydrogen ion decomposition of thiosulfate. *Can. J. Chem.* **43**, 2802-2811.

Dalton, H. and L. E. Mortenson, 1972: Dinitrogen (N_2) fixation (with a biochemical emphasis). *Bacteriol. Rev.* **36**, 231-260.

Hoering, T. C. and H. T. Ford, 1960: The isotope effect in the fixation of nitrogen by *Azotobacter*. *J. Amer. Chem. Soc.* **82**, 376-378.

Jungermann, K., R. K. Thauer, G. Leimenstoll and K. Decker, 1973: Function of reduced pyridine nucleotide-ferredoxin oxidoreductases in saccharolytic *Clostridia*. *Biochim. Biophys. Acta* **305**, 268-280.

Kaplan, I. R. and S. C. Rittenberg, 1964: Carbon isotope fractionation during metabolism of lactate by *Desulfovibrio desulfuricans*. *J. Gen. Microbiol.* **34**, 213-217.

Kobayashi, K., S. Tachibana and M. Ishimoto, 1969: Intermediary formation of trithionate in sulfite deduction by sulfate-reducing bacterium. *J. Biochem. (Tokyo)* **65**, 155-157.

Krouse, H. R., R. G. L. McCready, S. A. Husain and J. N. Campbell, 1967: Sulfur isotope fractionation by salmonella species. *Can. J. Microbiol.* **13**, 21-25.

Krouse, H. R., R. G. L. McCready, J. N. Campbell, S. A. Husain and A. Sasaki, 1967: Isotope fractionation by Salmonella sp. *Studia Biophysica* **4**, 169-177.

Krouse, H. R., R. G. L. McCready, S. A. Husain and J. N. Campbell, 1968: Sulfur isotope fractionation and kinetic studies of sulfide reduction in growing cells of *Salmonella heidelberg*. *Biophysical. J.* **8**, 109-124.

Krouse, H. R., and A. Sasaki, 1968: Sulfur and carbon isotope fractionation by *Salmonella heidelberg* during anaerobic $SO_3^=$ reduction in Trypticase soy broth medium. *Can. J. Microbiol.* **14**, 416-422.

Laishley, E. J., R. G. L. McCready and J. W. Costerton, 1974: Ultrastructural modification of cells of *Clostridium pasteurianum* caused by growth on sulphite. *Can. Soc. Microbiol.* **24**, 32.

Laishley, E. J., P. M. Lin, and H. D. Peck, Jr., 1971: A ferredoxin-linked sulfite reductase from *Clostridium pasteurianum*. *Can. J. Microbiol.* 17, 889-895.

Lee, J-P., J. LeGall and H. D. Peck, Jr., 1973: Isolation of assimilatory and dissimilatory-type sulfite reductases from *Desulfovibrio vulgaris*. *J. Bacteriol.* 115, 529-542.

McCready, R. G. L., E. J. Laishley and H. R. Krouse, 1975: Stable isotope fractionation by *Clostridium pasteurianum* I. S^{34}/S^{32}: Inverse isotope effects during $SO_4^=$ and $SO_3^=$ reduction. *Can. J. Microbiol.* 21, 235-244.

Munro, H. N. and A. Fleck, 1969: Analysis of tissue and body fluids for nitrogenous constituents. In: *Mammalian Protein Metabolism*, Vol. III, H. N. Monro, ed. (New York: Academic Press), pp. 423-525.

Rittenberg, D., 1946: The preparation of gas samples for mass spectrographic isotope analysis. Part of a symposium on preparation and measurement of isotopic tracer. (Ann Arbor: J. W. Edwards)pp. 31-39.

Seto, B. and L. E. Mortenson, 1974: In vivo kinetics of nitrogenase formation in *Clostridium pasteurianum*. *J. Bacteriol.* 120, 822-830.

Šmejkal, V., F. D. Cook and H. R. Krouse, 1971: Studies of sulfur and carbon isotope fractionation with microorganisms isolated from springs of Western Canada. *Geochim. Cosmochim. Acta* 35, 787-800.

Streicher, S. L. and R. C. Valentine, 1973: Comparative biochemistry of nitrogen fixation. *Ann. Rev. Biochem.* 42, 279-302.

CHAPTER 24

MICROBIOLOGICAL CONTRIBUTIONS TO THE
ATMOSPHERIC LOAD OF PARTICULATE SULFATE

DIAN R. HITCHCOCK

Norton Lane, Farmington, Connecticut 06032

INTRODUCTION

Studies of the global sulfur budget show that biogenic sulfur enters the atmosphere at rates which are large compared to the emissions of anthropogenic pollutant sulfur. Estimates in the literature range from about 100 to 230 million tons per year in contrast to about 70 million tons of pollutant sulfur at the present time (Conway, 1943; Eriksson, 1960; Robinson and Robbins, 1968; Kellogg et al., 1972; Friend, 1974).
Although there is good evidence that most of this sulfur is redeposited in precipitation and as dry deposition, there is little information about the form or distribution of the sulfur while it remains in the atmosphere. Examination of the current literature reveals uncertainty regarding the nature of the biological source (Lovelock et al., 1972; Rasmussen, 1974; Friend, 1974).
Some organic sulfur derived from decomposing plant tissue enters the atmosphere (Lovelock et al., 1972; Rasmussen, 1974), but extrapolations of available emission rate data indicate that this source accounts for only about 2 to 5 million tons of sulfur per year (Hitchcock, 1975). These global extrapolations are shown in Table 24.1.
The only other biological source of atmospheric sulfur is H_2S produced metabolically by the sulfate-reducing bacteria (such as *Desulfovibrio desulfuricans*) which inhabit anoxic muds and water-logged soils. Recent evidence indicates

Table 24.1
Annual Global Organic Sulfur Emissions[a]

Material	Average Standing Crop (g)	Reported DMS-S Emissions (10^{-12} g gm^{-1} hr^{-1})	Assumed Mean S Emissions (10^{-12} g gm^{-1} hr^{-1})	Calculated Global S emissions (10^{12} g S yr^{-1})
Marine Algae	5×10^{15}	1060	1060	0.05
Intact Leaves	1×10^{17}	1 to 23	12	0.011
Senescent Leaves	5×10^{16}	"10 to 100 times greater"	1200	0.53
Soils	6×10^{18} to 2×10^{19}	11 to 45	28	1.5 to 4.9
			TOTAL	2.1 to 5.5

[a]Source: Hitchcock (1975)

that H_2S released from aquatic muds may survive oxidation by
by O_2 in the water column for many hours (Chen, 1970; Chen
and Morris, 1972), and consequently we may expect that, in
aquatic environments where turbulent diffusion occurs,
H_2S released from bottom sediments can be carried to the
surface and lost to the atmosphere. That this occurs near
coastal mudflats, salt marshes and estuaries as well as
fresh-water swamps, irrigated farmlands and some lakes and
rivers is clearly evident from reports of the odor of H_2S
near such sites. Measurements of this phenomenon are rare,
since the odor detection threshold—0.5 ppb—is below the
level which can be reliably measured with readily available
instruments (Bethea, 1973). The concentration of dissolved
H_2S in water which gives rise to this atmospheric concentration also defies measurement, as it can be as low as
$10^{-10}M$ to $10^{-8}M$ at the pH values characteristic of most
natural waters.

The fate of H_2S in the atmosphere is not well understood. Gas-phase reaction of H_2S with O_2 is slow, but H_2S
may react fairly rapidly with ozone, and its oxidation may
be enhanced by photochemical reactions (Cadle and Ledford,
1966; Cox and Sandalls, 1974). Kellogg *et al.* (1972)
have speculated that H_2S may be rapidly converted to sulfate
after absorption in atmospheric aerosols, and this has
received partial experimental confirmation by the work of
Penkett (1972).

We report here two studies which, taken together, support the hypothesis that H_2S derived from the metabolism
of bacterial sulfate reducers enters the atmosphere and
gives rise to local concentrations of particulate sulfate
which are comparable to those observed in many cities in
winter when pollutant sulfate sources are large.

EVIDENCE OF H_2S EMISSIONS
FROM POLLUTED COASTAL WATERS

Experimental evidence supporting the conclusion that
H_2S is emitted in large quantities from polluted coastal
waters is provided by measurements of the abundance and sulfur isotope distribution of gaseous and particulate sulfur
collected from tradewinds in Hawaii by Castleman and
Munklewitz in February, 1972.*

---

*Personal communication from H. W. Munklewitz and A. W.
Castleman, Brookhaven National Laboratory, Upton, Long
Island, New York.

Data

The samples were collected on two successive days at the University of Hawaii sampling tower at Waimanalo Bay, on the northeast coast of Oahu. Northeast tradewinds had been steady for several days; under these conditions, the closest land downwind is the Pacific coast of North America, approximately 3800 km distant. On both days the sampling duration was about 6 hr, the temperature varied between 21 and 25°C and the windspeed averaged 15.3 msec^{-1} (30 knots). On the first day, relative humidity averaged 66% and the surface of the bay was choppy; on the second, relative humidity averaged 77% and there were more whitecaps, a higher surf, and some brief rain squalls. The odor of H_2S was not detected on either day.

The total gaseous and particulate sulfur and the sulfur isotope distribution in each fraction were measured on each day and are shown in Table 24.2. Sampling and analysis procedures used have been described by Tucker (1969). The observed sulfur isotope ratio is expressed as the δS^{34} value, where

$$\delta^{34}S\ (\permil) = \frac{(S^{34}/S^{32})_{sample} - (S^{34}/S^{32})_{standard}}{(S^{34}/S^{32})_{standard}} \times 1000$$

δS^{34} values are expressed in parts per mil (‰) and are referred to a meteoritic standard. Analytical errors in the determination of the δS^{34} values are estimated as less than 2%, while those in the concentration measurements are less than 10%.

Sources of Gaseous Sulfur

These measurements, though fragmentary, are interesting because they demonstrate the presence of gaseous sulfur at this site which is extremely distant from anthropogenic sulfur sources—so distant that we may rule out these pollutant sulfur sources.

Although pollutant sulfur dioxide concentrations in surface air in cities can be quite high, especially in winter when ground-level emissions from home heating are large, concentrations decline very rapidly with distance from major population centers, due to dilution with clean air, absorption of the SO_2 on soil, vegetation, and other surfaces, absorption in cloud and fog droplets, and gas-phase reactions which form particles. Atmospheric oxidation and absorption in aerosols and cloud and fog droplets are estimated to reduce pollutant SO_2 concentrations by about 3% per hour over land (Junge, 1972). Over the ocean, SO_2 is also absorbed at the surface. The data of Liss and Slater (1974) imply a loss of about 0.5% per hour to the ocean, calculated by assuming that the air under the 2 km inversion is a well-mixed reservoir. Absorption in aerosols and atmospheric gas-phase reactions occur in the marine atmosphere, and consequently we may infer a total SO_2 loss rate considerably in excess of 0.5% per hour, and perhaps up to 4% per hour.

Table 24.2
Atmospheric Sulfur at Waimanalo Bay

	Sample 1	Sample 2
Observations		
Abundance		
Gaseous S	0.64 µg m^{-3}	0.64 µg m^{-3}
Particulate S	1.93 µg m^{-3}	2.56 µg m^{-3}
$\delta^{34}S$		
Gaseous S	+5.9 ‰	+6.0 ‰
Particulate S	+15.0 ‰	+16.2 ‰
Calculated Sources[a]		
Biogenic Gaseous S	0.64 µg m^{-3}	0.64 µg m^{-3}
Biogenic Particulate S	0.68 µg m^{-3}	0.7 µg m^{-3}
Total Biogenic S	1.32 µg m^{-3}	1.34 µg m^{-3}
Seasalt Particulate S	1.25 µg m^{-3}	1.86 µg m^{-3}

[a] Assuming δS^{34} of seasalt sulfur is +20‰ and all non-seasalt sulfur in particulate samples is derived from the same source as the gaseous S.

If we assume that the air reaching Waimanalo Bay is well-mixed and derived from the entire 2000 km Pacific coast of the United States, we may crudely estimate an upper limit for its anthropogenic SO_2 content, employing measurements of the ground-level concentration of SO_2 in polluted cities on this coast, together with generous assumptions regarding the anthropogenic SO_2 concentrations in air over nonurban sites. Geometric mean ground-level SO_2 concentrations in 15 cities monitored by the U.S.

Environmental Protection Agency (EPA) ranged from 4.6 to 30.5 μg m^{-3} in 1972,* and averaged 10 μg m^{-3}. Assuming this concentration is typical for air over 10% of the entire coast, and anthropogenic SO_2 concentrations in the remaining air average 1 μg m^{-3}, we may infer an average ground-level SO_2 concentration of 2 μg m^{-3}, or 1 μg m^{-3} as sulfur. If the average air speed over the ocean is 10 km hr^{-1} (Broecker and Peng, 1974) we may infer a concentration reduction by a factor of 50 (at 1%/hr) to 10^7 at 4%/hr) during the 380-hr journey. It follows that none of the gaseous sulfur observed at Waimanalo Bay can be anthropogenic sulfur from North America, and that it must, therefore, be biogenic sulfur of marine origin.

The global extrapolations of organic sulfide productivity indicate that H_2S from bacterial sulfate reduction is the principal source of biogenic sulfur in the atmosphere, and it seems reasonable to suppose that the gaseous sulfur at Waimanalo Bay has a similar source, especially in view of the fact that regions of intensive sulfate reduction leading to high concentrations of H_2S in bottom waters exist in Kaneohe Bay immediately to the north (Smith, et al., 1973), as well as in many sites on the southern coast of Oahu.

This conclusion is supported by the limited evidence available regarding sulfur isotope fractionation in the dissimilatory reduction of sulfate in the marine environment and in the decomposition of organic sulfur. Kaplan, Emery, and Rittenberg (1963) have shown that organic sulfur in marine plant and animal tissue exhibits a mean δS^{34} value of -1‰ relative to sulfate in the ocean, while Kaplan and Rittenberg (1964) have shown that hydrolysis of cysteine by *Proteous vulgaris* produces H_2S with a mean δS^{34} value of -5.1‰ relative to the substrate sulfur at 40°C and -4.3‰ at 25°C. If we assume that these results are typical of the fractionation effects occurring during the production of organic sulfides in the marine environment, then we may infer an average δS^{34} value of about +15‰ or more for organic sulfides of marine origin, assuming a sea water sulfate δS^{34} value of +20‰.

If these assumptions are correct, then the gaseous sulfur observed at Waimanalo Bay cannot be derived from an organic sulfide source, since its δS^{34} value is only about +6‰ (Table 24.2). This sulfur isotope distribution is consistent with an origin as H_2S produced by bacterial sulfate reduction. Although much more extreme fractionation

---

*Personal communication from N. H. Frank, EPA, Research Triangle Park, North Carolina.

has been observed in laboratories (Goldhaber and Kaplan, 1974), values in the range of -10 to -20‰ seem to be common in nature (Holser and Kaplan, 1966; Sangster, 1968).

Sources of Particulate Sulfur

The δS^{34} values of the particulate sulfur samples are +15.0 and +16.2; these demonstrate the presence of isotopically "light" non-seasalt sulfur, but the available evidence is not sufficient to identify its origin with certainty.

Continental sources of particulate sulfur at this site cannot be confidently ruled out, although available evidence suggests that continental aerosols are unlikely to be abundant. The upper limit for the average pollutant sulfur concentration in the air mass near the Pacific coast has been estimated above as 1 µg m^{-3}, and most of this will have been removed as SO_2, rather than converted to particulate form.

Studies of the abundance of trace metals in aerosols at this site have shown that lead averages 3 x 10^{-9} g m^{-3} (Hoffman et al., 1972), while analyses of the elemental composition of aerosols in a typical polluted city indicate that the S/Pb ratio ranges from about 5 to 38 (Tanner et al., 1974). Assuming that all the lead observed by Hoffman et al. (1972) is of continental origin, all the sulfur observed by Tanner et al. (1974) is derived from pollutant sulfur sources, and that these ratios are typical of the pollutant sources on the Pacific coast of the United States, we may infer a probable mean continental pollutant particulate sulfur abundance in Waimanalo Bay aerosols to be between 0.015 and 0.11 µg m^{-3}.

Particulate sulfur could also be derived from the upper troposphere during the formation of the tradewinds. Concentrations in the range of 0.15 to 0.5 µg S m^{-3} have been observed in the upper troposphere of the Pacific.* We have no information regarding the sulfur isotope distribution of these particulates, although Castleman et al. (1974) provide evidence that the background sulfate aerosol in the lower stratosphere which predominates at times of volcanic inactivity has an average δS^{34} value of about +2‰ (which they interpret as signifying a biogenic source for H_2S; such a source is probably located in tropic regions which are the known sources of most stratospheric air.)

---

*Personal communication from R. D. Cadle, National Center for Atmospheric Research, Boulder, Colorado.

Seasalt sulfur in aerosols has the same δS^{34} value as that in the ocean, or about +20‰. If organic sulfides of marine origin have a δS^{34} value of +15‰ or more, then we may infer that little of the sulfur in the particulate samples at Waimanalo Bay is derived from this source, because this would imply that these samples contain no seasalt sulfur— a finding at variance with the results of many studies of aerosols at this site (see, for example, Duce and Woodcock, 1971).

We are inclined to believe that most of the non-seasalt sulfur in the particulate samples is derived, like the gaseous sulfur, from H_2S produced in local muds, partly for reasons described in the next section. If so, we may calculate the total biogenic sulfur contribution by assuming its δS^{34} value is the same as that of the gaseous sulfur, while that of seasalt sulfur is +20‰ (Table 24.2).

PARTICULATE ATMOSPHERIC SULFATE IN URBAN AND NONURBAN SITES

Atmospheric scientists concerned with sulfur in polluted city atmospheres have long assumed that virtually all particulate sulfate in the air is derived directly or indirectly from anthropogenic pollutant sulfur emissions. Relatively high annual geometric mean atmospheric sulfate levels in nonurban sites monitored by the EPA in northeastern states have been explained as due to the presence of a widespread background sulfate level derived by atmospheric reactions from pollutant sulfur dioxide emissions over an extremely wide geographical region (Altshuller, 1973; Frank, 1974).

The Waimanalo Bay studies suggest that bacterial sulfate reduction contributes gaseous sulfur to the atmosphere and may contribute particulate sulfur. If so, we may expect similar H_2S emissions in urban and nonurban sites near coasts and other regions where sulfate reduction is active. In relatively cold climates, these emissions will peak in the warm months of the year, and could give rise to measurable levels of sulfate which may also peak in warm months.

These considerations lead us to study the atmospheric sulfate measurements collected by the EPA at urban and nonurban coastal and inland sites in the northeastern United States, to look for evidence of biogenic contributions.

Routine measurements of 24-hr mean sulfate levels obtained from hi-vol particulate samples usually collected every 13 or 6 days have been performed in a number of sites by federal or state agencies for several years. Sulfate

levels show extreme day-to-day variation at all sites in
this region, and we limited our attention to those for which
long records were obtainable. We found suitable data for
only seven nonurban sites in New England and middle Atlantic
states, and one of these—Cape Hatteras in North Carolina—
has an undesirably mild climate. These seven sites include
four "coastal" sites located on the shores of extended
bodies of water—Lake Ontario, the Chesapeake Bay, and the
Atlantic Ocean—and three sites in inland forested locations.

These nonurban sites were roughly matched with urban
sites in the same region, of which eight are "coastal" sites
(including three on Great Lakes), and ten are inland sites
(all in upstate New York). Local pollutant sulfur dioxide
sources are nonexistent or negligible near all the nonurban
sites and significant in all the urban sites, although as
a group the inland urban sites have lower average local
pollutant sulfur emissions than the coastal urban sites.

The questions considered in our study were: Do atmospheric sulfate levels in nonurban sites peak in warm months
of the year? Is there evidence that all of the sulfate in
nonurban sites is derived from a widespread anthropogenic
sulfate background in summer? Are sulfate levels in warm
months of the year higher in locations near extended bodies
of polluted water than in other locations?

RESULTS AND DISCUSSION

The results of our study strongly support the hypothesis that biogenic sources contribute significant amounts
of sulfate to the atmosphere, and that these biogenic contributions are higher in coastal locations than in inland
sites, although they exist in both.

Analyses of seasonal differences in geometric mean sulfate levels at each site considered show that summertime
sulfate levels are higher than wintertime levels in all
seven nonurban sites, and in ten of the eighteen urban sites
(Table 24.3). These summertime elevations are significant
in five of the nonurban sites, and in three of the urban
sites. In contrast, summertime sulfate levels are significantly higher than the group mean wintertime level, but
there is no significant difference between these seasonal
means for any grouping of the urban sites (Table 24.3).

The hypothesis that sulfate in these nonurban sites is
anthropogenic background sulfate while that in the urban
sites consists of the background common to all sites plus
a specifically urban local contribution was explored in
a number of ways. Comparisons of seasonal means of the

Table 24.3
Seasonal and Annual Geometric Sulfate Means[a]

Site	Years	N	Annual	DJF	MAM	JJA	SON	JJA-DJF[b]	Significance[c]
A. Nonurban Sites									
Jefferson County, NY	1962; 65-70	154	6.9	5.7	5.6	11.9	5.8	+6.2	0.00001
Acadia Nat. Pk., ME	1962; 65-70	168	4.9	4.5	4.3	7.4	4.2	+2.9	0.0002
Calvert County, MD	1965-69	147	8.6	7.9	7.5	12.0	7.6	+4.1	0.0002
Cape Hatteras, NC	1965-70	143	7.3	6.9	6.8	8.5	8.6	+1.6	ns
Coos County, NH	1965-70	144	5.0	4.7	4.3	5.9	5.1	+1.2	0.09
Washington County, RI	1965-70	114	7.9	7.3	8.0	8.6	6.7	+1.3	ns
Shenandoah Nat. Pk., VA	1962; 65-70	178	7.3	5.2	6.6	9.6	6.9	+4.4	0.001
Group Mean		7	6.7	5.9	6.0	8.9	6.3	+3.0	0.007
B. Coastal Urban Sites in New York									
Buffalo	1968-72	278	9.5	8.0	10.0	11.2	8.8	+3.2	0.0009
Niagara Falls	1968-72	283	11.3	11.5	11.6	12.6	9.7	+1.1	ns
Rochester	1969-72	189	8.1	8.0	8.1	9.9	6.8	+1.9	0.03
New Rochelle	1969-72	179	10.6	11.3	9.9	10.5	11.0	-0.8	ns
Group Mean		4	9.8	9.5	9.8	11.0	8.9	+1.5	ns
C. Other Coastal Urban Sites									
Portland, ME	1959-65	165	10.7	13.2	10.4	9.8	10.0	-3.4	0.003
Providence, RI	1965-70	151	12.2	13.8	11.8	12.2	11.0	-1.6	ns
Washington, D.C.	1965-70	148	11.5	11.7	10.2	13.7	11.0	+2.0	ns
Norfolk, VA	1964-70	178	11.3	10.7	10.6	12.4	9.7	+1.7	ns
Group Mean		4	11.3	12.2	10.6	12.0	9.6	-0.2	ns

D. *All Coastal Urban*

		n	Annual	DJF	MAM	JJA	SON	JJA−DJF	Sig.
Group Mean		8	10.5	10.8	10.2	11.5	9.6	+0.7	ns

E. *Inland Urban Sites (New York)*

		n	Annual	DJF	MAM	JJA	SON	JJA−DJF	Sig.
Albany	1965-67;69-72	325	10.3	10.7	9.9	10.3	10.5	−0.4	ns
Troy	1965-67;69-72	345	8.6	10.0	7.1	9.8	7.7	−0.2	ns
Syracuse	1969-72	185	7.9	9.9	8.1	7.4	6.9	−2.5	0.01
Schenectady	1969-72	164	8.3	9.0	8.5	9.2	6.7	+0.2	ns
Massena	1964-68	221	7.2	8.8	7.2	6.6	6.3	−2.2	0.01
Utica	1969-72	180	7.4	8.6	7.3	7.1	6.5	−1.5	0.07
Poughkeepsie	1968-72	225	8.7	8.4	8.8	10.7	6.7	+2.3	0.015
Jamestown	1969-72	185	8.5	8.0	9.4	9.3	7.1	+1.3	ns
Binghamton	1969-72	175	6.8	7.6	6.8	6.9	6.3	−0.7	ns
Elmira	1969-72	188	6.0	6.1	5.9	7.1	5.2	+1.0	ns
Group Mean		10	7.9	8.6	7.8	8.3	6.9	−0.3	ns

[a] Sulfate levels expressed as micrograms per standard cubic meter.
[b] JJA level minus DJF level.
[c] Significance of the difference between JJA and DJF means, as determined by Student's *t* test

DJF = December, January and February; or winter; MAM = March, April and May, or spring; JJA = June, July and August, or summer; SON = September, October and November or fall.

urban and nonurban groups of sites show that while the nonurban sites have significantly less sulfate in their atmospheres in winter, spring and fall, and for the year as a whole, the two groups do not differ significantly in summer (Table 24.4). This indicates that if sulfate in nonurban sites is background at all times of the year, then in summer, the local urban contributions present in cities are negligible in comparison to the background.

A high summertime background sulfate level is not physically implausible. It could be due to more efficient conversion of sulfur dioxide to sulfate, promoted by such seasonally varying factors as atmospheric water content, temperature, solar insolation, or the presence of microbiologically produced trace gases such as ammonia.

Alternatively, it is possible that most of the atmospheric sulfate in these nonurban sites in summer is not background sulfate, but sulfate derived from locally generated biogenic H_2S. If so, the similarity in the levels in urban and nonurban sites could be due to the fact that biogenic sources exist in the urban sites, too.

Examination of the summertime mean sulfate levels in the 15 sites in New York state indicates, that sulfate at the one nonurban site there (Jefferson County, located at Cape Vincent on the eastern shore of Lake Ontario) cannot all be derived from a widespread background common to all sites. At Jefferson County the summer mean is 11.9 $\mu g\ m^{-3}$, which is 68% to 80% higher than that in the three closest urban sites (Utica, Syracuse, and Massena). The summertime means for the fourteen urban sites range from 6.6 to 12.6 $\mu g\ m^{-3}$. We must infer that a minimum of nearly half the sulfate at Cape Vincent in summer is local and, therefore, biogenic.

The influence of biogenic sulfur sources on atmospheric sulfate levels is further demonstrated by comparing the levels in a group of four New York cities located on the shores of seriously polluted bodies of water (Buffalo on Lake Erie, Niagara Falls and Rochester on Lake Ontario, and New Rochelle on Long Island Sound) with those in the ten inland cities in New York (Table 24.4). These comparisons show that sulfate levels are significantly higher in the coastal cities in spring, summer and fall, but do not differ significantly in winter, when pollutant sulfur emissions are highest in all cities in this state, in spite of the fact that pollutant sulfur sources are far larger in the coastal cities than in the inland cities.

In addition to confirming a "coastal" effect, these comparisons suggest that biogenic sulfur sources may contribute to atmospheric levels in fall and spring, as well as

Table 24.4
Comparisons of Group Mean Sulfate Levels[a]

		N	Annual	DJF	MAM	JJA	SON
A.	All Urban Sites	18	9.0	9.4	8.8	9.6	8.0
	All Nonurban Sites	7	6.7	5.9	6.0	8.9	6.3
	Significance[b]		0.004	0.00002	0.0005	ns	0.03
B.	NY coastal Urban Sites	4	9.8	9.5	9.8	11.0	8.9
	NY Inland Urban Sites	10	7.8	8.6	7.8	8.3	6.9
	Significance		0.03	ns	0.03	0.06	0.003
C.	Coastal Urban Sites	8	10.5	10.8	10.2	11.5	9.6
	Coastal Nonurban Sites	4	6.8	6.1	5.9	9.8	8.1
	Significance		0.002	0.002	0.0003	ns	ns
D.	Inland Urban Sites	10	7.8	8.6	7.8	8.3	6.9
	Inland Nonurban Sites	3	6.5	5.6	6.1	7.9	6.2
	Significance		ns	0.004	0.09	ns	ns

[a] Sulfate levels expressed as micrograms per standard cubic meter.

[b] As determined by Student's t test.

in summer. Comparisons of urban and nonurban sites after grouping with respect to proximity to coasts provide evidence of major biogenic contributions in fall as well as in summer (Table 24.4). During these two seasons, mean levels in each set of paired groups do not differ significantly. (These are also the seasons when the sulfate levels in the nonurban sites are highest.) In winter, levels in the urban sites are significantly higher than in the nonurban sites, and this is also true in spring in the coastal pair of groups. However, the difference between the means of the inland pair of groups in spring is only marginally significant.

These comparisons between urban and nonurban sites indicate that urban centers have little effect on atmospheric sulfate levels in warm months when microbiological metabolism is most active, but we regard these results with caution because of the small number of sites available for comparison, even though the seasonal variation in the differences indicates that the results are not wholly due to small values for n. In winter and spring, the average geometric mean sulfate level in the coastal urban sites exceeds that in the coastal nonurban sites by 77% and 73%, while in summer and fall the figures are 17% and 19%, respectively. For inland sites, the corresponding figures are 45% and 28% for winter and spring, and 5% and 11% for summer and fall. It should also be noted that while waters local to all coastal urban sites are seriously polluted, there is no local water pollution near two of the coastal nonurban sites and comparatively little near two others (Jefferson County, New York, and Calvert County, Maryland). There appears to be no local water pollution near any of the inland nonurban sites, but there is significant local water pollution near some of the inland urban sites.

CONCLUSIONS

The results of the two studies presented above suggest that biological sulfur sources give rise to measurable levels of particulate sulfate in the urban and nonurban atmospheres, and that these sources are most active in summer and fall, when they appear to exceed local anthropogenic sources. These results confirm and extend the studies of many workers who have presented evidence to show that bacterial sulfate reduction in estuaries or lakes gives rise to atmospheric sulfur. (See especially Koyama et al., 1965; Berner, 1971; Nakai and Jensen, 1967; and Grey and Jensen, 1972.)

Although our results do not demonstrate that water pollution leads indirectly to atmospheric sulfate pollution, this seems very likely in view of the well known fact that organic and nutrient water pollution promotes sulfate reduction in local muds, and also in view of the results of Koyama et al. (1965) and Berner (1971), who associated increases in the amount of sulfur cycling through the atmosphere with water pollution. The results also may explain the observations of Frank (1974), who found that while sulfur dioxide levels in cities have generally decreased in recent years, particulate sulfate levels have either remained the same or increased.

There remain many other questions which must be answered before the influence of microbiological processes on the atmospheric load of sulfate will be fully understood. Subjects requiring further research include details of the atmospheric processes which convert H_2S to particulate sulfate, the factors which influence the emission of H_2S from surface water, and the total magnitude of biogenic contributions in cities where sulfate levels are undesirably high. Recent studies have shown that many respiratory diseases are associated with atmospheric particulate sulfate (EPA, 1974) and consequently we may expect that these and other related topics will receive attention in the near future.

REFERENCES

Altshuller, A. P., 1973: Atmospheric sulfur dioxide and sulfates. *Environ. Sci. Tech.* **7**, 709-712.

Bethea, R. M., 1973: Comparisons of hydrogen sulfide analyses techniques. *J. Air Poll. Control Assoc.* **23**, 710-713.

Berner, R. A., 1971: Worldwide sulfur pollution of rivers. *J. Geophys. Res.* **76**, 6597-6600.

Broecker, W. S. and T. H. Peng, 1974: Gas exchange rates between air and sea. *Tellus* **26**, 21-35.

Cadle, R. D., and M. Ledford, 1966: Reaction of ozone with hydrogen sulfide. *Int. J. Air and Water Pollution* **10**, 25-30.

Castelman, A. W., Jr., H. R. Munkelwitz and B. Manowitz, 1974: Isotopic studies of the sulfur component of the stratospheric aerosol layer. *Tellus* **26**, 222-234.

Chen, K. Y., 1970: Oxidation of Aqueous Sulfide by O_2. PhD Dissertation, Harvard University.

Chen, K. Y., and J. C. Morris, 1972: Kinetics of aqueous oxidation of sulfide in O_2. *Environ. Sci. Tech.* **6**, 529-537.

Conway, E. J., 1943: Mean geochemical data in relation to oceanic evolution. *Proc. Roy. Irish Acad.* **A48**, 119-159.

Cox, R. A. and F. J. Sandalls, 1974: The photo-oxidation of hydrogen sulphide and dimethyl sulphide in air. *Atmos. Environ.* **8**, 1269-1281.

Duce, R. A. and A. H. Woodcock, 1971: Difference in chemical composition of atmospheric sea salt particles produced in the surf zone and on the open sea in Hawaii. *Tellus* **23**, 428-435.

Eriksson, E., 1960: The yearly circulation of chloride and sulfur in nature; meterological, geochemical and pedological implications. Part II. *Tellus* **12**, 63-109.

Environmental Protection Agency, 1974: Health Consequences of Sulfur Oxides; A Report from CHESS, 1970-1971. EPA-650/1-74-004 USEPA Research Triangle Park, North Carolina.

Frank, N. H., 1974: Temporal and spatial relationships of sulfate, total suspended particulates, and sulfur dioxide. Presented at Air Pollution Control Association annual meeting, Denver, Colorado, June, 1974.

Friend, J. P. 1974: The global sulfur cycle. In: *Chemistry of the Lower Atmosphere*, S. I. Rasool, ed. (New York: Plenum Press), pp. 177-201.

Grey, D. C., and M. L. Jensen, 1972: Bacteriogenic sulfur in air pollution. *Science* **177**, 1099-1100.

Goldhaber, M. B. and I. R. Kaplan, 1974: The sulfur cycle. In: *The Sea*, Vol. 5, E. D. Goldberg, ed. (New York: John Wiley & Sons) pp. 569-655.

Hitchcock, D. R., 1975: Biogenic contributions to atmospheric sulfate levels. Paper presented at 2nd Annual Conference on WateReuse, Chicago, Illinois, May 1975.

Holser, W. T. and I. R. Kaplan, 1966: Isotope geochemistry of sedimentary sulfates. *Chem. Geol.* **1**, 93-135.

Hoffman, G. L., R. A. Duce and E. J. Hoffman, 1972: Trace metals in the Hawaiian marine atmosphere. *J. Geophys. Res.* **77**, 5322-5329.

Junge, C. E., 1972: The cycle of atmospheric gases—natural and man-made. *Quart. J. Roy. Met. Soc.* **98**, 711-729.

Kaplan, R. R., K. O. Emery and S. C. Rittenberg, 1963: The distribution and isotopic abundance of sulphur in recent marine sediments off southern California. *Geochim. Cosmochim. Acta* 27, 297-321.

Kaplan, R. R. and S. C. Rittenberg, 1964: Microbiological fractionation of sulfur isotopes. *J. Gen. Microbiol.* 34, 196-212.

Kellogg, W. W., R. D. Cadle, E. R. Allen, A. L. Lazrus and A. E. Martell, 1972: The sulfur cycle. *Science* 175, 587-596.

Koyama, T., N. Nakai and E. Kamata, 1965: Possible discharge rate of hydrogen sulfide from polluted coastal belts in Japan. *J. Earth Sci.* 13, 1-11.

Liss, P. S. and P. G. Slater, 1974: Flux of gases across the air-sea interface. *Nature* 247, 181-184.

Lovelock, J. E., R. J. Maggs and R. A. Rasmussen, 1972: Atmospheric dimethyl sulphide and the natural sulphur cycle. *Nature* 239, 252-253.

Nakai, N. and M. L. Jensen, 1967: Sources of atmospheric sulfur compounds. *Geochim. J.* 1, 200-210.

Penkett, S. A., 1972: Oxidation of SO_2 and other atmospheric gases by ozone in aqueous solution. *Nature Phys. Sci.* 240, 105-106.

Rasmussen, R. A., 1974: Emission of biogenic hydrogen sulfide. *Tellus* 26, 254-260.

Robinson, E. and R. C. Robbins, 1968: Sources, Abundance, and Fate of Gaseous Atmospheric Pollutants. SRI Project Report PR-6755, American Petroleum Institute, New York.

Sangster, D. F., 1968: Relative sulphur abundances of ancient seas and strata-bound sulphide deposits. *Geol. Assoc. Can. Proc.* 19, 79-91.

Smith, S. V., M. K. Chave and D. Kan, 1973: Atlas of Kaneohe Bay. Technical Report UNIHI Sea Grant TR-72-01 University of Hawaii, Honolulu.

Tanner, T. M., J. A. Young and J. A. Cooper, 1974: Multi-element analysis of St. Louis aerosols by nondestructive techniques. *Chemosphere* 5, 211-220.

Tucker, E. D. (ed.), 1969: The atmospheric diagnostic program at Brookhaven National Laobratory. BNL-50206. Brookhaven National Laboratory, Upton, Long Island, New York.

CHAPTER 25

A METHOD FOR THE DIRECT DETERMINATION OF
SULFIDE IN WATER AND SEDIMENTS

PAUL GIAMMATTEO

Department of Zoology, University of Georgia
Athens, Georgia 30602; Present Address: 509
Woodlawn Avenue, Newark, Delaware 19711

INTRODUCTION

Measurement of sulfide in the natural environment is generally a tedious process. Only two methods for the direct determination of sulfide are known—the silver/silver sulfide electrode and direct UV spectrophotometry. The latter is applicable only to water samples and is subject to strong interferences from some oxidation products of sulfide, notably thiosulfate, tetrathionate and sulfite. The sulfide electrode, though sensitive, measures concentrations in a microenvironment. This is inappropriate for the determination of sulfide in many cases because of the patchy distribution patterns often found in sediments. The methods enjoying the most widespread use are the iodometric titration and methylene blue methods. The titration is generally used for large samples and is suitable for detection of sulfide in concentrations higher than 1 mg/liter (Standard Methods, 1971). The methylene blue method is used for the determination of small concentrations (in the ppb range) of sulfide. Reported errors vary from 10% to 1% (Standard Methods, 1971; Chen and Morris, 1972), with the latter generally achieved after proper manipulation of reagents and when the sulfide range is known. This method is time-consuming, requiring approximately an hour and a half per determination (Budd and Bewick, 1952). Neither method can be used directly, mainly because of

interferences, and both are subject to some interferences
even when used as prescribed (Gustafsson, 1960a,b). Sulfite and thiosulfate interfere with methylene blue if
their concentrations are greater than 10 mg/l (Standard
Methods, 1971); nitrite interferes at 0.5 mg/l. Copper and
mercury are also reported as interferences (Beaton, Burns
and Platou, 1968). Organic sulfur compounds may interfere
by liberating various amounts of sulfide during the extraction (Gustafsson, 1960a,b). Furthermore, the methylene
blue reaction must be run under a nitrogen atmosphere
with strict temperature control.

Sulfide can be measured using a carbon tetrachloride
solution of silver dithizonate. The technique offers some
advantages over other techniques in that it is sensitive
and accurate on a level comparable to the methylene blue
method, but easier to use. Furthermore, it can be used to
measure sulfide directly, without additional acid extractions. The reagent appears to be selective and is relatively
immune to interferences. It is also capable of distinguishing between free and bound sulfide.

MATERIALS AND METHODS

Preparation and Calibration of the Reagent

The reagent is prepared by adding 0.2 g of dithizone
(Eastman) to 400 ml of carbon tetrachloride (reagent grade)
in a liter separatory funnel. Shake to dissolve the dithizone. Add a solution of silver nitrate (5 g $AgNO_3$ in
100 ml distilled water) followed immediately by 5 ml of
6 M H_2SO_4. Shake vigorously. The green dithizone solution
turns orange in the presence of the silver ion. When the
orange color appears, vent the funnel. Continue shaking
for about 30 sec, then allow the layers to separate. At
this point a thick, orange floc forms in the upper aqueous
layer and the carbon tetrachloride layer appears somewhat
cloudy. Collect the carbon tetrachloride layer, wash it
twice with about 200 ml distilled water, then store in a
foil-covered bottle. Kunkel, Buckley and Gorin (1959)
prefer to filter the washed solution of silver dithizonate,
but this seems to be unnecessary as the washed solution is
quite clear.

Preliminaries to the actual measurement of sulfide
include preparing the buffer and finding the appropriate
concentration of silver dithizonate. Because of variations
in ionic strength, a dilution factor for each batch of dithizonate reagent must be determined empirically. This

is done by measuring absorption at 460 nm (ABS_{460}), the
wavelength at which silver dithizonate shows maximum absorbance. (The reacted product is read at 618 nm.)
   To measure sulfide concentrations in the range of 20
to 100 µg, it is necessary to prepare a reagent solution
such that a 1:10 dilution results in an ABS_{460} of 0.62.
The 1:10 dilution is necessary because the concentration of
reagent actually used to measure sulfide is too opaque to
read on the spectrophotometer at 460 nm. The reagent stock,
when diluted 0.45:10 (with CCl_4) results in an ABS_{460} of
0.60, with a deviation of approximately 0.15. This can be
corrected to read 0.62 by addition of CCl_4 or silver dithizonate, whichever is appropriate. When the proper dilution
has been found, record the percentage silver dithizonate, then
multiply by 10 to find the percentage silver dithizonate
to be used in the reagent solution. Once the reagent
solution has been prepared, it is wise to make a 1:10 dilution and check the absorbance at 460 nm. The procedure for
preparing the adjusted reagent solution involves placing
carbon tetrachloride in the reference cell (1 cm) of a
Beckman DB-G spectrophotometer, and the diluted stock solution in the sample cell. The instrument is zeroed at 618
nm, then absorbance of silver dithizonate read at 460 nm.
The instrument is zeroed at 618 nm because that is the
wavelength at which the reaction product is measured.
This also provides an internal correction for any absorbance due to silver dithizonate at this wavelength.

Preparation of the Buffer

   Preparation of the buffer solution employs the method
described in the *Handbook of Chemistry and Physics* (42nd
ed., p. 1718). It is necessary to substitute 0.1 M H_2SO_4
for 0.2 M HCl. Also, K_2SO_4 is used in place of KCl. This
should result in a solution of pH 2.2. Buffer can be stored
for considerable periods of time in polyethylene bottles
with no measurable change in pH.

The Standard Sulfide Solution

   One liter of distilled water was added to a liter flask
with side arm. The side arm of the flask had been previously
filled with aquarium sealer (Silicone Rubber Aquarium Sealer,
Dow Corning Corp.), then capped with a rubber septum. This
flask, or one similarly prepared, was used for all subsequent determinations requiring a known amount of sulfide.
The water was purged with commercial N_2 gas for 2 hr prior
to use to decrease the concentration of dissolved O_2.

A standard sulfide solution, containing 20 μg sulfide-
sulfur per liter, was prepared by dissolving 0.15 g $Na_2S \cdot 9H_2O$ (Fisher Chem.) in the water. Aliquots were withdrawn
from the standard solution through the enclosed side arm.
This minimizes introduction of air. Transfers involving
sulfide solutions were done with a 5-ml Hamilton syringe
equipped with a Chaney adapter.

Measurement of Sulfide in Aqueous Solutions

   The sulfide ion can be detected in water, sediments, and
possibly air. Add 15 ml of reagent using 10-ml disposable
pipettes, then 10 ml of buffer (pH 2.2) to a 125-ml pyrex
flask. Cap the flask with a rubber stopper. The liquid
in the reaction vessel separates into two layers—the buffered aqueous portion settling on the surface of the carbon
tetrachloride solution. Add the sample to the aqueous layer,
replace the cap, and mix on a vortex mixer at low speed
(3 on a scale of 1 to 10) for 30 sec. Do not shake. Let
the solution stand for about an hour to allow for complete
color development. The orange silver dithizonate layer
turns green in the presence of a sufficient amount of
sulfide. Withdraw 1 ml of the dithizone solution volu-
metrically, dilute with 9 ml of carbon tetrachloride, and
read absorbance at 618 nm against a carbon tetrachloride
reference. Silver dithizonate has a negligible absorbance
at 618 nm. The reaction product, presumably free dithizone,
absorbs strongly at this wavelength.

Preparation of the Standard Curve

   Reaction vessels were prepared with the appropriate
amount of silver dithizonate and buffer. Sulfide-containing
solutions were withdrawn from the standard solution as
described above, in most cases using integral aliquots from
1 to 5 ml. It was unnecessary to acid-wash the syringe
between samples. Standards were introduced into a reaction
vessel by removing the stopper and injecting the sample
into the aqueous layer. The reaction vessel was recapped
immediately and mixed on the Vortex mixer as described
above. Determination of sulfide by absorbance at 618 nm
has already been described. Results of the standard curve
determinations and the effect of varying sample size will
be discussed in the next section.

Measurement of Sulfide in Mud

Measuring sulfide in a mud sample is essentially the same as for a water sample. Reaction vessels are prepared in the same manner and the mud can be introduced in whatever way is feasible. For the work described here, 1- to 2-g aliquots (wet weight) were introduced into the acid layer via weighing spatulas, then mixed. Before reading absorbance, it is necessary to withdraw approximately 4 to 6 ml of the carbon tetrachloride solution from the reaction flask and centrifuge it for 1 min at 2000 rpm. This removes any particles that may have settled into the carbon tetrachloride layer.

In the course of this work, it was necessary to determine the efficiency of extraction of sulfide from mud. Description of this method is more appropriately considered with discussion of the results.

Cleaning

Throughout the course of experimentation the glassware was rinsed in acetone, then washed in hot, soapy water, rinsed well with tap water, and finally with distilled water. Flasks used for sulfide standards were acid-washed.

RESULTS

The Standard Curve

A concentration of silver dithizonate resulting in an ABS_{460} of 0.62 was chosen as a reasonable concentration with which to prepare a standard curve. At this concentration silver dithizonate is capable of detecting between 20 and 100 µg of sulfide. The standard curve (Figure 25.1) was prepared from three sets of data collected over a period of three days. The actual concentration of reagent as measured at 460 nm ranged from 0.63 to 0.655. All the determination was done from the same batch of reagent. The response indicated by the standard curve did not follow Beer's Law. Thus, as might be expected, a simple linear regression to determine the equation of the line proved to be inappropriate. Similar results were obtained after applying log-log and semi-log transformations. The line best fitting the data proved to be a second degree polynomial. The equation

$$y = -0.280333 + 0.1551(x) - 0.000075(x^2)$$

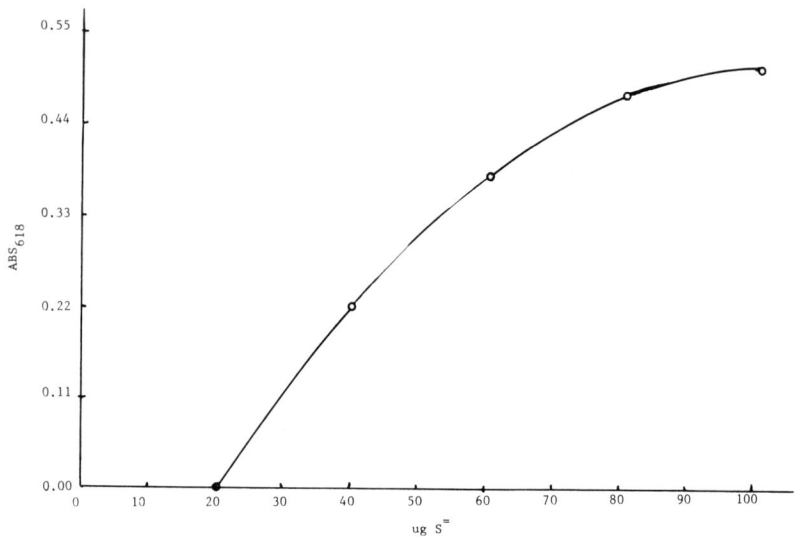

Figure 25.1 The standard curve. Each point is an average of six readings, taken over a period of three days, using reagent solution whose ABS_{460} ranged from 0.63 to 0.655.

was generated by the Polynomial Regression Program (Hewlett Packard, Model 20 Stat Pac I-10) using the Hewlett Packard 9820A calculator. Degree of fit, R_{xy}, equaled 0.961. Statistical computations on a line of this nature involve nonparametric procedures. In view of the above difficulties computation of error was handled as follows: Confidence Intervals (CI) for absorbance were determined at each measured concentration of sulfide. Percent error was calculated using the formula

$$\% \text{ error} = \left\{ [CI/(dy/dx)]/\mu g\ S^= \right\} \times 100$$

where dy/dx = slope of the tangent at a given concentration of sulfide. This formulation seemed reasonable in view of the fact that there was little change in slope over any 10-μg range in sulfide concentrations. The resulting error was in all cases less than ±6%, and in most cases less than ±3% (Table 25.1).

Color Development and Stability

Determination of the absorption characteristics of the product over a period of time is essential for any spectro-

Table 25.1
Calculation of Error of the Standard Curve.
CI = Confidence Interval; % Error = $[(CI/Slope)/\mu g\ S^=] \times 100$

$S^=\mu g$	Slope	Confidence Interval	% Error
20	0.0125	0.003	1.04
30	0.0110	0.010	
40	0.0095	0.010	2.62
50	0.0080		
60	0.0065	0.010	2.31
70	0.0050		
80	0.0035	0.015	5.36
90	0.0020		
100	0.0005	0.013	2.60

photometric technique. For this reagent, it seems that full color development occurs after 30 min and may remain unchanged for up to nine days (Table 25.2). For added assurance, no measurement reported in this work, other than those specifically intended, was taken within less than 1 hr, or more than 3 hr, after reaction. Data indicate that the reaction product, when left at room temperature, is stable for up to 24 hr in the light and up to 9 days in the dark. An unusual feature, not tabluated, is the response of the reaction product to chilling (-15°C)—an increase in absorbance relative to a nonchilled control can be seen. Tests indicating this response were conducted at a reagent concentration of approximately 0.07 absorbance units at 460 nm.

Mixing Intensity and Time

All determinations of sulfide, except where indicated, involved the use of a variable-speed Vortex Genie Mixer (Fisher Sci. Co.) as the dispersive agent initiating contact between the aqueous and organic layers in the reaction vessel. Shaking the solution by hand resulted in a much larger variance than that found when the sample was mixed using the Vortex mixer. Further study of the mixing regime was conducted to determine the effect of mixing intensity and time on the absorbance characteristics of the final product. There is insufficient data to make a definitive statement, but t-tests on a few samples indicated that there is no difference in ability to detect sulfide due to mixing intensity. However the F test for variance

Table 25.2
Stability of the Product[a]

Sample Number	Time (hr)	ABS_{618}	Illumination	Sample Number	Time (hr)	ABS_{618}	Illumination
2	0.5	0.243	Light	3	0.5	0.252	Dark
	24.0	0.237	Light		24.0	0.268	Dark
	72.0	0.193	Light		432.0	0.337	Dark
5	1.0	0.246	Light	6	1.0	0.262	Dark
	24.0	0.239	Light		24.0	0.269	Dark
7	2.0	0.253	Light	12	4.0	0.252	Dark
	24.0	0.236	Light		24.0	0.258	Dark
	72.0	0.212	Light		72.0	0.260	Dark
10	4.0	0.231	Light	15	6.0	0.246	Dark
	24.0	0.235	Light		24.0	0.246	Dark
13	6.0	0.241	Light		216.0	0.249	Dark
	24.0	0.232	Light	18	8.0	0.252	Dark
	72.0	0.173	Light		24.0	0.263	Dark
17	8.0	0.251	Light		72.0	0.268	Dark
	24.0	0.240	Light				

[a] All samples used for stability determinations were reacted with 40 µg of sulfide.

was significant when the samples compared were mixed at intensities of 3 and 6, indicating that a greater variance may be associated with those samples mixed at the lower intensity. There seemed to be a slight increase in absorbance of the product if mixing occurred for more than 30 sec, but this was not subjected to statistical testing because it was felt that more data were needed (Table 25.3).

Variations Due to Sample Size

Sample size is a legitimate, well-recognized variable in any study. Two aspects of concern in this work were the variabilities associated with small samples (1 to 5 ml) and large samples (greater than 25 ml). The standard curve was generated from a standard sulfide solution by taking aliquots ranging in size from 1 to 5 ml. Since there is an inherent potential for error in subsamples this small, this may have been a source of variation in the standard curve. To test this possibility a number of sulfide solutions were prepared at different concentrations. Concentrations were manipulated so that a 1-ml sample from one solution contained the same amount of sulfide as a 5-ml sample from another solution, and so on. Results of these

Table 25.3
Mixing Intensity and Time[a]

µg $S^=$	Mixing Intensity	Mixing Time (sec)	ABS_{618}	F 0.05	(sig)	t 0.975	(sig)
40	3	15	0.278				
		30	0.298				
		60	0.327				
		120	0.327				
		180	0.318				
	6	15	0.292				
		30	0.288				
		60	0.292				
		120	0.303				
		180	0.309	5.82	(−)	1.243	(−)
60	3	30	0.380				
			0.362				
		60	0.418				
			0.398				
		120	0.400				
			0.405				
	6	30	0.394				
			0.400				
		60	0.391				
			0.388				
		120	0.395				
			0.385	13.78	(+)	0.198	(−)
40	3	30	0.266				
		60	0.260				
		180	0.258				
	6	30	0.243				
		60	0.250				
		180	0.250	22.28	(+)	0.029	(−)

[a] The statistics compare mixing intensities.

tests are presented in Table 25.4. There seems to be no difference in the mean absorbance due to sample size at any given concentration of sulfide.

Another concern involving sample size was determining the maximum volume that the reagent could handle effectively. Clearly, the larger the sample size, the smaller the error will be. It was found that sample volumes up to 53 ml resulted in an error of less than 10%. When converted to ppm this is equivalent to a lower limit of detection in water of 0.4 ±0.04 mg/l of sulfide. This

Table 25.4
Variation Due to Sample Size[a]

	μg S⁼ Added	ml Sample (Total)	ABS_{618}
Large Sample Size	60	3	0.335
			0.352
		53	0.312
			0.315
Small Sample Size	60	1	0.365
			0.332
		3	0.372
			0.352
		5	0.351
			0.359
	30	1.5	0.085
			0.083
		2.5	0.091
			0.086

[a] Student's t-test indicates a significant difference in the means of the large (53-ml) and small (3-ml) samples. See text for discussion.

limit can be lowered further by reducing the concentration of reagent. Sample volumes larger than 50 ml are probably best handled in larger flasks, but the problem was not extensively pursued beyond this level. It is also advisable to increase the quantity of buffer to 15 ml if samples of large size are to be taken.

Additive effects of sulfide on the reagent were also studied. It was found possible to add sulfide to the reagent in increments, within limits of the sample size, with seemingly little effect on its ability to detect sulfide (Table 25.5). This may prove practicable in field situations where no indicators of sulfide concentration, such as color or smell, are evident.

Determinations of Sulfide in Sediments

Silver dithizonate can also be used to determine sulfide concentration in mud. Laboratory work regarding this aspect of the reagent was directed toward determining the extraction efficiency from the mud milieu. The procedure involved adding a known quantity of sulfide to a sediment

Table 25.5
Sulfide Addition in Increments

μg $S^=$ Added		Mixing Intensity	Mixing Time (sec)		ABS_{618}
Initial Sample	Increment		Initial Sample	Increment	
20.3	20.3	3	30	15	0.275
20.3	20.3	3	30	15	0.270
40.6		3	30		0.268
20.3	40.6	3	30	15	0.364
20.3	40.6	3	30	15	0.369
60.9		3	30		0.378

sample, performing the analysis, then comparing the results to controls. The actual manipulations were carried out by placing an aliquot (0.1 to 0.2 g wet weight) of highly reduced mud into an empty reaction vessel. A small volume (1 to 2 ml) of known sulfide solution was then added to the mud and mixed to form a slurry. Reagent and buffer were immediately added from a previously prepared flask and the normal procedure then followed. Highly reduced mud was used in preference to a more oxidized sediment to minimize the possible error associated with oxidation of sulfide by the ferric ion. Results are presented in Table 25.6, together with computed efficiency. Disregarding the two extremes, the recovery ranges from 85.1% to 109.1%. There is no statistical difference between the actual and expected recovery; the wide range noted is most probably due to the variance associated with the controls. This indicates essentially complete recovery of added sulfide from mud. This result is supported by qualitative observation. The black color of the mud sample disappears over the course of the reaction to be replaced by the brownish red color characteristic of low-sulfide sediments. These results are interpreted to indicate essentially complete recovery of sulfide in the monosulfidic forms—greigite, mackinawite, and noncrystalline iron sulfide. In all probability, the disulfidic forms—pyrite and marcasite—are not detected by this method due to the extreme conditions needed to liberate sulfide from these minerals (Berner, 1970).

During the course of experimentation with mud, duplicate sets of sediment were reacted at pH 2 and pH 6. In all cases, the samples determined at pH 2 showed the characteristic green color indicating the presence of sulfide. No green was seen in any sample reacted at pH 6, even though some samples reacted at pH 2 indicated up to 300 μg sulfide

Table 25.6
Spiking Experiments[a]

S= Added (µg)	Dry Weight (g)	S= Recovered (µg/g Dry Weight)	S= Expected (µg/g Dry Weight)	% Recovery
--	0.1153	407.6		
--	0.1763	330.6		
--	0.0937	345.8		
--	0.1004	305.8		
--	0.1675	323.0		
16.7	0.0850	443.5	521.2	85.1
	0.1415	382.3	443.1	86.3
	0.1458	414.3	439.6	94.3
25.1	0.1225	578.0	529.8	109.1
	0.1217	491.4	530.8	92.6
	0.0778	602.8	647.8	93.1
25.6	0.2070	468.6	492.8	95.1
30.7	0.1178	565.4	629.6	89.8
33.6	0.1241	415.8	595.5	69.8
	0.0504	1059.5	992.1	106.8
	0.1343	545.1	574.8	94.8
41.0	0.0878	1070.6	836.0	128.1
	0.1355	649.5	671.9	96.7

[a] Statistics comparing S= expected *vs* S= recovered indicate no significant difference between means or variances associated with each grouping.

per g dry weight. Furthermore, aqueous samples of sulfide, when reacted at pH 6, responded similarly to those reacted at pH 2. At pH 6 the sulfide bound to heavy metals is not likely to be liberated, and empirically may not be detected by silver dithizonate. These results indicate that the method is capable of distinguishing between free and bound sulfide, if so desired, simply by changing the pH regime at which the reaction is run.

Interferences

Development of the method was pursued with the intent of using it as a direct measure of sulfide in the environment. Specificity is thus an extremely important parameter. The ions tested included the common ions of salt and fresh water, reduced compounds, and other sulfur compounds. Results are presented in Table 25.7. The values reported for each ion in the table are the actual quantities that

Table 25.7
Interferences[a]

Ion	µg Added	Interference	Ion	µg Added	Interference
Na^+	50705	−	NO_2^-	1973	+
Mg^{++}	6100	−		493	+
Ca^{++}	2300	−	NO_3^-	2656	−
K^+	1845	−	Fe^{+++}	600	−
Sr^{++}	30	−	NH_4^+	52	−
Mn^{++}	5	−	SO_3^{--}	1003	−
Cl^-	92525	−	$S_2O_3^{--}$	1391	+
SO_4^{--}	12510	−		69.5	−
HCO_3^-	700	−	Hg^{++}	22200	−
Br^-	85	−	Cys(e)	305	−
$BO_3^{\equiv}$	110	−	Met	600	−
			β-Mercaptoenthanol	620	+

[a] (−), no detectable interference; (+), detectable interference. See text for discussion.

were tested. For the discussion following, the levels of nitrite and thiosulfate have been converted to ppm. This conversion assumes use of a 1-ml sample size. However, it is expected that the absolute quantity of an ion, rather than its concentration, is more important in determining interfering activity.

Of those ions tested, only two potential interferences were found, with the interfering activity occurring at uncommonly high levels in both cases. A concentration of 1973 ppm nitrite discolored the reagent immediately; whereas at 493 ppm discoloration did not occur before 2-3 hr. The above trend is expected to continue. Thiosulfate interfered at a concentration of 1391 ppm, but not at 69.5 ppm. It is interesting to note that even at the higher concentration, there was no interference from thiosulfate if tested at pH 6. Hence, interference by this ion may be due to formation of H_2S from thiosulfate at low pH. Alternatively, thiosulfate may interfere by removing silver from the dithizonate complex (Welcher, 1947). Even though nitrite and thiosulfate interfere with the reagent, their concentrations in the environment are so low that the problem can be disregarded (Hutchinson, 1957; Raymont, 1965; and Sorokin, 1972). Cysteine and methionine did not interfere at the concentrations tested. This was somewhat surprising considering that the technique was designed to measure organic

sulfides, though it should be noted that the compounds tested
originally were all alkane thiols. The explanation for
these results may involve effects related to the activity
of amino and carboxy groups of the two amino acids, or,
perhaps in the case of methionine, the placement of the
sulfide moiety. β-mercaptoethanol (0.01 mmole) was found
to interfere, but this compound is not to be expected in
the environment. It should be noted that all interference
tests were conducted at a reagent concentrations (ABS_{460})
of 0.07. This is sufficiently dilute to detect 3 µg of
sulfide.

Field Work

Measurements were made on a number of sediment cores
collected from two dissimilar impoundments. Lake Jackson,
near Atlanta, Georgia was created as a hydroelectric source
in 1907. It is fed by three major streams, at least one
of which serves as a carrier for effluent from Atlanta's
sewage plants. Par Pond, in South Carolina, has had a quite
different history, being a cooling pond for heated wastewater from the Savannah River Plant. Cores were taken with
a gravity coring device fitted with polyethylene tubes from
several locations in Lake Jackson and one location in Par
Pond. These were transported to the laboratory and either
analyzed within 12 hr after collection or stored in the
cold (10°C) until needed. Cores were analyzed for various
cations and for sulfide.

Cores were subsampled by extruding the contents onto
a cutting board calibrated in cm. A subsample was taken
at 2- or 5-cm intervals and immediately analyzed for sulfide.
A second subsample was stored at 10°C in Whirl Paks (Nasco)
for subsequent analysis of cations.

Cations were analyzed for both acid and base extractable fractions. The subsample was partitioned into two
aliquots. One was used for duplicate acid extractable determination, and the other for duplicate base extractable
determination. Approximately 1-g wet weight aliquots were
placed in glass scintillation vials, to which 10 ml of
6 N HCl or 0.1 N KOH were added; the acid extract was
refluxed overnight. Refluxing was done in a scintillation
vial capped with a marble. Heat was supplied by a continuous flow of hot water. The extracts were collected after
24 hr following centrifugation. Base extract was centrifuged
at 14000 rpm for 20 min in a Sorvall RC 2 centrifuge.
Acid extracts were collected after centrifuging for 15 min
at 2500 rpm in an International centrifuge. Samples were
then brought up to a common volume of 25 ml and analyzed

on a Perkin Elmer Atomic Absorption Spectrophotometer Model 303. Results are presented in Figures 25.2, 25.3 and 25.4.

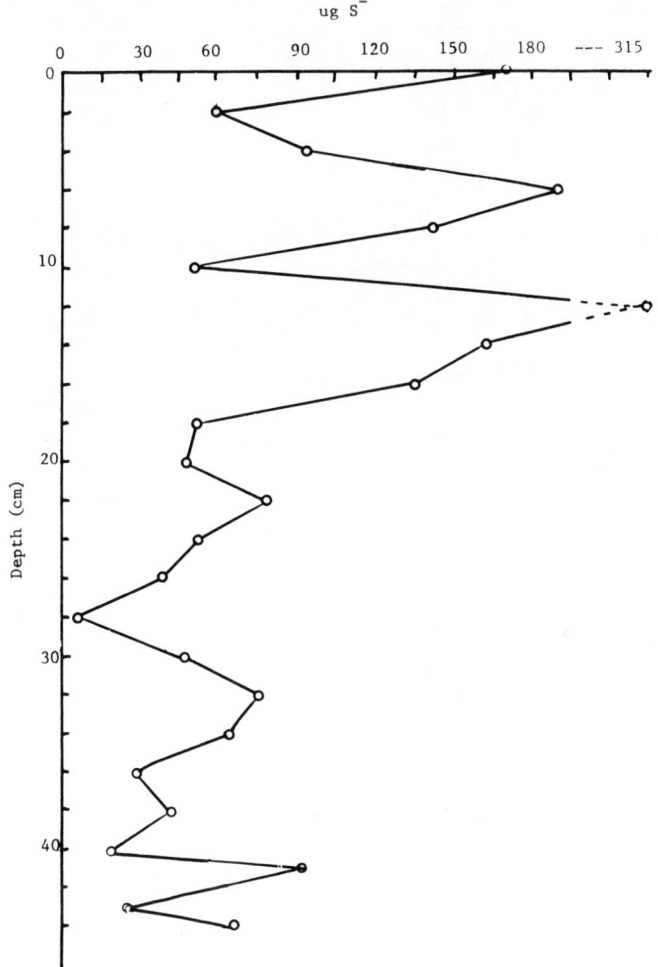

Figure 25.2   Sulfide determined on a core from Lake Jackson at 2-cm intervals.

The purpose of these experiments was to determine the feasibility of the sulfide technique for field work and also to determine whether or not metal deposition could be explained as a function of sulfide concentration. It is sufficient for the purposes intended to note that the quantity of sulfide determined varied with qualitative

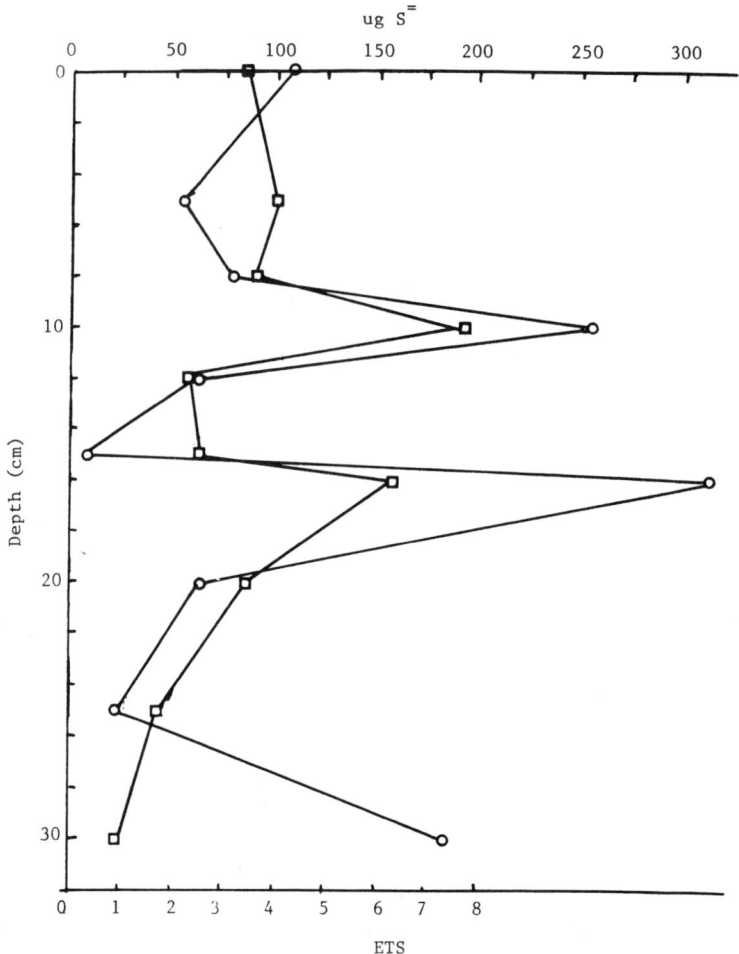

Figure 25.3 Sulfide and ETS determined on a core from Lake Jackson. ○———○ $S^=$; □———□ ETS.

coloring of sediments—blacker zones yielding higher concentrations of sulfide. Sulfide patterns also correlate with patterns of other parameters in these lake sediments— notably ETS, chlorophyll and pollen (Zimmerman, in press). With respect to metal deposition, it is seen (Figure 25.4) that sulfide deposition alone cannot explain the prodigious quantities of metal, especially iron, present in the Par Pond sediments. The Lake Jackson results are similar. Further discussion of the dynamics of these lakes is inappropriate for this paper.

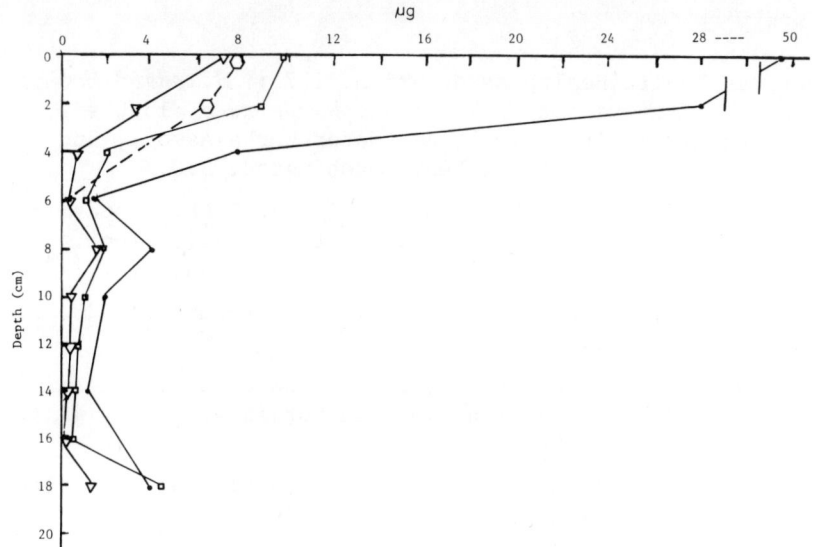

Figure 25.4 Sulfide and metals determined on a core from Par Pond.
Fe x 10^{-1} ●—●; Mn x 10^0 □—□
Zn x 10^1 ▽—▽; $S^=$ x 10^0 ⬡—⬡

DISCUSSION

A method for the measurement of sulfide using a carbon tetrachloride solution of silver dithizonate has been presented. The technique is accurate, reproducible, sensitive, and practically immune to interference. Furthermore, the method can distinguish between free and bound sulfide by changing the pH regime at which the reaction is run. Also, the self-extracting properties of the reagent allow a considerable saving in analysis time. For these reasons the method appears to be useful for measuring sulfide in mud and water.

The mechanism of the reaction is unknown but is suspected to involve the stripping of silver by sulfide from the silver dithizonate complex (Kunkel, Buckley and Gorin, 1959). The effect of variables such as pH, temperature, volume of reagent, and concentration of reagent have not yet been fully investigated. It was necessary first to develop a working technique. A more lengthy discussion of these topics can be found in Giammatteo (1975).

REFERENCES

American Public Health Association, 1971: *Standard Methods for the Examination of Water and Wastewater*, 13th ed., Am. Public Health Assoc., Am. Water Works Assoc., and Water Pollution Control Fed., Washington, D.C.

Beaton, J. D., G. R. Burns and J. Platou, 1968: Determination of Sulphur in Soils and Plant Material. Tech. Bull. #14, The Sulphur Institute, Washington, D.C.

Berner, R. A., 1970: Sedimentary pyrite formation, *Am. J. Sci.* $\underline{268}$, 1-23.

Budd, M. S. and H. A. Bewick, 1952: Photometric determination of sulfide and reducible sulfur in alkalies. *Anal. Chem.* $\underline{24}$, 1536-1540.

Chen, K. Y. and J. C. Morris, 1972: Kinetics of oxidation of aqueous sulfide by O_2. *Environ. Sci. Technol.* $\underline{6}$, 529-537.

Giammatteo, P. A., 1975: Silver dithizonate method to measure sulfide. M. S. Thesis, University of Georgia, Athens, Georgia.

Gustafsson, L., 1960a: Determination of ultramicro amounts of sulphate as methylene blue--I. The Color Reaction. *Talanta* $\underline{4}$, 227-235.

Gustafsson, L., 1960b: Determination of ultramicro amounts of sulphate as methylene blue--II. The Reduction. *Talanta* $\underline{4}$, 236-243.

Hodgman, C. D., R. C. Weast, and S. M. Selby (eds.), 1960: *Handbook of Chemistry and Physics*. (Cleveland, Ohio: The Chemical Rubber Publishing Co.).

Hutchinson, G. E., 1957: *A Treatise on Limnology*. Vol. 1. (New York: John Wiley and Sons, Inc.), 1015 pp.

Kunkel, R. K., J. E. Buckley, and G. Gorin, 1959: Determination of alkanethiols in hydrocarbons with silver ion and dithizone. *Anal. Chem.* $\underline{31}$, 1098-1099.

Raymont, J. E. G., 1963: *Plankton and Productivity in the Oceans*. (Oxford: Pergamon Press Ltd.), 660 pp.

Sorokin, YU. I., 1972: The bacterial population and the processes of hydrogen sulfide oxidation in the Black Sea. *J. Cons. Int. Explor. Mer* $\underline{34}$, 423-454.

Welcher, F. J., 1947: *Organic Analytical Reagents*, Vol. III. (D. Van Nostrand Co., Inc.), 593 pp.

Zimmerman, A. P., 1974: Electron transport analysis as an indicator of biological oxidation in freshwater sediments. *Verh. Internat. Verein. Limnol.* (in press).

CHAPTER 26

SELENIUM IN BIOLOGICAL SYSTEMS, AND PATHWAYS
FOR ITS VOLATILIZATION IN HIGHER PLANTS*

BARBARA-ANN GAMBOA LEWIS

Argonne National Laboratory, 9700 South Cass Avenue
Argonne, Illinois 60439

Selenium is a nonmetallic element often included in the
group of "trace metals" of environmental concern. It occurs
naturally in the biosphere (see Table 26.1), and is also
released to the environment via industrial waste and burning
of fossil fuels. Prior to the mid-1950s, selenium was considered to be toxic to most biological systems even at
relatively low concentrations. Work on liver necrosis
(Schwarz and Foltz, 1957), however, and continual improvements in analytical methods for selenium have led to the
present-day concept that this element is essential to animal well-being. The nutritional requirement for selenium
is in the range of 0.1-0.3 ppm in the diet.

Most of the work implicating selenium in animal nutrition has centered around the interrelationships among four
compounds, namely α-tocopherol or Vitamin E (a fat-soluble
vitamin), Coenzyme Q or Ubiquinone (an aromatic isoprenoid
lipid), the sulfur amino acids cystine and methionine, and
the element selenium (Olson et $al.$, 1965). The members
of this group can replace each other to various degrees
in preventing signs of given deficiency symptoms such as
unthriftiness ($i.e.$, lack of vigor, often associated with
infertility and periodontal disease) in rats and sheep;
muscle dystrophy in rabbits, guinea pigs, and lambs; exu-
dative diathesis and encephalomalacia in chicks; hepatic

---
*Paper prepared under the auspices of the U.S. Nuclear
Regulatory Commission.

necrosis in rats, pigs, and other animals. Olson reported the turnover of these nutrients by weight, setting Se = 1: Coenzyme Q = 10^2, α-tocopherol = 10^3, and total sulfur amino acids = 10^5, based on rate of dietary intake. The ratio of average tissue concentration in micrograms/gram is Se = 1, Coenzyme Q = 40, α-tocopherol = 25, and organic sulfur = 2000. These ratios suggested catalytic roles for all the group members except the sulfur amino acids (Olson, 1965).

An antioxidant role for selenium has also been postulated. Vitamin E and other antioxidants have been found to retard deteriorative processes, and the close relationship of selenium and Vitamin E in animal nutrition has implicated selenium also as an antioxidant (Olson et al., 1965; Tappel and Caldwell, 1967; Rotruck et al., 1973).

There is indirect evidence that selenium deficiency may play a role in SID (Sudden Infant Death) (Money, 1971; Rhead et al., 1972) and carcinogenesis (see the review by Shapiro, 1972). In the latter case, low doses of selenium may protect against malignant tumors and/or high doses may be carcinogenic. If these effects are confirmed, addition of selenium to deficient diets will require some relaxation of the "Delaney clause" and may present a serious dilemma (Scott, 1973). Presently, certain rates of feed supplementation for the prevention of selenium deficiencies in swine and nonlaying poultry are permitted by the FDA (Schmidt, 1974).

There is also some evidence that certain compounds of selenium may provide protection against the toxicity of Hg and Cd to animals (Allaway, 1973; Ganther et al., 1972).

Additional aspects of selenium requirement and toxicity in animal and human nutrition are reviewed elsewhere (Muth et al., 1967; Rosenfeld and Beath, 1964).

Trace element research with higher plants has lagged behind animal research, and the essentiality of selenium for plants is an unresolved question. An apparent selenium requirement was demonstrated for ten of the so-called selenium-accumulator* species of *Astragalus* (Trelease and Trelease, 1938), but the results may have been confounded by a selenium-phosphate interaction whereby the increase in growth in the presence of selenium could have been

---

*Defined as plants which can accumulate selenium to levels 100 to 10,000 times greater than the levels in most other native and crop plants. These species have also been termed selenium indicators, since their growth is restricted to seleniferous areas. In this paper, the words "accumulator" and "non-accumulator" refer to this definition.

partially due to a depression of phosphate toxicity by selenium (Broyer et al., 1972). One of the difficulties in showing a plant requirement for selenium is that plants apparently can absorb detectable amounts of selenium from the ambient air. For example, alfalfa plants in a greenhouse, cultured with purified nutrient solutions with a total of 0.0395 µg Se, contained 1.5 µg Se upon harvest (Broyer et al., 1966).

Much of the work on selenium in biological systems has been concerned with the ability or inability of selenium to form analogs of sulfur compounds. Selenium and sulfur are Group VI elements of the periodic table. Like sulfur, selenium can exist in several allotropic forms; "metallic" selenium is considered to be the stable form at 25°C. Ordinarily, it is a poor conductor of electricity, but its conductivity can be increased a thousandfold in the light over that in the dark. Light of 700 nm appears to be the most effective in increasing the conductivity. This property is made use of in photocells and rectifiers. A role for selenium in animal vision has also been suggested, based on this property (Aberg, 1966; McFarland et al., 1970).

The selenium analogs of sulfurous and sulfuric acids are selenious acid (H_2SeO_3) and selenic acid (H_2SeO_4), respectively. Standard free energy differences between selenate and selenite are much smaller than the corresponding differences for sulfate and sulfite; consequently, selenates are much stronger oxidizing agents than are sulfates. Aqueous solubilities of selenate salts are, in general, greater than those of sulfate salts, up to a factor of ten. The structure of the sulfate and selenate ions in crystals of their salts is tetrahedral, the sulfur or selenium atom occupying the center, and the four oxygen atoms the corners of a regular tetrahedron. The approximate interatomic distance of S-O is 0.151 nm, while that of Se-O is 0.161 nm (Yost and Russell, 1946). This similarity in structure of selenate and sulfate has been used to explain the competition between these salts observed in ion absorption experiments with excised barley roots (Leggett and Epstein, 1956). Similarities in the crystal structures of sulfur and selenium isologs are discussed elsewhere (Mautner, 1972).

Selenium and sulfur can exist in oxidation states of +6, +4 and -2, and both have the six-electron system of valence orbitals. Selenium is more electronegative than sulfur, however, and this property has been used to account for some of the differences in the biochemical behavior of the two elements.

Table 26.1

Occurrence of Selenium in Nature

Material	Selenium Concentration (ppm Se)	Reference
Meteorites	3-15	Byers, 1938
Lunar Basalts	0.1-0.2	Anders et al., 1971
Volcanic Rocks	<1-2	Rosenfeld and Beath, 1964
Shales	<1-675	Rosenfeld and Beath, 1964
Phosphate Rock	1-300	Rosenfeld and Beath, 1964
Limestone	0.1-6	Rosenfeld and Beath, 1964
Vanadium-Uranium Ores	500-2600	Rosenfeld and Beath, 1964
Coal	0.5-3.9	Magee et al., 1973
Soils (Nonseleniferous)	<1-2	Rosenfeld and Beath, 1964
(Seleniferous)	0.1-80	Rosenfeld and Beath, 1964
Vegetation:		
Primary Selenium Accumulators	100-10,000	Rosenfeld and Beath, 1964
Astragalus bisulcatus[a]	5530	Rosenfeld and Beath, 1964
Stanleya pinnata[a]	1190	Rosenfeld and Beath, 1964
Atriplex nuttallii[a]	300	Rosenfeld and Beath, 1964
Grasses[a]	23	Rosenfeld and Beath, 1964
Grains (Seleniferous Areas)	0.1-30	Rosenfeld and Beath, 1964
Vegetables (Seleniferous Areas)	0.1-4.5	Rosenfeld and Beath, 1964
Waters:		
Oceans	0-0.001	Rosenfeld and Beath, 1964; Bowen, 1966
Colorado River	0.001	Rosenfeld and Beath, 1964

Irrigation Drainage from		
Seleniferous Lands	2.7	Rosenfeld and Beath, 1964
Canada (Surface Waters)	0.010	Goulden and Brooksbank, 1974
(Ground Water)	0.0001	Goulden and Brooksbank, 1974
(Note: EPA recommendation for public water supply)	(not to exceed 0.01)	(EPA, 1972)
Blood:		
Animals with acute Se poisoning	7–27	Rosenfeld and Beath, 1964
Animals with chronic Se poisoning	1–4	Rosenfeld and Beath, 1964
Normal human placental blood	0.18	Rosenfeld and Beath, 1964
Normal fetal cord blood	0.12	Rosenfeld and Beath, 1964
Milk:		
Normal human	0.020	Hadjimarkos and Shearer, 1964
Cows and pigs with alkali disease	0.02–3.0	Rosenfeld and Beath, 1964

^{a}Collected from same outcrop of the (seleniferous) Niobrara formation, Wyoming.

BIOCHEMICAL DIFFERENCES BETWEEN SELENIUM AND SULFUR

Differences in the biochemistry of sulfur and selenium have been demonstrated. For example, no selenium analogs have been found for glutathione in *E. coli* (Cowie and Cohen, 1958) and in *Astragalus* (Virupaksha and Shrift, 1963), choline sulfate in *Aspergillus* and plant sulfolipid in *Chlorella* and *Euglena* (Nissen and Benson, 1964), or 3'-phosphoadenosine 5'-phosphosulfate in yeast (Wilson and Bandurski, 1958). It was suggested that selenate could be metabolized to amino acids through activation as adenosine phosphoselenate, rather than as 3'-phosphoadenosine 5'-phosphoselenate (Nissen and Benson, 1964).

Although selenium incorporation into the proteins of various animal tissues has been reported (McConnell and Roth, 1966), there was little incorporation of selenium into the sulfur amino acids in plant proteins of the selenium accumulator *Neptunia amplexicaulis* (Peterson and Butler, 1967). However, extensive protein incorporation of selenoamino acids occurred in ryegrass, wheat, and clover (Peterson and Butler, 1967). It was suggested that selenium accumulator species have evolved a detoxification mechanism whereby selenium is excluded from protein incorporation, while nonaccumulator species do not have this mechanism. Selenium incorporated into proteins could result in alteration of the protein structure, inactivation of the protein, and eventual poisoning of the plant (Peterson and Butler, 1967).

In a comparison of the chemical properties of selenocysteine and selenocystine with their sulfur analogs it was concluded that the chemical differences between these analogs are sufficiently great that it seems unlikely that the selenium analog could replace cysteine or cystine in a protein molecule without marked alteration of the protein structure and function (Huber and Criddle, 1967). This should be particularly true for the "sulfhydryl proteins" whose activity seems directly dependent on the presence of a free cysteine side chain. It was argued that near pH 7 the cysteine side chain exists predominantly in the sulfhydryl form, while selenocysteine is found almost exclusively as the selenide ion. The pH dependence of activity of a selenocysteine-containing "sulfhydryl protein" would therefore be expected to be greatly changed, if not completely lost (Huber and Criddle, 1967).

There are also reactions involving selenium for which sulfur analogs have not been found (Shrift, 1969). One

of the more interesting cases is selenocystathionine, which
has been found in the accumulators *Stanleya pinnata* (Virupaksha and Shrift, 1963) and *Neptunia amplexicaulis* (Peterson and Butler, 1967). The sulfur analog cystathionine
has not been found to any large extent in higher plants,
although it is a major metabolite of sulfur in animals and
microorganisms (Fowden, 1965).

BIOCHEMICAL SIMILARITIES BETWEEN
SELENIUM AND SULFUR

A number of selenium analogs of sulfur compounds have
been found in biological systems in addition to those mentioned above. The incorporation of selenium into various
animal proteins such as hemoglobin, cytochrome c, and myo-
globin, the muscle enzymes myosin and aldolase, plasma
proteins, lipoproteins, and the total proteins of milk and
egg has been reported (McConnell and Roth, 1966). Certain
selenium compounds have been shown to function as well as,
or better than, their sulfur analogs. For example, when
selenomethionine was incubated with methionine-activating
enzyme prepared from rabbit liver or yeast, it was utilized
as well as or, in the case of the yeast enzyme, better
than methionine (Mudd and Cantoni, 1957). When Se-
adenosylselenomethionine* was incubated with guanidoacetate
and creatine methylpherase from pig liver, creatine formed
in high yield by transmethylation (Mudd and Cantoni, 1957).
Se-adenosylselenomethionine was as efficient a methyl donor
as S-adenosylmethionine in choline biosynthesis.

There is also evidence for the existence of selenothiamine in
E. coli which apparently adapted to growth on media containing
selenate instead of sulfate (Shrift, 1956). Selenomethionine
could completely replace methionine for the exponential growth
of a methionine-requiring mutant of *E. coli* (Cowie and Cohen, 1957).
Selenium, however, could not entirely replace sulfur for the growth
of *E. coli*, and as was mentioned above, no selenoglutathione
was formed.

Recently, compounds identified as 4-selenouridine were
found in tRNA isolated from *E. coli* grown on media containing
$^{75}SeO_3^=$ (Hoffman and McConnell, 1974) or sodium ^{75}Se-
selenosulfate (Rao et al., 1974).

In *Lactobacillus helveticus*, an organism which required
preformed pantethine, selenopantethine could completely
replace the common metabolite, the selenopantethine apparently being a true functional analog of pantethine (Mautner

---

*
Se-adenosylselenomethionine is the selenium analog of
S-adenosylmethionine, the form of methionine considered to
be "active" in transmethylations.

and Gunther, 1959). Selenocoenzyme A was later found in rat liver (Lam et al., 1961).

A selenium-containing ferredoxin, putidaredoxin (an iron-sulfide protein), in which the acid-labile sulfur was replaced with selenium, was found to have biological activity comparable to that of the sulfide enzyme (Tsibris et al., 1968). When thioNADP (nicotinamide adenine dinucleotide phosphate) and selenoNADP were compared as cofactors in certain reactions catalyzed by NADP-linked oxido-reductases, SeNADP was fully accepted as a cofactor by glucose-6-phosphate dehydrogenase, in the latter two cases, SeNADP was a competitive inhibitor of the reactions (Christ et al., 1970).

Other aspects of selenium biochemistry are reviewed by Stadtman (1974) and Klayman et al. (1973).

SELENIUM-SULFUR EXCHANGE

The question has been raised whether selenium does actively replace sulfur in its biological pathway or merely exchanges and/or binds to sulfur in a particular compound. Evidence for both types of reactions is reported in the literature. Evidence for nonenzymatic binding and/or exchange between selenite and certain sulfur compounds has been demonstrated (Schwarz and Sweeney, 1964; Cummins and Martin, 1967; Jenkins, 1968). On the other hand, there is an extensive literature that suggests true biochemical incorporation of selenium into sulfur compounds.

BIOLOGICAL PRODUCTION OF VOLATILE SELENIUM

Workers at the University of California became interested in the assumption that selenium accumulator plants give off volatile selenium in some form (Lewis et al., 1966). The assumption was based on the characteristic "garlicky" odor of *Astragalus* species growing on seleniferous soils (Rosenfeld and Beath, 1964). Biological production of volatile selenium has been acknowledged for decades, primarily as a result of experiments with animals (Schultz and Lewis, 1940; Petersen et al., 1951; Kamstra and Bonhorst, 1953; Ganther and Baumann, 1962; Olson et al., 1963; Ganther et al., 1966; and others), and microorganisms (Challenger, 1935). In the latter case, dimethyl selenide, $(CH_3)_2Se$, was identified as the compound given off by *Penicillium* sp. (Challenger, 1935; Fleming and Alexander, 1972), and a scheme by which inorganic selenium compounds are methylated by fungi was presented by Challenger (see Figure 26.1).

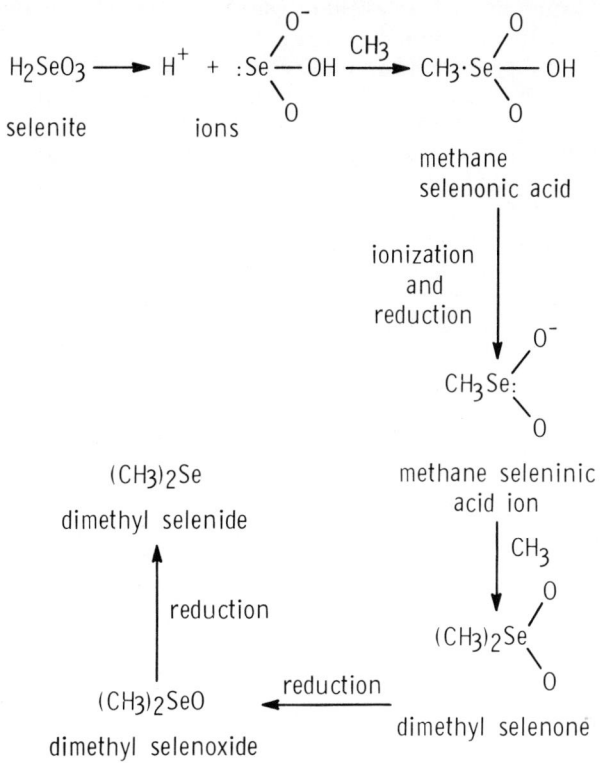

Figure 26.1 Pathway for the microbial production of dimethyl selenide (after Challenger, 1935).

Interestingly, the compound given off by rats was also identified as dimethyl selenide (McConnell and Portman, 1952; Sternberg et al., 1968). [As early as 1894, Hofmeister proposed that selenium in the animal is detoxified by methylation to form dimethyl selenide, which is excreted by way of the lungs. He had no direct evidence of this, but based his proposal on the fact that the odor of dimethyl telluride was detected in the breath of dogs injected with sodium tellurite, and that selenium would presumably behave like tellurium. Due to its greater toxicity, the injection of sufficiently large amounts of sodium selenite was not expedient (Schultz and Lewis, 1940; Byard, 1969).]

The production of dimethyl selenide by animals has been viewed as a detoxification mechanism which becomes of major importance when doses in excess of about 1 mg Se/kg body weight are given. At lower levels of selenium in the body, the urine appears to be the major excretory route for the element (Ganther et al., 1966; Hopkins et al., 1966; Byard and Baumann, 1967; Byard, 1969; Palmer et al., 1969). A model system for enzymic synthesis of dimethyl selenide from sodium selenite, using mouse liver extracts, has been described. The crude system had a specific requirement for glutathione. Reduced triphosphopyridine nucleotide, coenzyme A, adenosine-5-triphosphate, and magnesium were also required for optimal activity. S-adenosylmethionine was the probable methyl donor. These results were taken as confirmation that the reductive utilization of inorganic selenium occurs directly in mammalian tissues without the involvement of intestinal flora (Ganther, 1966).

VOLATILE SELENIUM FROM HIGHER PLANTS

As mentioned above, the characteristic odor of *Astragalus* species growing on seleniferous soils is an indication that some higher plants may produce volatile compounds of selenium. Additionally, losses of selenium have been observed during storage or drying of accumulator and nonaccumulator plant tissues containing selenium (Moxon and Rhian, 1938; Beath and Eppson, 1947; Asher et al., 1967; Ehlig et al., 1968; Gissel-Nielsen, 1970). Volatile selenium was first collected from growing plants, including the nonaccumulator alfalfa (*Medicago sativa* L.), in 1966 (Lewis et al., 1966). Subsequently, dimethyl diselenide, $(CH_3)_2Se_2$, was identified in volatile material collected from the accumulator *Astragalus racemosus*. Dimethyl selenide was not found in the volatiles from this species (Evans et al., 1968).

Studies with nonaccumulator plant species indicated that sunflower (*Helianthus anuus*), spinach (*Spinacia olerarea*), cabbage (*Brassica oleracea* var. Capitata), and four other cruciferous species, gave off volatile selenium during growth on nutrient media containing up to 3 mg/l selenium as selenite or selenate. The rate of volatile selenium production varied from less than 0.05% of leaf selenium per hour to 0.40% per hour, the proportion having little correlation with the concentration of selenium in the plant tissues or with selenium concentration in the growth medium. Sunlight intensity appeared to be the primary influence on rate of volatile selenium production and the effect was later shown to be due to increased temperatures of the leaf tissues, rather than to an effect of light per se (Lewis, 1971a). Experiments with fresh leaf homogenates demonstrated that tissues from plants cultured in solutions containing selenite released 10 to 16 times as

much volatile selenium as leaves from plants cultured in
solutions containing selenate. A pH of 7.8 and a temperature
of 40°C were optimum for the production of the volatile
compound from fresh leaf homogenates (Lewis et al., 1974).
 Using cold trap techniques and gas chromatography,
dimethyl selenide was subsequently identified as the single
selenium volatile compound produced by fresh cabbage leaves.
The sulfur analog, dimethyl sulfide, was also released by
these leaves (Lewis, 1971; Lewis et al., 1974). A crude
enzyme fraction was isolated from fresh cabbage leaves,
and shown to cleave Se-methylselenomethionine selenonium
salt to dimethyl selenide and homoserine. No additional
cofactor was apparently required for the cleavage reaction.
The cabbage enzyme fraction also cleaved S-methyl methionine sulfonium salt to dimethyl sulfide and homoserine
(Lewis et al., 1971). Similar cleavage of the sulfonium
compound has been demonstrated with enzyme fractions separated from soil bacteria (Tanaka and Nakamura, 1964;
Mazelis et al., 1965). The dimethyl selenide precursor,
Se-methyl selenomethionine selenonium compound, was isolated from cabbage plants cultivated in nutrient solutions
containing up to 3 mg/l selenium as selenite or selenate
(Lewis, 1971a; Lewis et al., 1974). As mentioned above,
the selenonium compound has been found in other species of
nonaccumulators, and has not been found in species of accumulator plants. Its presence or absence has been suggested as a biochemical method for distinguishing selenium
nonaccumulators from accumulators (Shrift, 1969).
 It is of some interest that no dimethyl disulfide was
detected in the volatile material collected from freshly
harvested cabbage leaves in the experiments (Lewis, 1971a).
The characteristic odor of crucifers has been partly attributed to the production of dimethyl disulfide (Thompson,
1967), and the compound has been reported to occur in sulfur
volatile compounds from cabbage (Bailey et al., 1961). In
the experiments discussed above, dimethyl disulfide was
found only in volatile material collected from cabbage
leaves that had been harvested a day previously, and it
is suggested that this compound is a decomposition product
that is not produced by growing cabbage plants or freshly
harvested leaves from these plants. This conclusion
remains to be confirmed.

PATHWAYS FOR SELENIUM VOLATILIZATION IN HIGHER PLANTS

 Based on a review of the literature and on results of
the experiments with cabbage, the schemes outlined in
Figures 26.2 and 26.3 are proposed for the production of

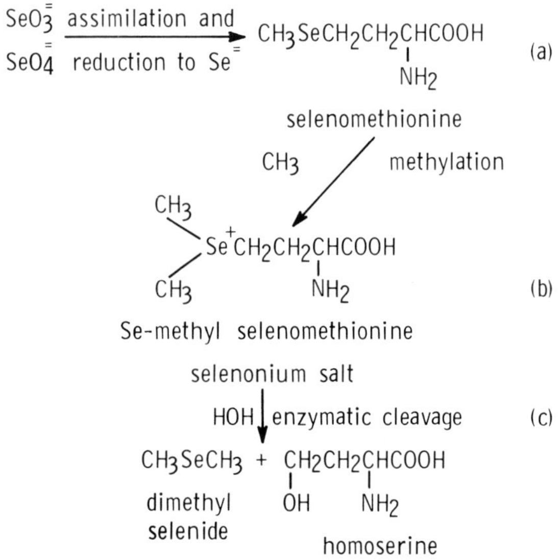

$SeO_3^=$ assimilation and
$SeO_4^=$ reduction to $Se^=$ → $CH_3SeCH_2CH_2\underset{NH_2}{CHCOOH}$ (a)

selenomethionine

$\underset{CH_3}{\overset{CH_3}{>}}\overset{+}{Se}CH_2CH_2\underset{NH_2}{CHCOOH}$ $\overset{CH_3}{\underset{methylation}{\nearrow}}$ (b)

Se-methyl selenomethionine
selenonium salt

HOH ↓ enzymatic cleavage   (c)

CH_3SeCH_3 + $CH_2CH_2\underset{NH_2}{CHCOOH}$
dimethyl $\overset{|}{OH}$
selenide
                    homoserine

Figure 26.2 *Formation of dimethyl selenide in nonaccumulator species of higher plants (letters in parentheses refer to text explanations).*

$SeO_3^=$
$SeO_4^=$ → $\begin{array}{c}CH_2-Se-CH_2\\ | \quad\quad\quad | \\ CH_2 \quad\quad CHNH_2\\ | \quad\quad\quad | \\ CHNH_2 \quad COOH\\ | \\ COOH\end{array}$ (a)

selenocystathionine
(b)

$CH_3SeCH_2\underset{NH_2}{CHCOOH}$ (c)

Se-methyl selenocysteine

$CH_3Se CH_2\underset{NH_2}{CHCOOH}$
$\downarrow$
O (d)

Se-methyl selenocysteine selenoxide

$CH_3SeSeCH_3$ (e)
dimethyl diselenide

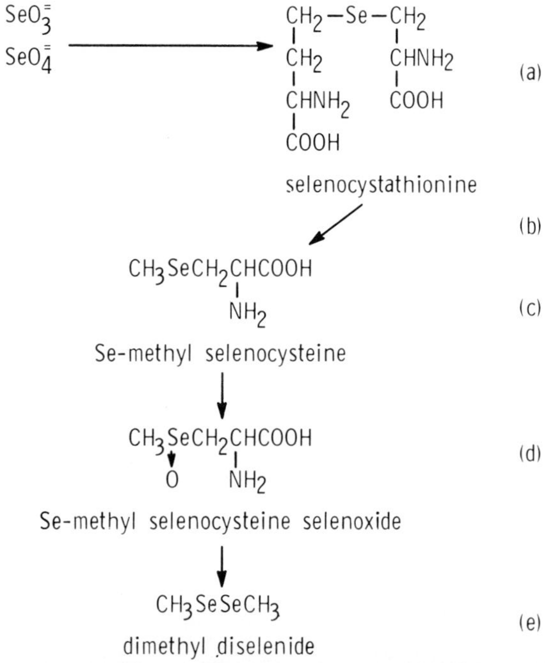

Figure 26.3 *Formation of dimethyl diselenide in accumulator species of higher plants (letters in parentheses refer to text explanations).*

volatile selenium by nonaccumulator and accumulator species of plants, respectively, under normal physiological conditions when nontoxic levels of selenium are present in the growth medium. Evidence for the individual steps is summarized below, where the letters in parentheses refer to the corresponding steps in the figures.

Figure 26.2, Nonaccumulator Species

(a) Selenomethionine has been found in rye grass, wheat, and clover, all nonaccumulators. No selenomethionine was found in the accumulator *Neptunia amplexicaulis* (Peterson and Butler, 1962, 1967). Similarily, no selenocystine or selenomethionine was found in the hydrolysate of protein isolated from the accumulator *Astragalus bisulcatus* grown on soil containing selenate and selenite (Martin and Lee, 1967).

(b) Se-methyl selenomethionine selenonium compound has been identified as the predominant soluble organic selenium compound in four nonaccumulator species of *Astragalus*. The compound could not be detected, or occurred only in traces, in accumulator species (Virupaksha and Shrift, 1965). This compound was also found in cabbage, a nonaccumulator, grown on media containing selenite or selenate (Lewis *et al.*, 1971, 1974).

(c) Evidence that this reaction takes place in intact plants has been inferred from the studies discussed above (Lewis *et al.*, 1971, 1971a, 1974).

Figure 26.3, Accumulator Species

(a) The formation of selenocystathionine from selenate and selenite in accumulator species has been reported (Virupaksha and Shrift, 1962; Peterson and Butler, 1967). Selenocystathionine has also been found in selenate-treated *A. bisulcatus* (Chow *et al.*, 1972).

(b) The formation of S-methyl cysteine in plants is usually considered to result from the methylation of cysteine (Thompson, 1967). An analogous pathway for the formation of Se-methyl selenocysteine in *A. bisulcatus* has been proposed, based on experiments with ^{75}Se-selenomethionine in the growth medium (Chow *et al.*, 1972). It has not been shown that this pathway occurs in the absence of selenomethionine in the growth medium. The formation of Se-methylselenocysteine from selenocystathionine in accumulator plants fed selenite or selenate needs elucidation.

(c) This compound has been found extensively in accumulator plant species (Virupaksha and Shrift, 1965), and has been suggested as a biochemical basis for distinguishing accumulators from nonaccumulators.

(d) There is no direct evidence that this process occurs in plants, other than that for the sulfur analog. The natural (+) isomer of S-methyl-L-cysteine sulfoxide was formed preferentially from labeled S-methyl-L-cysteine in broccoli leaves and turnip leaf discs. The stereospeicficity of the oxidation in intact cells indicated to the authors that the conversion was probably enzymatic (Arnold and Thompson, 1962).

(e) It has demonstrated (Ostermayer and Tarbell, 1959) that acid hydrolysis of S-methyl-L-cysteine sulfoxide produces dimethyl disulfide (see Figure 26.4). Enzymes which degrade this sulfoxide compound are known to occur in plants (Thompson, 1967). The identification of dimethyl diselenide in volatile material from the accumulator *A. racemosus* (Evans et al., 1968) suggests that an analogous reaction for selenium may be taking place in accumulator plant species.

$$CH_3 \overset{O}{\underset{O}{S}} SCH_3$$

methylmethane thiosulfonate

+

$$4\ CH_3 \underset{O}{S} CH_2 \underset{NH_2}{C} HCOOH \xrightarrow{2H_2O}$$

S-methyl-L-cysteine sulfoxide

CH_3SSCH_3
dimethyl disulfide

+

$4\ CH_3COCOOH$
pyruvate

+

$4\ NH_3$

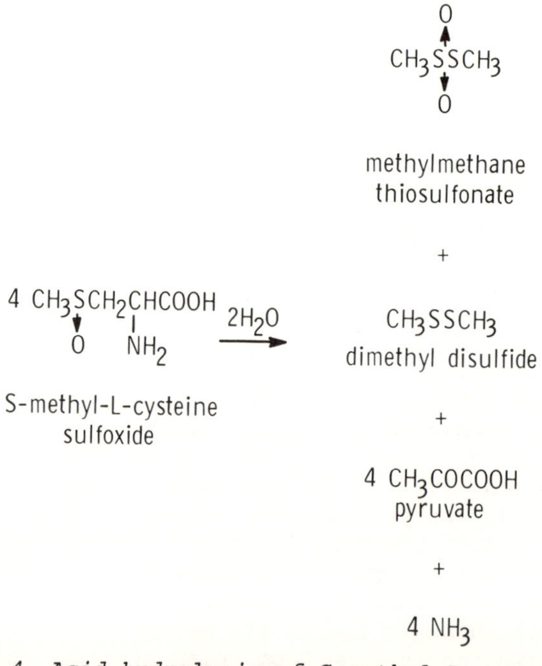

Figure 26.4 Acid hydrolysis of S-methyl-L-cysteine sulfoxide (after Ostermeyer and Tarbell, 1959).

SIGNIFICANCE OF SELENIUM VOLATILIZATION IN PLANTS

The release of volatile selenium compounds occurs in apparently healthy plants, under physiological conditions, and does not require the presence of high or toxic levels of selenium in the growth medium. Unlike animal exhalation of dimethyl selenide (which is apparently a detoxification mechanism), the dimethyl selenide released by nonaccumulator species of higher plants seems to be simply a by-product of normal, biochemical reactions within the plant, where selenium is metabolized along the pathways of sulfur.

Production of dimethyl diselenide by accumulator species of higher plants is twice as efficient for removal of selenium as the reaction in nonaccumulator species; that is, two atoms of selenium are removed per two methyl groups in accumulator species, compared to one atom of selenium removed per two methyl groups in nonaccumulator species. However, rather than a detoxification mechanism, the release of dimethyl diselenide by accumulator plants is, again, more likely the result of reactions in which selenium and sulfur behave analogously. Volatilization of selenium appears to play only a minor role, if any, in the relative tolerance of accumulator species for seleniferous growth media.

The release of volatile selenium compounds by intact plants and the possibility of subsequent assimilation of the volatile compound(s) by adjacent plants* are factors requiring consideration when planning controlled experiments on selenium requirement.

REFERENCES

Aberg, B., 1966: Selenium as a trace element. *Chem. Abstr.* 65, 20579h 19 [*Nord. Med.* 75 (21), 589 Swed.].

Allaway, W. H., 1973: Selenium in the food chain. *Cornell Vet.* 63, 151-170.

Anders, E., R. Ganapathy, R. R. Keays, J. C. Laul and J. W. Morgan, 1971: Volatile and siderophile elements in lunar rocks: comparison with terrestrial and meteoritic basalts. *Proc. 2nd Lunar Sci. Conf.*, 2, (Cambridge, Mass: MIT Press) pp. 1021-1036.

---

*Transfer of ^{75}Se between separate alfalfa plants in a greenhouse has been reported (Asher *et al.*, 1967).

Arnold, W. and J. Thompson, 1962: The formation of (+)-S-methyl-L-cysteine sulfoxide from S-methyl-L-cysteine in crucifers. *Biochim. Biophys. Acta* 57, 604-606.

Asher, C., C. Evans and C. M. Johnson, 1967: Collection and partial characterization of volatile selenium compounds from *Medicago sativa* L. *Aust. J. Biol. Sci.* 20, 737-748.

Bailey, S., M. Bazinet, J. Driscoll and A. McCarthy, 1961: The volatile sulfur components of cabbage. *J. Food Sci.* 26, 163-170.

Beath, O. and H. Eppson, 1947: The forms of selenium in some vegetation. Who. Agr. Exp. Sta. Bul. 278.

Bowen, H., 1966: *Trace Elements in Biochemistry*, (New York: Academic Press) p. 19.

Broyer, T. C., D. Lee and C. Asher, 1966: Selenium nutrition of green plants: effect of selenite supply on growth and selenium content of alfalfa and subterranean clover. *Plant Physiol.* 41, 1425-1428.

Broyer, T. C., C. M. Johnson and R. P. Huston, 1972: Selenium and nutrition of *Astragalus:* I. Effects of selenite or selenate supply on growth and selenium content. *Plant and Soil* 36, 635-649.

Byard, J. and C. Baumann, 1967: Selenium metabolites in the urine of rats given a subacute dose of selenite. *Federation Proc.* 26, 476.

Byard, J., 1969: Trimethyl selenide: a urinary metabolite of selenite. *Arch. Biochem. Biophys.* 130, 556-560.

Byers, H., 1938: Selenium in meteorites. *Ind. Chem. Eng., News Ed.* 16, 459.

Challenger, F., 1935: The biological methylation of compounds of arsenic and selenium. *Chem. and Ind.* 54, 657.

Chow, C., S. Nigam and W. McConnell, 1972: Biosynthesis of Se-methyl-selenocysteine and S-methyl cysteine in *Astragalus bisulcatus*. *Biochem. Biophys. Acta* 273, 91-96.

Christ, W., D. Schmidt and H. Coper, 1970: Comparison of thio-NADP and seleno-NADP in NADP-dependent oxidoreductases. *Z. Physiol. Chem.* 351, 427-434.

Cowie, D. and G. Cohen, 1957: Biosynthesis by *E. coli* of active altered proteins containing Se instead of S. *Biochim. Biophys. Acta* 26, 252-261.

Cummins, L. and J. Martin, 1967: Are selenocystine and selenomethionine synthesized *in vivo* from sodium selenite in mammals? *Biochem.* 6, 3162-3168.

Evans, C., C. Asher and C. M. Johnson, 1968: Isolation of dimethyl diselenide and other volatile selenium compounds from *Astragalus racemosus* (Pursh.). *Aust. J. Biol. Sci.* 21, 13-20.

EPA, 1972: Water quality criteria. Environmental Protection Agency, Washington, D.C.

Fleming, R. and M. Alexander, 1972: Dimethyl selenide and dimethyl telluride formation by a strain of *Penicillium*. *Appl. Microbiol.* 24, 424-429.

Fowden, L., 1965: Origins of the amino acids. In: *Plant biochemistry*, J. Bonner and J. Varner, eds., pp. 361-388.

Ganther, H. and C. Baumann, 1962: Selenium metabolism. I. Effects of diet, arsenic, and cadmium. *J. Nutr.* 77, 210-216.

Ganther, H., 1966: Enzymic synthesis of dimethyl selenide from sodium selenite. *Biochem.* 5, 1089-1098.

Ganther, H., O. Levander and C. Baumann, 1966: Dietary control of selenium volatilization in the rat. *J. Nutr.* 88, 55-60.

Ganther, H., C. Goudie, L. Sunde, M. Kopecky, Oh Sang-Hwan and W. Hoekstra, 1972: Selenium: relation to decreased toxicity of methyl-mercury added to diets containing tuna. *Science* 175, 1122-1124.

Gissel-Nielsen, G., 1970: Loss of selenium in drying and storage of agronomic plant species. *Plant and Soil* 32, 242-245.

Goulden, P. and P. Brooksbank, 1974: Automated atomic absorption determination of arsenic, antimony, and selenium in natural waters. *Anal. Chem.* 46, 1431-1436.

Hadjimarkos, D. and T. Shearer, 1973: Selenium in mature human milk. *Am. J. Clin. Nutr.* 26, 583-585.

Hoffman, J., and K. McConnell, 1974: The presence of 4-selenouridine in *Escherichia coli* tRNA. *Biophys. Acta* 366, 109-113.

Hopkins, L., A. Pope and C. Baumann, 1966: Distribution of microgram quantities of selenium in tissues of the rat, and effects of previous selenium intake. *J. Nutr.* 88, 61-65.

Huber, R. and R. Criddle, 1967: Comparison of the chemical properties of selenocysteine and selenocystine with their sulfur analogs. *Arch. Biochem. Biophys.* 122, 164-173.

Jenkins, K., 1968: Evidence for the absence of selenocystine and selenomethionine in the serum proteins of checks administered selenite. *Can. J. Biochem.* 46, 1417-1425.

Kamstra, L. and C. Bonhorst, 1953: Effect of arsenic on the expiration of volatile selenium compounds by rats. *Proc. S. Dak. Acad. Sci.* 32, 72-76.

Klayman, D. and W. Gunther, (eds.), 1973: *Organic selenium compounds, their chemistry and biology*. (New York: Wiley-Interscience).

Lam, K., M. Riegl and R. Olson, 1961: Biosynthesis of selenocoenzyme A in the rat. *Federation Proc.* 20, 229.

Leggett, J. and E. Epstein, 1956: Kinetics of sulfate absorption by barley roots. *Plant Physiol.* 31, 222-226.

Lewis, B., C. M. Johnson and C. C. Delwiche, 1966: Release of volatile selenium compounds by plants: collection procedures and preliminary observations. *J. Agr. Food Chem.* 14, 638-640.

Lewis, B., C. M. Johnson, and T. C. Broyer, 1971: Cleavage of Se-methyl-selenomethionine selenonium salt by a cabbage leaf enzyme fraction. *Biochim. Biophys. Acta* 237, 603-607.

Lewis, B., 1971a: Ph.D. dissertation, University of California, Berkeley.

Lewis, B., C. M. Johnson and T. C. Broyer, 1974: Volatile selenium in higher plants. *Plant and Soil* 40, 107-118.

Magee, E., H. Hall and G. Varga, Jr., 1973: Potential pollutants in fossil fuels. EPA-R2-249, EPA Office of Research and Monitoring, NERC, RTP, Control Systems Laboratory, Research Triangle Park, North Carolina.

Mautner, H. and W. Gunther, 1959: Selenopantethine, a functional analog of pantethine in the *Lactobacillus helveticus* system. *Biochim. Biophys. Acta* 36, 561-562.

Mautner, H., 1972: Sulfur and selenium isologs as probes of active sites. *Ann. N. Y. Acad. Sci.* 192, 167-174.

Martin, J. and J. Lee, 1967: The absence of selenocystine and selenomethionine in the hydrolysate of protein isolated from *Astragalus bisulcatus*. *Proc. Am. Soc. Exptl. Biol.* 26, 476.

Mazelis, M., B. Levin and N. Mallinson, 1965: Decomposition of methyl-methionine sulfonium salts by a bacterial enzyme. *Biochim. Biophys. Acta* **105**, 106-114.

McConnell, K. and O. Portman, 1952: Excretion of dimethyl selenide by the rat. *J. Biol. Chem.* **195**, 277-282.

McConnell, K. and D. Roth, 1966: Incorporation of selenium into rat liver robosomes. *Arch. Biochem. Biophys.* **117**, 366-374.

McFarland, I., C. Winget, W. Wilson and C. M. Johnson, 1970: Role of selenium in neural physiology of avian species. *Poultry Sci.* XLIX, 216-221.

Money, D., 1971: Cot deaths and deficiency of Vitamin E and selenium. *Br. Med. J.* **4**, 559.

Moxon, A. and M. Rhian, 1938: Loss of selenium by various grains during storage. *Proc. S. Dak. Acad. Sci.* **18**, 20.

Mudd, S. and G. Cantoni, 1957: Selenomethionine in enzymatic transmethylations. *Nature* **180**, 1052.

Muth, O., J. Oldfield and P. Weswig, (eds), 1967: *Selenium in Biomedicine.* (Westport, Conn.: The AVI Publishing Co., Inc.).

Nissen, P. and A. Benson, 1964: Absence of selenate esters and "selenolipid" in plants. *Biochem. Biophys. Acta* **83**, 400-402.

Olson, R., K. Schwarz, M. Horwitt, A. L. Tappell *et al.*, 1965: Nutrition symposium: interrelationships among Vitamin E, Coenzyme Q, and selenium *Federation Proc.* **24**, 55-92.

Ostermayer, F. and D. Tarbell, 1959: Products of acidic hydrolysis of S-methyl-L-cysteine sulfoxide; the isolation of methyl methanethiosulfonate, and mechanism of hydrolysis. *J. Am. Chem. Soc.* **82**, 3752-3755.

Palmer, I., D. Fischer, A. Halverson and O. Olson, 1969: Identification of a major selenium excretory product in rat urine. *Biochim. Biophys. Acta* **177**, 336-342.

Petersen, D., H. Klug and R. Harshfield, 1951: Expiration of volatile selenium compounds from selenized rats. *Proc. S. Dak. Acad. Sci.* **3**, 73.

Peterson, P. and G. Butler, 1962: The uptake and assimilation of selenite by higher plants. *Aust. J. Biol. Sci.* **15**, 126-146.

Rao, Y. and J. Cherayil, 1974: Number and proportion of selenonucleosides in the transfer RNA of *Escherichia coli*. *Life Sci.* 14, 2051-2059.

Rhead, W., G. Schrauzer, S. Saltzstein, E. Cary and W. Allaway, 1972: Vitamin E, selenium, and the sudden infant death syndrome. *J. Pediatr.* 81, 415-416.

Rosenfeld, I. and O. Beath, 1964: *Selenium*. (New York: Academic Press).

Rotruck, J., A. Pope, H. Ganther, A. Swanson, D. Hafeman and W. Hoekstra, 1973: Selenium: biochemical role as a component of glutathione peroxidase. *Science* 179, 588-590.

Schmidt, A., 1974: Selenium in animal feed. Federal Register 39, 1355-1358.

Schultz, J. and H. Lewis, 1940: The excretion of volatile selenium compounds after administration of sodium selenite to white rats. *J. Biol. Chem.* 133, 199-207.

Schwarz, K. and C. Foltz, 1957: Selenium as an integral part of Factor 3 against dietary necrotic liver degeneration. *J. Am. Chem. Soc.* 79, 3292.

Schwarz, K. and E. Sweeney, 1964: Selenite binding to sulfur amino acids. *Federation Proc.* 23, 421.

Scott, M., 1973: The selenium dilemma. *J. Nutr.* 103, 803-810.

Shapiro, J., 1972: Selenium and carcinogenesis: a review. *Ann. N. Y. Acad. Sci.* 192, 215-219.

Shrift, A., 1967: Microbial research with selenium. In: *Selenium in Biomedicine*. O. Muth, J. Oldfield and P. Weswig, eds. (Westport, Conn.: The AVI Publishing Co., Inc.).

Shrift, A., 1969: Aspects of selenium metabolism in higher plants. *Ann. Rev. Plant Physiol.* 20, 475-494.

Stadtman, T., 1974: Selenium biochemistry. *Science* 183, 915-922.

Sternberg, J., J. Brodeur, A. Imbach and A. Mercier, 1968: Metabolic studies of seleniated compounds. III. Lung excretion of ^{75}Se and liver function. *Int. J. Appl. Radiat. Isotopes* 19, 669-684.

Tanaka, K. and J. Nakamura, 1964: Metabolism of S-methylmethionine. 1. Bacterial degradation of S-methylmethionine. *J. Biochem.* (Tokyo) 56, 172-176.

Tappel, A. and K. Caldwell, 1967: Redox properties of selenium compounds related to biochemical function. In: *Selenium in Biomedicine.* O. Muth, J. Oldfield and P. Weswig, eds, (Westport, Conn.: The AVI Publishing Co., Inc.).

Thompson, J., 1967: Sulfur metabolism in plants. *Ann. Rev. Plant Physiol.* 18, 59-84.

Trelease, S. and H. Trelease, 1938: Selenium as a stimulating and possibly essential element for indicator plants. *Am. J. Botany* 25, 372-380.

Tsibris, J., M. Namtvedt and I. Gunsalus, 1968: Selenium as an acid labile sulfur replacement in putidaredoxin. *Biochem. Biophys. Res. Comm.* 30, 323-327.

Virupaksha, T. and A. Shrift, 1963: Biosynthesis of selenocystathionine from selenate in *Stanleya pinnata.* *Biochim. Biophys. Acta* 74, 791-793.

Virupaksha, T. and A. Shrift, 1965: Biochemical diffusion between selenium accumulator and nonaccumulator *Astragalus* species. *Biochim. Biophys. Acta* 107, 69-80.

Wilson, L. and R. Bandurski, 1958: Enzymatic reactions involving sulfate, sulfite, selenate, and molybdate. *J. Biol. Chem.* 233, 975-981.

Yost, D. and H. Russell, 1946: Systematic inorganic chemistry of the fifth and sixth group non-metallic elements. (New York: Prentice Hall, Inc.).

INDEX TO VOLUME 1

INDEX TO VOLUME 1

acetylation of fatty acids 153
adenine 79,177,179
alcohols 159,165
  in lake sediments 84
  isoprenoid 84
algae
  blue-green 133,162-164, 230
  coccoid 133
  green 311
  growth rate 291
  phytoplanktonic 77
  prokaryotic 133
  storage of phosphorus in 290
algal mats 133-137
  alcohols in 161,168
  alkanes in 156
  alkanoic acids in 154,166
  alkenes in 160,167
  diatoms in 152,155
  elementary analysis of 139
  fatty acids in 141,152
  formation of by blue-green algae 133-135
  kerogen structures in 131
  lipids in 149,151,153
  phytanic acid in 161
  sterols in 161,168
algal ooze 135
  alcohols in 161,168
  alkenes in 160,167
  diagenesis of 142

algal ooze, *continued*
  ether extracts of 140-142
  fatty acid distributions in 152
  heptane extracts of 140-142
  oxidation of 139
  sterene 160
alkanes 89,92
  branched/cyclic 81,154, 157,163,165
  cyclic 81
  in algal mats 154,156
  in algal ooze 152,154-172
  in blue-green algae 165
  isolation from humic substances 99-102
  isoprenoid 81
  See also n-alkanes
alkanoic acids in algal mats and ooze 156-159,168
alkenes in algal ooze 160,167
alkylmercury derivatives 113,114
allochthonous sources of organic material 75,85
amino acids 77,90,105,233
  hydrolizable 77
  in lake sediments 76
amino sugars
  in lake sediments 78
  nitrogen 233
ammonia 300
  absorption 227

ammonia, *continued*
  adsorption by ion exchange 255
  concentrations near feedlots 227
  effect of illumination on 301
  effect of organic matter on release of 300
  effect on *Nitrobacter* growth 266
  growth conditions for clostridium 337
  in the atmosphere 227
  oxidation rate 251
  volatilization of 118
ammonium oxidation 252
  oxidizers 256
anaerobic ecology 51
ancient sediments 131,149, 150
andesitic agglomerate 118
  ash 118
Ando soil 192,198
anthropogenic sulfur 354, 356,358
anthropogenic sulfur dioxide concentrations 355,356
arylmercury derivatives 113,114
Ashitaka Loam 191,192,198, 199
atmosphere 351,353,359,362
autochthonous sources 75, 80,82,85,188

bacteria 232
  fractionation of isotopes by 60
  isotope composition of cell mass 58
bacterial sulfur cycle 304
Beaufort Sea 176,177
bicarbonate 310,314,322
  See also Dead Sea
biomass 246,252

branched/cyclic acids 76
branched/cyclic alkanoic acids 83,154,157,166

calcium nitrate fertilizer 274
carbohydrates 76,90
  composition of 188
  determination of by phenol-sulfuric acid method 186
  in lake sediments 76,185
  origin of 188
carbon
  cycle 3-10
  dating 238
  in humic substances 91
  inorganic, in Lake Ontario sediments 190
  isotopes 4,51,52,309,316, 341,343
  organic 190,195,311,323
carbon-13 313,316,323
carbon-14 192,236
  age dating 195
carbon dioxide
  air-sea exchange 3,8,18
  atmospheric 3-10
  exchange 13,14
  isotopic concentration of 54,66
  isotopic fractionation of 64
  isotopic variations 4
  production by yeast 341-343
  production in anaerobic environments 51
  reduction 341
  removal by photosynthesis 324
  technogenic 3
  up-take by the oceans 14
carbon monoxide
  anthropogenic production 26,29,33,34
  atmospheric cycle 25

carbon monoxide, *continued*
  concentrations of 39,40
  distribution of 29-36
  natural production of 27
  oxidation of by hydroxyl radicals 39,41
  residence time 39
  sources and sinks 26,28, 39,41,43,44
  steady-state analysis of 39,40
carbonate 309,310,316,324
  Phanerozoic 8
  Precambrian 9
carotenoid pigments 76,80
chlorophyll 79
  diagenesis 80
  epilimnetic 80
  in Bay of Quinte 285
clostridium 334-339

DDT
  in North Atlantic sediment 213
  in North Atlantic water 203
Dead Sea
  bicarbonate in 314,323
  carbon isotopes of 316,324
  carbonate in 316
  hydrogen sulfide 321
  isotopic variations 314,323
  major ions in 311
  sediments of 311
  sulfur cycle in 309-324
  sulfur isotopes of 315
denitrification 228,231,238,271
  in laboratory soil columns 250
  rates of 250-253
denitrifier population 250,252
deposition 105

differential thermal analysis of metal organic complexes 122,128
diffusion coefficient for isotopic exchange 19
diffusion model for oceanic mixing 13
dimethyl selenide 396,397,399
dissolved carbonate species 323

eddy diffusion coefficient 15
electron microscopic investigation of fulvic acids 128
esters 102,137,141
eutrophic lakes 76-80,85,180
eutrophication 287
exchange capacity of nitrifying organisms 255

fats 90
fatty acids 89,92,105,153
  acetylation of 155
  distributions 152
  in algal mats 141,154
  in algal ooze 154
  in lake sediments 82
  isolation from humic substances 99-102
  methylation of 155
  See also n-fatty acids
fertilizer nitrogen 233
field plots 245,247
flowing systems
  See perfusion
fractionation techniques 234
fulvic acids 89-100,125,195
  chemical structure of 103
  electron microscopic investigation of 128
  in Ashitaka Loam formation 194
  in plant residues 238

gases
 analysis of 56
 exchange between ocean and
  atmosphere 3
 in landfill leachate 55
 released by sewage sludge
  55
glucose 77,195-197,201
gravitational drainage 275
Great Lakes 175,359
growth rate constant 248
guanine 79,177
gypsum 310,311,319

halophylic bacteria 311
humic acids 90,120,195,236
 extracellular electron
  transport by 109
 hydrolysis of 99
 in Ashitaka Loam formation
  198-201
 metabollic oxidation 111
 saponification of 100
humic materials
 aromaticity of 99
 methoxyl content of 91
humic substances 89
 analysis of 91,92
 chemical degradation 93
 chemical structure of 102
 degradation products of
  94-99
 extraction of 90
 fractionation of 90
 major elements in 91
 methods for characterizing
  92
hydrocarbons 105
 in lake sediments 81,82
hydrodynamic dispersion
 coefficient 246
hydrogen bonding 103,104
hydrogen sulfide 309,319,
 320
 buffering effect on pH
  304

hydrogen sulfide, *continued*
 emissions from polluted
  coastal waters 353
 kinetics of production in
  sediment-water systems
  302
 rate of atmospheric oxidation 355
hydrolysates 76,77
hydrophobic organic compounds
 89
hydroxyl groups, alcoholic
 91
hydroxyl radicals 39,47
 concentration of 39,41,
  42,44
 See also carbon monoxide
hypoxanthine 79,177,179

inceptisols 119
infra-red spectra
 of metal-fulvic complexes
  123
 of metal-organic complexes
  125-128
inhibition constant 265
inorganic-organic associations
 117
isoprenoid acids 82
isotope
 analysis 20,56
 exchange 16
 fractionation 19,51-70,
  344,346
isotope variations
 caused by methane-producing
  bacteria 55
 in anaerobic environments
  54
 of nitrogen in organic
  matter 234

Jordan River 309,310,317,
 323,324
 sulfur isotopes in sulfate
  318

kerogen
  elementary composition of 139
  extraction of 135
  fractionation 135
  in algal mats and ancient sediments 131
  oxidation of 135,137-142
  structures of 143
ketones in lake sediments 84
ketonic groups 91
kinetic isotopic effect 331-339
kinetics of bacterial sulfate reduction 302
Kuroboku soil 192,198

lactobacillic acid 82
lacustrine environments 75
  sediments 83,84
Laguna Mormona, Baja California 131,149
landfill leachate gases 55
Lake Erie 175,180,362
Lake Huron 175,180
Lake Ontario 175-177,179, 180,185-189,359,362
lake sediments 75
  alcohols in 84
  amino acids in 76
  amino sugars in 78
  fatty acids in 82
  hydrocarbons in 81
  ketones in 84
  pigments in 80
  sterols in 84
leaching 271
ligands 123
limnocorals 283,285,287
lipids
  diagenesis of 162-170
  in algal mats 151-170
  in blue-green algae 163
  in recent sediments 151-170

lipids, *continued*
  urea adduction of 155
loam 192
  See also Ashitaka Loam, and sandy loam
low-molecular organic acids 90
Lyngbya 133,153
lysimeter investigation of leaching and denitrification losses 271,274,276

marine food chains 206
marine sediments
  extraction of 219
  rate of sulfate reduction in 302-305
mercury
  as an electron acceptor 110-112
  evolution from anoxic sediment inocula 112
  reduction in presence of humic acids 113
metal ions 89,122
metal-organic complexes
  ash content of 120
  bromine oxidation of 127
  characteristics of 118, 122
  differential thermal analysis of 123
  extraction of 118
  infra-red spectra of 123,124
  metal contents of 123,124
  purification of 118,120, 121
  silica in 127
  ultimate analysis of 125
methane
  adsorption 254
  bacteria 56-63
  concentration of 44
  in marine sediments 54
  isotopic content of 54,66

methane, *continued*
    isotopic fractionation of
        64
    oxidation by hydroxyl
        radicals 45
    oxidation of 28,247,249
    -producing bacteria 51,
        55,63
    production in anaerobic
        environments 51
methanol-insoluble and
    -soluble materials 120
methyl hydroperoxide 45,47
methylation 99
    of fatty acids 155
    of selenium compounds 397
methyl esters 153
microbial biomass 247
    contributions to particu-
        late sulfate 351-364
    counts 249,250
    degradation of purines and
        pyrimidines 177
    ecology of 68
    hydrocarbons 99
    population 246
microbial reaction rate 246
microorganisms
    energetics of 253
    eukaryotic 84
    growth rates for 254
    yield coefficients 254
microsites, anoxic 253
modeling of nitrogen flow
    235
molar yield coefficient 253
monocarboxylic acids 83
Monod's growth equation 261
    bacterial constants of
        268

n-alkanes 81,94,99,154,162
n-alkanoic acids 82,154
n-alkanols 84
n-fatty acids 94,99,162
Nares Abyssal Plain 205

nitrate 230
    in soil solutions 278
    oxidizers 248
    production by *Nitrobacter*
        261-266
    reducers 250
nitrification 246,247,250,
    254
    environmental factors on
        259
    in mixed bacterial cultures
        266
    in relation to substrate
        concentrations 261
    in soils 251
    rates of 249,251,255
    steady-state 246
nitrifiers 250,252,254,255,
    259
nitrite
    assimilation 261
    oxidation 255
    oxidizers 248,252,256
Nitrobacter 255,259,263,
    266,268
nitrogen
    balance in lysimeter ex-
        periments 277
    biological fixation 228
    cycling 225,232
    diffusion control of ion
        uptake 257
    fixation 226
    fluxes 230,233
    hydrolizable 233
    in the biomass 231
    in humic substances 91
    in sediments 298
    in soils 225-240,271-279
    interecosystem transfers
        225
    isotopes 235,344,346
    isotopic fractionation
        51-70,344,346
    isotopic variations 234
    -labeled fertilizers
        272

nitrogen, *continued*
  modeling of transformations in soils 247-258
  oxides 45-47
  oxidizing enzyme systems 255
  rates of chemical transformation 245
  release in cropped soils 239
  role in eutrophication 287
  transfers within ecosystems 228
nitrogen-15 232,236,279
  content in effluent gas stream 277
  immobilized 234
  -labeled fertilizer 274
nitrogen dioxide, oxidation rate of 251
nitrous oxide, occurrence of 271,272
nonhumic substances 90
North Atlantic 203
nucleic acid bases 175-183
  extraction from soils and sediments 176
  identification of 176
nutrient enrichment 75
  metabolism 69
  regeneration 188

oligotrophic lakes 76-78, 80-82,85,180
oil shale 131,134,135
organic geochemistry of lake sediments 75-88
organic material
  carbon isotopic composition 53
  in anaerobic environments 52
  in recent algal mats 131-149

organic matter
  anaerobic biodegradation 298
  effect on release of ammonia 300
  humification of 197
  in soils and sediments 117
  loss 250
  nitrogen content 234
  of grasslands 232-239
  oxidation 252
  turnover rate in soils 236
organic waste decomposition 51
oxidation by alkaline solution 126
oxidation reagents 134
oxygen
  containing functional groups 91
  diurnal variation 302
  in humic substances 91
  isotopes 316
  isotopic fractionation 51-70

particulate sulfur sources 357,358
PCB 203
  codistillation of 210
  decline of 206,207
  degradation of 212
  determination in seawater 212
  extraction from seawater 218
  fluxes 216
  food chain magnification 204
  in North Atlantic water and sediments 205-217
  loss from Sargasso Sea 208
  partitioning between seawater and particulates 209

PCB, continued
  rate of deposition on sea
    bed 216
peptides 90
perfusion 90
pesticides 89
  See also DDT and PCB
petroleum source materials
  102
pH of the ocean 304,306
phenol-sulfuric acid method
  for measuring carbohy-
  drate content 186
phenolic acids 94
phosphate 298
  -organic matter relation-
    ships 297
phosphorus 283-293,298
photochemical reactions 47
photosynthetic production
  302,323
phytane 81
phytanic acids 82
phytol 81-84
phytoplankton 82,85
pigments in lake sediments
  80
  See also carotenoid
plastic hemisphere 277
pollution in rivers, mathematical modeling of 259
Polychlorobiphenyls
  See PCB
polysaccharides 186,191,
  195-200
pore waters 319,321
productivity
  effect of nitrite concen-
    tration on 262-266
  effect of pH on 261
  effect of temperature on
    261
  in limnocorals 287
  in the oceans 210
  in varied light 260
  measurement 292

purines and pyrimidines 79,
  175,177-180
  concentrations in sediments
    178-181
  concentrations in zoo-
    plankton 177
  diagenesis of 178
  extraction from soils and
    sediments 177
  microbial degradation of
    179
pyrite 317,321,379
pyruvate, decarboxylation of
  347

sandy loam soil 274,277
saponification 99,101
Sargasso Sea 207,208
  PCB loss in 208
  sedimentation rate in 210
saturation constant 247,263
sediment
  anaerobic 321
  as sinks 285,298
  biological activity in 175
  carbohydrates in 185
  extraction of pyrimidines
    from 177
  humic acids from 109
  kerogen structures in
    134-149
  lipids in 151
  marine 177,302-305
  metals in 385
  nitrogen source in 298
  organic geochemistry of
    75-88
  oxygen isotopic composition
    of 313
  phosphorus regeneration
    from 283-286
  release of sediments from
    296
  sulfide in 369,383,385
  -water interface 295

sediment, *continued*
  -water system simulation
    294
  See also lake sediments
sedimentary rocks 105
sedimentation in soils 102
selenate and sulfur 391
selenium
  accumulator 390,401
  analogs 391,394
  and sulfur 394
  assimilation in plants
    400
  human health effects 390
  in animal nutrition 389
  in biological systems
    389,391
  methylation of compounds
    397
  occurrence of in nature
    392
  oxidation states 391
  phosphate interaction 390
  physical properties 391
  plant requirements 390
  toxicity in animals 389,
    390
  volatilization 389,396,
    399,403
sewage sludge gases 55
silica, in metal-organic
  complexes 127
silicic acid 127
silver dithizonate 370-373,
  378,385
sitosterols 84
soils 177,178,195
  analysis of 121
  and sediments 89,90,102
  atmosphere 272,273
  biological factors 245
  carbon contents of 238
  cycling of nitrogen in
    225
  columns 245,247
  decomposition in 239

soils, *continued*
  diffusion control of ion
    uptake 255
  extraction of polysacchar-
    ides from 195
  extraction of pyrimidines
    from 177
  glucose in 195-198
  humic acids from 89
  Lower Lias clay 275
  nitrates in 280
  nitrification in 249
  nitrogen contents of 238
  nitrogen losses from 271-
    279
  nitrogen transformation
    modeling 245-256
  nitrous oxide in soil
    atmosphere 273
  organic carbon in 195-197
  polysaccharides in 191
  purines and pyrimidines in
    177-181
  tropical volcanic 117
  See also Kuroboku soils,
    and loam
soil organic matter
  availability 252
  half lives of constituents
    234
  in the activity of denitri-
    fiers 256
  transfers 233
  uronic acid in 198
spectrophotometry 122
spodosols 119
stanols 84
steady state
  conditions of nitrification
    246,247
  in perfusion 255
  kinetics 250
  model of carbon monoxide in
    the troposphere 40,42
sterenes 158,165
sterols 159,165

sterols, continued
  in lake sediments 84,85
stigmatasterol 84
straw residue nitrogen 239
sugars, hydrolyzable 76,77
sulfate 302,309,313,315,
    320,340
  effect on isotope frac-
    tionation 342,344
  effect on sulfite reduc-
    tion 340
  in rain water 318
  isotopic composition of
    318
  particulate in the atmos-
    phere 351
  pore water 315,321
  reducers 298,321,351,
    353,356,358
  reduction 302-305,317,
    319,321,330
  See also Dead Sea
sulfide 302,315,319,320,
    347
  acid volatile 315,321
  determination 371-374,
    378,380
  in sediments 383,385
  measurement of 369-374,
    378
  methods for direct deter-
    mination of 369
  -oxidizing bacteria 304
  production by yeast
    330-347
sulfite
  assimilatory and dissimu-
    latory reductase 347
  effects of concentration
    340
  effects on isotope frac-
    tionation 341,344
  reduction 330,333
sulfonium compound
    cleavage of 399
sulfur 309,321
  analogs 394

sulfur, continued
  anthropogenic pollution
    351
  biogenic 351,356,358,362
  cycle 304-306,309-322
  isotopes of 309,315,329,
    330,344,347
  isotope distribution
    353,354
  isotope fractionation
    51-70,319,356
  isotopic variations in
    marine plants 356
  organic 315,352
  oxidation states of 391
  particulate 351,355,357,
    359
  sources of gaseous 354
  See also Dead Sea
sulfur dioxide 356

tensioned drainage 277
thymine 79,177
toxic pollutants 102
transport dispersion 102
tree-ring dating 4,18,20
tri-carboxylic acids 94
triterpanoic acids 83
troposphere 357
  photochemical models of
    39,45

uracil 79,177,178
urea
  adduction of lipids 155
  hydrolysis in soils 249,
    250
uronic acid 195-197,201
ultimate analysis of humic
    fractions 91
ultrasonic dispersion
    99

vivianite 298

volatilization 118,389,399, 403

X-ray fluorescence analysis 125

zero-order kinetics 250,255
zero-order reactions 303
zooplankton grazing 290
zooplankton in Lake Ontario 177